BRUCE B. MASON

BRUCE B. MASON

THE CONDUCT OF POLITICAL INQUIRY
Behavioral Political Analysis

THE CONDUCT OF POLITICAL INQUIRY

Behavioral Political Analysis

LOUIS D. HAYES *University of Montana*

RONALD D. HEDLUND *University of Wisconsin at Milwaukee*

Editors

Prentice-Hall, Inc., Englewood Cliffs, New Jersey

© 1970 by Prentice-Hall, Inc., Englewood Cliffs, New Jersey

All rights reserved. No part of this book may be reproduced in any form or by any means without permission in writing from the publisher.

13-167304-1

Library of Congress Catalog Card Number: 77-110491

Current printing (last digit):
10 9 8 7 6 5 4 3 2 1

Printed in the United States of America

Prentice-Hall International, Inc., London
Prentice-Hall of Australia, Pty. Ltd., Sydney
Prentice-Hall of Canada, Ltd., Toronto
Prentice-Hall of India Private Limited, New Delhi
Prentice-Hall of Japan, Inc., Tokyo

PREFACE

As the sophistication of the social sciences increases, in terms of more widespread use of complex methods and techniques for investigating and explaining social phenomena, it is becoming necessary to develop materials designed to acquaint students with some fundamentals of social science inquiry. Lack of an adequate grounding in at least the basic assumptions and approaches to social science research will prevent the student from making full use of the best knowledge in the field. In political science, especially, the gap between levels of research and levels of training results in part because students have not been exposed to the nature of inquiry, which may be broadly defined as the systematic search for answers to questions.

Before 1945 political science as a discipline was primarily occupied with attempting to answer practical or philosophical questions about politics. The results tended to take the forms of moralistic discussions, institutional descriptions, and political reports. Little attention was given to the alternate approaches a researcher might use when seeking answers to questions concerning the political world. Now, however, larger numbers of political scientists are concerned with the structure of inquiry as well as with the subject matter of politics. Investigators are giving greater attention to research strategies and techniques, as may be seen in the growing number of scholarly books and articles partially or entirely devoted to the structure and procedures of inquiry. As a result of this attention, the post-1945 development of political science methodology surpasses its entire evolution up to that time.

Many of these developments have been toward adopting more scientific approaches for political inquiry such as those used in the "exact sciences." A wide variety of adjectives has been used to describe the nature of these new approaches, including *behavioral, objective, scientific,* and *empirical.* But political scientists differ regarding which adjective most appropriately describes the outlooks, approaches, and structures of contemporary political inquiry. Since all of these adjectives accurately describe some aspect of current political inquiry, any one of them might be used. Many political scientists would no doubt prefer using empirical or scientific in referring to the inquiry described in this book. However, these terms have fairly narrow meanings and may not be completely descriptive of contemporary political inquiry. We have chosen, therefore, the term *behavioral* because of its more comprehensive meaning.

This greater emphasis upon behavioral inquiry encouraged many political scientists to consider its nature and its relationship to understanding political phenomena. Such considerations pose a variety of questions, for example: What are the goals of inquiry? What are the approaches to inquiry? What are the common elements among these approaches to inquiry? Without answers to these questions, understanding and utilization of the various forms of contemporary political inquiry will probably be more limited. The editors of this volume have selected excerpts from some basic source material on inquiry, which hopefully provide insights into the major concerns of contemporary political science.

No doubt some of our colleagues will find parts of this volume objectionable. We are not, however, advocating a particular methodological viewpoint here, but have consistently attempted to steer a middle course through a field about which there is considerable disagreement. Accordingly, the reader should keep in mind that this effort is intended for students, especially those in introductory courses, and is not directed to professional political scientists.

The readings and commentaries that follow are intended to broaden understanding of political research by presenting a basic introduction to the structures and procedures of inquiry. In an effort to lend coherence to the various topics covered in the readings, we have included an original essay in which we discuss the main issues of contemporary political inquiry. The reading selections which follow this essay are divided into four parts: (*1*) the nature and development of political science as a field of inquiry; (*2*) the logic of behavioral political inquiry, especially the elements within it; (*3*) the methods and techniques used in behavioral inquiry; and (*4*) some of the conceptual frames of reference employed in various behavioral studies of political phenomena.

In our endeavor we have drawn upon suggestions, comments, and encouragement of colleagues from all areas of the discipline. We wish to acknowledge the help of Professors John Wahlke, Cornelius P. Cotter, Roy G. Francis, Robert P. Boynton, Irvin L. White, Meredith W. Watts, Jr., Ira S. Rohter, David J. Koenig, and Robert Erikson. We are also indebted to William Ferris, Joyce Beilke, Gilda Malofsky, and Rose Karlsen for their research and secretarial assistance. We owe a special debt to our wives and families for their patience and support; but, of course, we alone assume full responsibility.

L.D.H.
R.D.H.

CONTENTS

I

THE CONDUCT OF POLITICAL INQUIRY: AN OVERVIEW 2

The Study of Politics *3*
Behavioral Political Inquiry *7*
Methods of Political Inquiry *12*
Conceptual Frames of Reference *17*

II

POLITICS AS A FIELD OF INQUIRY 20

The Scope of Political Science
The Scope of Political Science, Vernon Van Dyke *27*

The Purpose of Political Inquiry
Values in Political Inquiry, Charles S. Hyneman *36*

The Development of a Science of Politics
The "Behavioral-Traditional" Debate in Political Science, Albert Somit and Joseph Tanenhaus *45*

III

THE NATURE OF BEHAVIORAL POLITICAL INQUIRY   56

Scientific Methods of Inquiry
Scientific Operations and Scientific Justification, Arnold Brecht *60*

Implementing Scientific Inquiry
Problem Formation and Hypothesis Generation, Fred. N. Kerlinger *64*
Concepts in Empirical Research, Carlo Lastrucci *72*
The Translation of Concepts into Indices, Paul F. Lazarsfeld *78*

Goals of Inquiry: Explanation, Prediction, Causation, and Theory Building
Explanation and Scientific Investigation, Warren Weaver *81*
Predicting Future Events, Irwin D. Bross *86*
Causation in Social Research, R. M. MacIver *89*
Empirical Theory: Explanation in the Social Sciences, Eugene J. Meehan *93*
Research and Theory Building, Robert K. Merton *99*

IV

METHODS AND TECHNIQUES FOR BEHAVIORAL INQUIRY   104

Data Collection: Observation
Roles in Social Field Observation, Raymond L. Gold *109*

  Interviewing and Survey Research
The Interview: A Data Collection Technique, William J. Goode and Paul K. Hatt *115*
Survey Research in Political Science, Herbert McClosky *120*

  Experiments
The Social Science Experiment, Barry Anderson *127*
Data Generation Through Simulation, Richard E. Dawson *132*

Data Analysis: Measurement
Levels of Measurement, S. S. Stevens *141*

  Quantification
Quantitative Descriptions of Data, Clair Selltiz, Marie Jahoda, Morton Deutsch, and Stuart W. Cook *145*
Problems in Statistical Analysis, Leslie Kish *153*

  Machines in Data Analysis
Computers: Their Built-in Limitations, Max Gunter *158*

V
CONCEPTUAL FRAMES OF REFERENCE   168

Unit Focus: The Individual
Role: A Basic Unit for Analyzing Political Behavior, Heinz Eulau   173
Individual Socialization: Maintaining the Political System, William C. Mitchell   177

The Group
The Group in Political Science, Charles B. Hagan   187

The Society
An Approach to the Study of Politics: The Analysis of Political Systems, David Easton   197
The Political Culture Approach, Sidney Verba   208

Process Focus: Communications
Political Communications and the Political System, Karl W. Deutsch   216

Power
Power and Influence, Robert A. Dahl   222

Decision-Making
A Decision-Making Approach to the Study of Political Phenomena, Richard C. Snyder   233

Political Change
The Concept of Political Development, Lucian W. Pye   241

Functionalism
Functionalism in Political Science, William Flanigan and Edwin Fogelman   248

THE CONDUCT OF POLITICAL INQUIRY
Behavioral Political Analysis

I THE CONDUCT OF POLITICAL INQUIRY: AN OVERVIEW

THE STUDY OF POLITICS

The study of political phenomena is one of the oldest areas of inquiry. Although most disciplines, at one time or other, have attempted to trace their lineage back to ancient Greece, few have been able to establish quite as secure a claim for this heritage as has political science. That Plato and Aristotle were concerned with the Greek polity is not disputed. Many of the questions considered by the Greeks are discussed in political science today. One issue that has preoccupied students of politics throughout history is: Given the necessity for government, how can it best be accomplished? Plato sought to answer this question by exploring the nature of the ideal state, whereas medieval philosophers concerned themselves with a framework for establishing God's kingdom on earth. More recent discussions on the necessity for government have focused on the nature of political power.

Topics currently under investigation by political scientists indicate that students of politics are by no means in agreement over the definition of their subject matter. One thing is clear, however; traditional notions of what constitutes politics seem inadequate for describing the topics currently under investigation. For example, definitions of politics only in terms of governments, states, sovereignty, and authority appear too restrictive for characterizing contemporary investigations. Many political scientists seem more at ease with broader definitions of politics, such as the ones offered by David Easton (Politics is the authoritative allocation of values)[1] and Harold Lasswell (Politics

[1] David Easton, *The Political System: An Inquiry Into the State of Political Science* (New York: Alfred A. Knopf, Inc., 1953), pp. 129–34.

is who gets what, when, and how).[2] These definitions and others like them, though perhaps more vague and ambiguous than traditional definitions, indicate a reluctance by political scientists to restrict politics to narrow limits.

Further, contemporary definitions of politics take into account the impracticality of drawing distinct lines around the subject matter of the field in order to state that all "x" is political and all "non-x" is nonpolitical. Little can be gained by disciplinary provincialism, which, in effect, says to the sociologist or economist: "Do not Trespass—Property of Political Science!" Accordingly, political scientists consider it legitimate to study topics like "Psychopharmacology and Political Belief,"[3] and "Card Sorting, A Psychometrically 'Clean' Method for Survey Interviewing,"[4] in addition to more familiar topics like "Committee Characteristics and Legislative Oversight of Administration,"[5] "Ballot Forms and Voter Fatigue: An Analysis of the Office Block and Party Column Ballots,"[6] and "Transaction Flows in the International System."[7]

The appearance of new and less restrictive conceptualizations of politics has been accompanied by a somewhat different posture on the part of many political scientists regarding what one ought to study and how one ought to go about it. The older, established orientation, the traditional, has been faced with competition for prominence in the discipline by a new approach, the behavioral. These two orientations differ primarily in the manner by which inquiry is undertaken, the traditional being heavily philosophical and descriptive and the behavioral being more empirical and analytical.

Traditional political inquiry consists in general of three types of studies—political philosophy, institutional description, and primitive empiricism. Political philosophy has three identifiable features: a tendency to employ

---

[2]Harold Lasswell, *Politics: Who Gets What, When, How* (New York: McGraw-Hill Book Company, 1936).

[3]Albert Somit, "Psychopharmacology and Political Belief," paper delivered at the 39th annual meeting of the Southern Political Science Association, November 2-4, 1967.

[4]Lester W. Milbrath, Everett F. Cataldo, Richard M. Johnson, and Lyman A. Kellstedt, "Card Sorting, A Psychometrically 'Clean' Method for Survey Interviewing," paper delivered at the 39th annual meeting of the Southern Political Science Association, November 2-4, 1967.

[5]John F. Bibby, "Committee Characteristics and Legislative Oversight of Administration," *Midwest Journal of Political Science*, 10 (February, 1966), 78-98.

[6]Jack L. Walker, "Ballot Forms and Voter Fatigue: An Analysis of the Office Block and Party Column Ballots," *Midwest Journal of Political Science*, 10 (November, 1966), 448-63.

[7]Steven J. Brams, "Transaction Flows in the International System," *American Political Science Review*, 60 (December, 1966), 880-98.

deductive reasoning in deriving conclusions (Assuming the validity of a general proposition, what specific conclusions logically follow?), an emphasis on the normative element (a concern with what ought to exist), and to a lesser extent a concern with the nature of politics and its place in the human order. As a result, the major contribution of political philosophy is the development of political values and ideas. These studies are largely conjectural and speculative, their contributions being evaluated in terms of their logical coherence, insightfulness, and the relative moral desirability of the conclusions rather than by their objective accuracy.

Institutional descriptions are discussions of the formal properties of political organizations and processes. These accounts tend to be legalistic in that they draw upon constitutions and legal documents to describe structures and organizations. Their purpose is to depict the formal characteristics of some political institution or governmental structure. Implicit in these studies is the notion that political structures remain relatively unchanged over time. Numerous examples of this approach can be found in studies of organization and procedure in the British House of Commons, the American Congress, and bureaucratic agencies.

The third type of traditional political study can be labeled primitive empiricism. Research included in this category is concerned more than that of the others with political *phenomena*. There is also a greater tendency to use empirical methods at the expense of intuition, argumentation, and related methods. Concern for political events leads to a regard for the dynamic elements of politics as well as the institutional framework. These studies are generally not as systematic or objective in procedure as are more recent examples of empiricism. Although the rigor of observation and analysis may be minimal, the conclusions are in many cases extraordinarily accurate. For example, an analysis of political and social reality in several South American societies was undertaken by James Bryce early in the twentieth century. His conclusions were extremely accurate and were drawn from a sound empirical basis, but his methods for data collection and analysis were not always rigorous. In comparison, studies of contemporary political and social reality, such as those by Gabriel A. Almond and Sidney Verba reported in *The Civic Culture*, also have an empirical basis, but their data gathering methods were more rigorous.[8]

[8]See James Bryce, *South America, Observations and Impressions* (New York: The Macmillan Company, 1912); and Gabriel A. Almond and Sidney Verba, *The Civic Culture* (Princeton, N. J.: Princeton University Press, 1963). (We recognize that the *The Civic Culture* has certain methodological limitations. We cite it merely as an illustration of attempted rigor in data gathering.)

More recent trends in political inquiry have departed from the traditional approaches by seeking new modes for answering political science questions. The emphasis is upon studying observable behavior in order to make accurate statements about political phenomena. Thus the form of reasoning favored is basically inductive rather than deductive in its development of conclusions; that is, reasoning proceeds from specific observations to general conclusions. This reliance on inductive reasoning should not be interpreted to imply that deductive reasoning no longer appears in inquiry; it means that in deriving conclusions the behavioralist seeks specific information and from this he generalizes. Deductive reasoning is useful for behavioral inquiry in suggesting questions for the researcher to study and also in formulating possible answers. From existing knowledge deductive reasoning may be used to formulate specific questions, which when answered, are used inductively to generate more comprehensive explanations.

Another departure from traditional inquiry advocated by many behavioralists concerns the goals of inquiry. Behavioralists argue that questions concerning "what is" take precedence over questions concerning "what ought to be." Therefore, normative discussions are regarded as more meaningful when conducted in the light of empirical evidence. The distinctions outlined above are not always maintained in research. Exceptions can be pointed out concerning behavioral aspects of traditional research and traditional aspects of behavioral research. We feel that the basic difference lies in the distinctive research mood advocated by these two approaches. Behavioralists are inclined to use objective, systematic, and empirical research methods to offer explanations. A more precise differentiation between traditional and behavioral is presented in the readings that follow.

Political science experienced a growing interest in behavioral research around the end of World War II although traces of this approach were evident much earlier. This concern with behavioralism has sometimes been labeled a revolution because it advocated changes in the outlook of the discipline *qua* discipline. In spite of this change in mood, traditional approaches continue to be used for describing certain features of politics, especially in subject areas in which behavioral data are impractical—for example, political philosophy.

Part of the explanation for the behavioral revolution probably rests with the radical changes that have taken place in the world during the twentieth century. These changes have, among other things, broadened the sphere of politics. Throughout the world, more complex political institutions have emerged, particularly with more widespread acceptance of the welfare state concept. At one time governments had relatively few duties to perform, but with the emergence of welfare states they have become directly involved in almost every aspect of human activity. Further expansion of governmental action has been fostered by technological change. The industrial revolution, automation, and related technological changes have created more problems that the individual cannot solve alone. Such problems as air and water pollution, economic depression, and riots are beyond the effective control of individuals. The expansion of governmental activity in these areas has presented the political scientist with larger numbers of phenomena appropriate for behavioral study.

Developments in international relations have also contributed to the mood for behavioral inquiry. Since World War II, international relations have become increasingly complex and the involvement of the United States in international affairs has changed from relative isolation before the War to complete participation today. As a result, government officials and the general public want to know more about international political phenomena. This desire has spurred political scientists' investigations into the international scene.

The acceptance of behavioral approaches in other social science disciplines has also influenced their development in political science. Widespread application of such approaches alone would not have convinced political scientists of their usefulness; however, the success other disciplines have had in producing useful and reliable findings through the use of behavioral approaches has been a persuasive argument for their acceptance by political science. Interdisciplinary sharing of approaches, techniques, and concepts has been generally accepted when such sharing has noticeably contributed to the expansion of knowledge.

Finally, the inability of traditional methods to provide verifiable answers to certain questions about politics created a mood among many political scientists that favored the emergence of new approaches. This dissatisfaction is particularly evident with respect to the inability of traditional methods to explain individual political behavior. Traditional explanations tended to focus on public organizations, governmental institutions, and political processes rather than on individual behavior.

BEHAVIORAL POLITICAL INQUIRY

The growing use of empirical methods of inquiry in political research has fostered greater interest in the nature and structure of inquiry itself. Although such concerns are comparatively recent in political science, they have been of interest longer in other disciplines, particularly philosophy, sociology, and the physical sciences in general. The study of inquiry, sometimes called the philosophy of science, has an extensive literature. This literature has served as resource material in the movement toward behavioral political inquiry. Because much of the literature on the conduct of inquiry is addressed to the physical sciences, one initial task has been to establish its relevance for the study of political phenomena. Having accomplished this, political scientists have devoted considerable attention to the extension and refinement of behavioral methods in the conduct of political inquiry.

One major reason that political scientists are concerned with methodological questions stems from the tendency for the new types of inquiry to pose research problems that are relatively new to the discipline. Previously, when students of politics asked questions about the "good life" in politics, they did not direct much attention to questions of methodology. But once they began to investigate the nature of observed political phenomena they were unavoidably confronted with the task of deciding how meaningful inquiry into these topics was to proceed.

The Nature of Behavioral Political Inquiry In evaluating the interest of political scientists in behavioral forms of inquiry, one must consider the nature of these investigations. They are sometimes falsely said to reveal ultimate truth, but no type of inquiry can make this claim. Likewise, behavioral inquiry cannot be recommended because it is easier to undertake—such inquiry involves skills that are not easily acquired. Rather, the advantages of behavioral inquiry stem from the fact that findings are considered more reliable and precise than findings obtained by other means.

One quality of behavioral inquiry is its use of "scientific methods." In this context a scientific method of research means identifying a problem, hypothesizing the existence of certain relationships among factors, collecting pertinent data, and empirically testing the hypotheses advanced. Any one of several combinations of techniques may be exemplary of scientific methods. Behavioral inquiry relies heavily upon empirical techniques and procedures. Empiricism is a theory of knowledge which postulates that the most valid information about phenomena is that gathered through actual experience or observation. Thus, by advocating the use of empiricism, behavioralists show preference toward data gathered through one's senses rather than through intuition, extrasensory perception, or deductive reasoning.

Because of the greater use of scientific methods, the task of behavioral inquiry is viewed as making accurate statements about political phenomena which can be verified by further investigation. A preoccupation with discovering ultimate truths is considered irrelevant. Nothing is viewed as being "true" even if empirically valid; rather, one accepts statements as valid pending additional investigation, subject to refutation should contrary evidence be obtained. The political scientist using these forms of inquiry must be willing to adjust his findings if refuting evidence is offered. Also, it is possible to repeat the research processes used in many behavioral political studies. Repetition of this sort permits one investigator to examine the research procedures of another to determine if analytical mistakes were made. In contrast, statements about the world derived from intuition alone cannot be verified through replication, since the intuition of one person cannot be shown to be the intuition of another. The only basis for accepting conclusions arrived at in this manner is the investigator's reputation for having been correct in the past. Behavioral inquiry offers the opportunity to examine the reliability of research processes as well as to confirm conclusions.

Many social scientists are disdainful of research intended to reexamine existing knowledge and prefer to pursue unexplored lines of inquiry. However, the importance of research designed to evaluate the accuracy of earlier findings should not be underestimated, for it is actually an integral part of behavioral research. Indeed, when replicative investigations produce results that differ from earlier findings, the validity of the original is subject to reevaluation in order to determine the reasons for this difference. Thus, the behavioral investigator need not avoid studying topics that have already received attention. Moreover, he should not be preoccupied with the unique happening, as this precludes the possibility of replication.

In behavioral inquiry there is a tendency to formulate broad inclusive statements, called generalizations, from a few specific instances. These statements are derived from the inductive nature of behavioral inquiry. Observations of isolated, particularistic phenomena are only useful for describing what transpired at one point in time. When several specific observations are merged into a comprehensive conclusion the researcher can make meaningful statements about larger numbers of phenomena. The metaphysical view associated with behavioral inquiry (the assumptions made about the arrangement of the universe) is the belief that reality is ordered. Translated into operational terms, this means that phenomena are regular and consistent. If the investigator can identify these regularities and consistencies for several comparable phenomena, he can generalize for all other phenomena of the same type. Valid generalizations of this sort are of enormous practical consequence because the investigator need not examine every individual example in order to describe what principles govern the behavior of similar phenomena.

In behavioral inquiry, investigators have attempted to avoid questions about what they would prefer the findings from an investigation to be; rather, they have taken the problem as a given element and through investigation attempted to arrive at some conclusions about it. In other words, the investigator is expected to be objective in conducting his examinations. He is expected to "let the evidence speak for itself" rather than to impose his own subjective notions on the outcome. The extent to which this is possible when one is dealing with social rather than physical phenomena is a point of contention among social scientists. This dispute transcends social science: philosophers of science, too, have argued about the types of research and the disciplines which can appropriately undertake value-free inquiry. The most extreme statement against the entrance of personal values into inquiry is logical positivism. Some logical positivists argue that the only data useful for understanding reality are those derived from direct observation, especially experience. Values, preferences, beliefs, and metaphysical considerations must be excluded from the analysis process. It is difficult for the social scientist to accept this position, because values and preferences are deemed vital dimensions of human activity. Even if the social scientist is willing to accept a strictly determinist outlook on human behavior (that everyone acts in accordance with predetermined causes), he still must recognize the element of human caprice. The natural and physical sciences, on the other hand, do not face this same problem because natural phenomena do not behave in accordance with value preferences in spite of a certain element of unpredictability.

The problem of values in social inquiry must be recognized as such—a problem. The student of politics can probably never achieve totally value-free inquiry, but to the extent possible, the social researcher should strive to keep his own value preferences removed from the inquiry itself. When the researcher permits his own preferences to mix with reality so that it becomes impossible to distinguish one from the other, objectivity no longer exists.

In many instances, discussion about inquiry and scientific methods has created the faulty impression that there is a single set of rules which guarantee that an investigator's endeavors will be scientific. Analysis of behavioral inquiry is more realistic if one proceeds on the assumption that scientific methods establish a mood and guide inquiry rather than prescribe the operations one must follow.

The Goals of Behavioral Political Inquiry The purposes of political inquiry have been variously defined. Students of politics, from Aristotle to contemporary political scientists, have had distinctive notions about the goals of political inquiry. Some say that the purpose is understanding, others that it is the achievement of the good life, and still others that it is objective description of reality. There is no a priori reason to exclude any of these notions. None, however, adequately explains the goals of behavioral inquiry. Scientists, on occasion, tend to argue that their activities are self-justifying, but "knowledge for knowledge's sake" seems an insufficient justification in light of the social and financial demands research makes on society. Ultimately, all inquiry tends to be evaluated in terms of its contribution toward improving the human condition. Behavioralists in no way deny this contention but argue that man can direct his activities and control his destiny more meaningfully when he understands the nature of worldly phenomena. Such understanding is thought to depend in large part upon the amount of knowledge available about such phenomena.

In terms of methodology, one of the goals of expanding knowledge is what is called *theory building*. The term *theory* is used in a variety of ways by scientists and nonscientists alike, adding to the confusion of an already overworked concept. Theory is one of the most abstract and complex concepts in the lexicon of science. Moreover, the term is much misused, particularly in the social sciences. Political philosophy—speculation about the ideal state—is often referred to as theory. Research strategies, such as systems analysis and field techniques, are also sometimes regarded as theory. However, the term is used in behavioral inquiry to mean a body of knowledge fashioned in such a way that it draws together facts about reality and imparts to them meaning and significance not otherwise apparent. In other words, a theory identifies and describes the relationships among facts discovered through observation. Theories are provisionally acceptable statements. Initially they may be primarily speculative, tentative explanations, and several entirely different theories may seek to explain the same phenomenon. Eventually, it is thought, the preponderance of evidence will yield one theory, which will be widely accepted in the discipline as the most accurate explanation, that is, a "general theory."

Theory building, as it relates to the processes of inquiry, performs a variety of tasks. One of these is the development of explanatory schemes sufficiently general to account for large numbers of similar phenomena. A description and explanation of one specific observation, if accurate only for that one observation, is an explanation at a relatively low level of generality. On the other hand, an explanation that accounts for a large number of phenomena is considered to be at a high level of generality.

Prediction is sometimes viewed as a second general goal in theory building. Certain scholars have suggested that the ability to predict is directly proportional to the ability to explain. This has not always been the case. Often researchers have been able to offer rather complete explanations of phenomena without having been able to make predictions. For example, after any national election, the political scientist is able to explain why one candidate was elected without necessarily having been able to predict the outcome. On the other hand, man was able to predict the ebb and flow of ocean tides before he was able to account for them.

Related to explanation and prediction is the important concept of causation. Phenomena occur because of what are called "causal factors." The identification of these causal factors is thought to be a major part of explanation. If the researcher is able to say with confidence that certain factors produce a phenomenon, a causal statement has been made and the occurrence is thought to be explained. However, demonstrating causal relationships in the social sciences is more difficult than this formulation would indicate. The physical scientist is in the position of being able to isolate a few factors and manipulate these to investigate causal factors. The social scientist is not in this position, because the human being and his social interaction are frequently too complex for such manipulation and control. Moreover, factors that cause social events seem to be highly interrelated, transitory, and fugitive owing to the rapidly changing nature of the social environment. As a result, social scientists find it difficult to develop social theories capable of offering precise causal formulations.

Theory is also instrumental in hypothesis generation, a critical step in advancing knowledge. An hypothesis is a statement that specifies a suspected relationship among two or more factors. Thus, it may be viewed as a probable answer to the question being posed. The generation of hypotheses is the investigator's responsibility, and he may make use of deductive reasoning. His insight, knowledge of the facts, and familiarity with existing theories are the primary ingredients in generating hypotheses. The role of theory in generating hypotheses is particularly important, because theories explain observed relationships among factors and may suggest how other factors should be related. Moreover, the failure of existing theories to explain certain phenomena suggests an area in which investigation is needed.

Methods of Behavioral Inquiry

A critical element for behavioral inquiry is the actual research process employed by the investigator in seeking answers to questions. In essence this process consists of gathering and analyzing information (*data*) that can be used as evidence to support or reject proposed hypotheses, which in turn shed light on the validity of existing theories. Given the behavioralists' preferences that data employed in inquiry be drawn from observations of reality and be gathered through the use of one's senses, certain methods and techniques are better suited for data gathering than others. However, the use of these procedures is by no means a sufficient condition for achieving significant behavioral inquiry. These procedures seek to satisfy three basic questions: What data are relevant to the phenomenon under investigation? How can this information best be obtained? How must the data be studied and presented in order for valid and meaningful conclusions to be drawn?

The Nature of Data   Without a clear notion of which data are relevant to the purposes of the research problem and which are extraneous, the researcher is faced with the impossible task of obtaining information about everything; he must be able to limit the information collected to that which is relevant. Such delineation of pertinent data is based upon the hypotheses to be tested and the researcher's own hunches. Obviously, this process is open to error.

Data collected in any investigation provide information about some object, usually called a "unit of analysis." Since social scientists study people, the units of analysis they most generally employ are individuals or aggregates such as clubs, organizations, and nations. Once the unit of analysis has been established, the investigator needs to determine which data are relevant to the research problem. The unit of analysis itself also suggests which data are relevant. If, for example, one is studying judicial decision making, information concerning presidential decision making is probably irrelevant.

By specifying the information desired, the investigator is determining which variable(s) should be examined. As the name suggests, a variable is some aspect of the objects under study that can differ in quality or quantity.[9] When the individual is the unit of analysis, most variables are either demographic (age, sex, socioeconomic status, and so on), attitudinal (preferences regarding the selection of a public official, predisposition toward authoritarianism, inclination to interact with others, and so on) or behavioral (number of times individuals interact, who speaks to whom, how a person reacts in some situation, and so on). By analyzing relationships among different types of variables, the investigator either supports or negates the hypotheses he set out to examine.

Finally, the investigator should consider the form variables must assume in order for them to be useful. In other words, what kinds of values can be assigned to the variables and which of these is most appropriate for the task at hand? Frequently, several alternative schemes can be used for the same variable, and the investigator must select the most appropriate. Data about age, for example, can be specified by phrases denoting various categories, for example, "young adult," "middle-aged adult," and "older adult"; by numbers, for example, "under 30," "30–50," "over 50"; or by the number of years since birth. Thus, when dealing with the variable "age," the investigator would select one scheme for specifying age.

The behavioralist will consider all of these aspects of data when planning his approach for gathering and analyzing information, before beginning the investigation, to make the collection and analysis easier and more meaningful.

Data Collection For the purpose of obtaining objective information the most desirable procedures for data gathering are those that rely upon one's senses—sight, hearing, smell, taste, and touch. However, opportunities for proper use of one's senses in data collection vary greatly with the phenomenon under study. The physical scientist is usually in a better position to use his senses in gathering information than is the social scientist, because physical phenomena lend themselves more easily to this type of inquiry. In most instances the social scientist must rely on various kinds of records for reconstructing the many events he is unable to observe. In addition to such ex post facto research he may be forced to use data collection techniques

---

[9]For a more complete discussion of units of analysis, variables, and values see Johan Galtung, *Theory and Methods of Social Research* (New York: Columbia University Press, 1967), Chap. I.

appropriate for probing into the human psyche. Unfortunately, the psychology of behavior does not always lend itself to direct observation, and the data collection techniques that can be used are sometimes of limited reliability.

Social science techniques can be classified into document analysis, observation, interviewing, and experimentation-simulation. Political scientists have used all four types, but by far the greatest number of efforts in data collection have used document analysis or interviewing. Currently an increasing number of studies are being undertaken that use observation and experimentation-simulation for gathering political data.

Documents are important repositories of information. Data about individuals, groups, and nations are reported in a wide variety of public and private documents including legislative journals, census materials, and governmental reports. Several problems arise, however, in the use of these materials. If the desired information is not included in the documents, it probably can be obtained only with great difficulty, if at all. Collected data frequently will not be reported in the most useful form. Also, because many politically relevant documents were originally compiled by governments, they are not always objective or open for inspection. In many instances governments have reported only information favorable to them.

In recent years investigators have relied increasingly upon interviewing techniques. Interviewing offers an advantage in that the information is collected from the subjects themselves; however, it presents certain difficulties. Many people are reluctant to subject themselves to probing into their personal affairs; and the researcher cannot always be certain that the information they provide is accurate.

Sociological studies of interaction in small group situations have indicated the potential usefulness of systematic observation as a data gathering device. On the basis of data gathered in this manner, numerous important conclusions have been made about individual behavior in group situations. Unfortunately systematic observation is difficult to implement in large group situations. Since many political phenomena involve large numbers of persons, observation has been used infrequently.

Pure experimentation as developed in the physical sciences is an exceedingly powerful device for collecting information about causal relationships, and has greatly assisted in expanding the horizons of knowledge. Unfortunately for social science, experimentation can be applied only under the most restrictive conditions. These conditions include researcher control over factors affecting the outcome and over the environment surrounding the phenomenon. Such control has been difficult to achieve in studying complex social and political activity. Recent developments in experimentation under contrived conditions (simulation) have produced new interest in applying these techniques to the study of politics. Simulation involves creating artificial political situations, usually of a decision making nature, and manipulating various factors so as to study their consequences.

Data Analysis After the data have been gathered, the investigator must analyze them in order to interpret their meaning for the problem under study. At this stage of the research process, the investigator must continue to be objective, systematic, and precise. Meaningful data improperly analyzed are as great an impediment to the advancement of knowledge as are meaningless data. Valid data and proper analysis are both essential to the orderly expansion of knowledge.

Analysis is the process of imparting meaning to data by interpreting them. Data analysis consists of a number of individual steps, including data manipulation, significance evaluation, and data presentation. The specific operations used in analysis vary widely, from simple procedures like creating frequency distributions and percentage tables, to more complex ones like scalogram analysis and paired comparisons, and to highly technical ones like factor analysis and causal modeling. These techniques and operations may be thought of as objective sets of rules for interpreting facts and figures. The range and complexity of these various techniques is an ever-expanding subject of concern to political scientists.

Before beginning data analysis, the researcher must plan how he will interpret the data he has collected. Some guidelines for the selection of appropriate analysis techniques are offered by the hypotheses under investigation. Choice of techniques will be facilitated by the investigator's familiarity with the entire spectrum of techniques. Because many are available, no single individual is likely to be entirely familiar with all of them; most behavioralists strive to be knowledgeable about representative types. The fact that many techniques are interrelated and are built upon similar principles assists in acquiring basic knowledge about those most commonly used. In addition to being familiar with the basic purposes of the techniques, the researcher must also recognize that certain procedures govern the proper use of any technique.

One view of the data analysis stage of inquiry is that it is a purely mechanical process in which data are automatically submitted to objective examination, with the investigator watching the proceedings from the sidelines; the conclusions emerge from this analysis without outside assistance and without human intervention and human error. In fact, however, the human element appears time and again. The investigator must make decisions concerning which technique to use and how to interpret findings. Poor judgment or faulty interpretation by the investigator can impair or negate the meaningfulness of the inquiry. A more realistic interpretation of the data analysis stage recognizes the role of the investigator but points out that many of the decisions about the statistical significance of the facts can be removed from the investigator's discretion. In those instances where some discretion is left to the investigator, rules and principles can be established to guide him. Thus, the important consideration in data analysis is the recognition that interpretation of facts and figures should proceed in an objective manner.

Aids to Data Analysis Recent technological advances have fostered the use of high-speed, sophisticated mechanical and electronic equipment in data analysis. Such equipment has assisted in developing techniques for data manipulation and analysis that previously had been either impossible to perform or so time consuming as to be impractical. The net effect of these developments has been to free the investigator from many mundane and repetitive tasks. Making the computer an extension of the investigator has made possible more comprehensive and complex analyses, with increasing efficiency and accuracy. However, a great mythology has grown up concerning the use of computers and allied hardware in the social sciences. One source of distrust is the belief that machines will begin to "think" and tell people what to do. Secondly, distrust may stem from the problems that arise whenever mechanical innovation is tried. Stories of individuals inconvenienced by computerized administrative processes are legion. An additional source of uneasiness is the potential for undesirable applications, for example, the proposed location and use of all information about each individual in one central computer, a type of Big Brother à la Orwell. The computer technician answers that computers are only machines capable of extremely efficient storage and manipulation of data. The calculations a computer can complete in nanoseconds—billionths of a second—might take tens or hundreds of man-hours. The advantages of computers lie not in their ability to do things that individuals cannot, but in their ability to perform operations more rapidly and efficiently.

Conceptual Frames of Reference

Previous sections of this introduction have examined some very important questions related to behavioral research in political science: "How do we conduct inquiry?" "Why do we conduct it in this fashion?" and "What are the implications of this inquiry to the discipline of political science?" Remaining is one additional concern: "How do we integrate the findings from this inquiry into some over-all pattern useful in understanding politics?" We are referring here to the development of conceptual frames of reference that assist the researcher and his audience in understanding the relevance of certain conclusions arrived at through research. The framework itself provides and stipulates relationships and offers a basis for imparting meaning to findings about phenomena. Some of these frames of reference take the form of broadly based theories advanced in an attempt to explain general political phenomena; others are best considered as schemes suggested for describing a specific aspect of politics. Common to all frames of reference, however, seems to be the thought that use of these schemes assists in ordering one's knowledge for a clearer understanding of politics. Thus, the ease with which an investigator and his audience can understand the significance of an isolated finding in terms of a broader perspective depends in large part upon the adequacy of the conceptual frameworks available to the investigator.

Large numbers of conceptual frames of reference are available to the political scientist. Some of the more popular ones currently used in the discipline include systems analysis, structural-functional analysis, the group approach, and role analysis. Each of these has its own technique for ordering political reality and for integrating research into an over-all perspective on the nature and meaning of politics.

* * *

The following pages provide insights into the purposes, nature, methods, and frames of reference of behavioral inquiry with special reference to political science. Pertinent selections examine some of the important aspects of behavioral inquiry. These should provide students of politics with an understanding of behavioral inquiry so that they can evaluate its strong and weak points. The editors of this book are convinced that an understanding of these matters must precede any intelligent application of behavioral methods to the study of politics.

II POLITICS AS A FIELD OF INQUIRY

Since the emergence of organized societies, man has pondered questions about the nature of society and the authority it exercises over him. This examination has led him to delve into questions about the origins of states, the foundations of government, the sources of authority, and the nature of political power. The study of these and related topics generally has been called "political science." But the ambiguities inherent in this formulation fostered questions about what political science includes. Early students of politics tended to concentrate on topics related to the governing of men. This focus produced the idea of defining and categorizing political science according to the subject being examined, that is, as the study of the organization called government; however, as the discipline grew and expanded, the topics under study exceeded the narrow concern with government.

In recent years political science has undergone self-analysis, which manifested itself in a rebellion against restrictions imposed by a narrow definition of the subject matter. Many political scientists refused to accept the idea that they ought to confine themselves to topics such as governmental organization and political party structure and began to argue that government and politics were affected by activities that generally had been regarded as beyond the pale of the discipline. They took the position that a restricted outlook made the discipline narrow, sterile, and provincial, and they argued that political science should include all areas that might produce politically relevant findings. Although this trend generally has been welcomed, it has created problems in identifying the cohesive element in the discipline.

In the first selection, Vernon Van Dyke examines this problem by treating various definitions of the scope of the discipline. He concludes that contemporary political science cannot properly be defined in terms of a predetermined range of topics; rather, he argues that the scope of a discipline is set by the questions being asked by practitioneers in the field. This pragmatic point of view is shared by many political scientists because the topics currently being investigated in the discipline are so varied that they defy simple classification. When defining the scope of the discipline, Van Dyke suggests, the purposes of inquiry as well as the questions being posed should be taken into account.

In political science the question of purpose has been a vexing one for some years. Debate over this question has pervaded all parts of the discipline, and varying statements of purpose have been suggested. One source of conflict stems from the intellectual heritage provided by the humanities. This tie discourages acceptance of the argument prominent in the physical sciences that knowledge should be sought for its own sake, and promotes the idea that the purpose of political inquiry includes human improvement and advancement.

From the humanitarian perspective, H. B. Spencer compared the goal of political inquiry to that of the medical profession, postulating that diseases exist in the body politic as in the human anatomy. Further he indicated that the purpose of political inquiry should be to diagnose these illnesses and develop cures. The metaphors Spencer used on this point are illustrative of the analogy he sought to establish. "We political scientists are still baffled by the cancer of war, a marvelous energy gone wrong; also the common cold of voters' ignorance and apathy, which saps vital energy and leaves the system flabby and inert, an absolutely lethal effect upon democracy."[1]

Henry Dennison argued for a somewhat similar point of view; the model he selected was that of an engineer rather than a doctor. From this vantage point the political scientist would attempt to improve the social and political world by adjusting it, much like an engineer who modifies a machine in order to make it operate more efficiently. Thus political science should seek to imitate

> ... the engineering point of view, which focuses upon the natural material and psychological forces found in a given social group and the measures and structures of organization which can be

---

[1] H. B. Spencer, "Pathological Problems in Politics," *American Political Science Review*, 43 (February, 1949), 3.

applied to them in order to work toward its fundamental purposes. It is the approach, not of the historian or of the moralist, but of the student of applied science, the engineer.[2]

A more individualistic conception of the purpose of political inquiry was suggested by Clinton Rossiter, who argued that the goal should be to teach people about their government and to prepare them for their citizen roles.[3] This view stresses the humanitarian aspects of political knowledge, focusing on the importance of proper indoctrination. Only through such efforts, argues Rossiter, can the system maintain stability and the existing political values and beliefs be perpetuated. The role of political scientists in political indoctrination is to determine what information should be presented, what procedures are best suited for this purpose, and which persons can best serve as agents in this process.

A statement of a fourth purpose was made by Harold Lasswell when he spoke about the importance of policy science for the discipline. The goal of inquiry from this viewpoint is the discovery of substantive policy alternatives.[4]

Another outlook is articulated by several political scientists who consider their goal to be objective, value-free reporting about the political realm. They seek to examine and analyze political reality without permitting personal preferences or individual points of view to affect their perspective.

Implicit in all of these conceptions is one basic issue: Should political inquiry seek to provide a basis for attaining "the good life" politically or should it be limited to describing and analyzing political reality? Stated another way, the issue concerns the role of values in political research. During the 1940's this issue was debated by William F. Whyte, a sociologist, and John H. Hallowell, a political scientist, in *The American Political Science Review*. In his initial challenge to political science, Whyte criticized the discipline's reliance on philosophy and ethics as the core of the discipline and called for a greater interest in political reality. "Political scientists should take an interest in politics. They should leave ethics to the philosophers and concern themselves primarily with

---

[2]Henry Dennison, "The Need for the Development of Political Science Engineering," *American Political Science Review*, 42 (June, 1948), 242.

[3]Clinton Rossiter, "Political Science I and Political Indoctrination," *American Political Science Review*, 42 (June, 1948), 542–49.

[4]Harold D. Lasswell, "The Policy Orientation," in Daniel Lerner and Harold D. Lasswell (eds.), *The Policy Sciences: Recent Developments in Scope and Method* (Stanford, Calif.: Stanford University Press, 1951), pp. 3–15.

the description and analysis of political behavior."⁵ He further elaborated this point in a later essay.

> I believe that the only useful sort of conceptual scheme in the social sciences is one made up of elements which are subject to first-hand observation and/or experimentation. The ethical values of the researcher do not meet this requirement, and therefore have no place in the scheme.
>
> This does not mean . . . that I have no scientific interest in what people believe in. I recognize . . . that the values held by people are an important dynamic factor in their behavior. I look upon the values of the people I study (discovered through interviewing and observation) as important research data. I feel that my own values have no place in my analysis of human behavior, and I try to disregard them as much as I possibly can.
>
> That is a vital distinction to make. I can write objectively and scientifically about the values of the people I study. If I interpret other people's behavior in terms of my own values—the procedure followed by most political scientists—I am no longer a social scientist.
>
> I believe that the social scientist can be more useful if he confines his scholarly efforts entirely to observing and analyzing the functioning of the organizations of our society. . . . I also believe that, as a citizen, he has the same rights and obligations as other citizens, but that he should make special effort to keep his personal value judgments and advocacy of policy from intruding into his scientific research. . . . [He] can see the results of his work put into action, whereas the academic moralist can enjoy only the pleasures of self-expression.⁶

In a well-known reply to this challenge, Hallowell questioned the feasibility of undertaking "value-free, scientifically-oriented" inquiry.

---

⁵William Foote Whyte, "A Challenge to Political Scientists," *American Political Science Review*, 37 (August, 1943), 697. By permission of the author and the American Political Science Association.

⁶William Foote Whyte, "Politics and Ethics: A Reply to John H. Hallowell," *American Political Science Review*, 40 (April, 1946), 304 and 307. By permission of the author and the American Political Science Association.

> The recognition of facts requires not only sensory awareness but judgments as to value and significance. As a matter of fact, it is only by fitting the data made available to him by his senses into some preformulated conceptual scheme that the individual is able to perceive facts at all. Actually, then, when the positivist insists that to be properly scientific we must confine ourselves to a description of "positive facts" that can be observed without transcending our immediate sensory experience, he is insisting upon the impossible.[7]

Hallowell explored in unequivocal terms what he believed were the basic purposes of political inquiry.

> The only rational reason for seeking to understand political institutions and behavior is the possible human benefit that may be derived from such a study. If our interest in studying politics is unrelated to human welfare and to such concepts as justice and freedom, there is, of course, no reason why we should seek to describe political behavior in ethical terms. But if our interest in politics is motivated by an interest in human welfare, it becomes not only our right, but our responsibility, to detect corruption, to identify injustice and tyranny, to unmask sophistry, and to guide men along the paths most likely to yield happiness and freedom.
>
> To assist in the implementation of that good is, as I see it, the major function of political science.
>
> The social sciences are not so much in need of new research techniques as of convictions based upon principles. If they have become sterile . . . it is not because they are lacking in adequate research techniques, but because they are hesitant about affirming unequivocally the principles upon which a free and just society can be based. A social science that is not committed to the search for freedom and justice has already signed its own death warrant.[8]

---

[7] John H. Hallowell, "Politics and Ethics," *American Political Science Review*, 38 (August, 1944), 647. By permission of the author and the American Political Science Association.
[8] John H. Hallowell, "Politics and Ethics: A Rejoinder to William F. Whyte," *American Political Science Review*, 40 (April, 1946), 310–12. By permission of the author and the American Political Science Association.

This exchange illustrates the polarization of positions with regard to the role of values in political inquiry that had developed in the discipline of political science. After years of debate in journal articles and in books, most political scientists have rejected both extremes in favor of a middle position, which recognizes the pitfalls of subjective bias and at the same time acknowledges the central importance of values in political life.

A comprehensive statement of the role of values in current political science research was made by Charles S. Hyneman, in the second reading selection. The author points out that values have a legitimate place in inquiry, and he proceeds to discuss this place in a simple and straightforward manner. He offers several specific suggestions useful both to the neophyte and to the experienced political scientist for conducting inquiry while maintaining one's values. This selection illustrates ways for combining socially meaningful and behaviorally significant inquiry.

As the discussion above indicates, political science has experienced an evolution in the subjects being studied and in its approaches to inquiry. A thorough assessment of the status of the discipline and trends in contemporary political science has been undertaken by Albert Somit and Joseph Tanenhaus.[9] The selection included here, taken from their most recent book, evaluates the behavioral mood in the discipline since World War II. One theme in this selection is the evolution of the discipline toward a "paradigmatic" stage of development. This stage is characterized by the achievement of a state of knowledge, a level of theory, and an approach to problem solving that are sufficiently general to provide a basis for answering all the questions posed by practitioners in the discipline. Although the full achievement of this level of development has yet to be reached, Somit and Tanenhaus seem optimistic that political science is approaching this stage.

The three reading selections in Part II should give the reader a sense of the nature and status of contemporary political science. These selections depict, from the editors' point of view, current thought in the discipline with respect to the nature of the subject matter, the role of values, and trends affecting political science.

[9] See their two books, *The Development of American Political Science: From Burgess to Behavioralism* (Boston: Allyn & Bacon, Inc., 1967); and *American Political Science: A Profile of a Discipline* (New York: Atherton Press, 1964).

Vernon Van Dyke

The Scope of Political Science[1]

We can think of scope in terms of the kinds of questions we try to answer—the kinds of problems that we attack—and in terms of the kinds of data that we bring to bear. In these terms, no one would contend that the scope of political science is fixed very sharply or that any boundaries are at all sacrosanct. We have somewhat different conceptions of the subject of our inquiries, variously describing it as politics or government or the policy process or the political system; and each label carries with it some probable implications for the scope of our work. Moreover, we pursue various purposes, are guided by different theories, and employ various approaches and methods; and again the choices that we make are interrelated with the selection of questions and data, and so affect scope.[2]

Since so many factors are involved, a discussion of scope could be organized in various ways. I propose to focus mainly on conceptions of our field of inquiry and their probable implications. I might note that a very large volume—I think an ever increasing volume—of material on political science is being published, and that we do not

[1] For comments and suggestions elicited by an earlier draft of this paper I wish to thank Karl von Vorys of the University of Pennsylvania and my colleagues at Iowa, especially G. Robert Boynton, Lane Davis, Samuel C. Patterson, and Joseph Tanenhaus.
[2] Charles S. Hyneman, *The Study of Politics* (Urbana: University of Illinois Press, 1959), p. 21.

Reprinted from James C. Charlesworth, ed., *A Design for Political Science: Scope, Objectives, and Methods* (Philadelphia: The American Academy of Political and Social Science, December, 1966), pp. 1–17, where it appeared under the title, "The Optimum Scope of Political Science." By permission of the author and the publisher.

have any very reliable classificatory or mapping system by which to tell what terrain is being covered or left unexplored. This means that it is difficult to say what the actual scope of political science is; we have to rely on our impressions. Impressions that our coverage is inadequate have led to recommendations for expanding the scope of the field, and I propose to take up some of them.

THE IMPLICATIONS OF A FOCUS ON POLITICS

It is common to say that *politics* is the subject of political science. More to the point, perhaps, is the fact that it is often simply taken for granted that *politics* is the subject. I am among those who have done this. This means that a definition of politics at least suggests the outer limits of the field, whether or not everything within those limits actually gets covered.

Edward C. Banfield's definition of politics and one that I have advanced can both be cited to serve our present purposes. Banfield's focus is on "actors (both persons and formal organizations) who are oriented toward the attainment of ends." When actors pursue conflicting ends, an issue exists. And "politics is the activity (negotiation, argument, discussion, application of force, persuasion, etc.) by which an issue is agitated or settled."[3] Without at the time knowing of Banfield's definition, I advanced one in 1960 that is somewhat similar. In its short version it is that politics consists of struggle among actors pursuing conflicting desires on public issues.[4] The principal differences are that whereas Banfield's stress is on "activity," mine is on a particular kind of activity, struggle; and whereas he does not say what kind of issues are political, I specify that they must be "public"—that is, that they must have to do with group policy,

[3] Martin Myerson and Edward C. Banfield, *Politics, Planning and the Public Interest* (Glencoe, Ill.: Free Press, 1955), pp. 304–305.
[4] Vernon Van Dyke, *Political Science: A Philosophical Analysis* (Stanford, Calif.: Stanford University Press, 1960), p. 134.

group organization, group leadership, or the conduct or regulation of intergroup relationships.

These definitions seem to me to imply, with respect to the scope of political science, that we should try, among other things, to identify political actors; to identify and clarify the goals that they seek, and perhaps recommend alternatives or at least call attention to them; to analyze interrelationships among ends; to describe and assess the means that are or might be employed in the pursuit of ends; to determine the reasons and causes influencing the choice of ends and means; and, in general, to find out why political issues arise, how the struggle over them is conducted, what governs the outcome, and what the effects of the struggle are or may be. Studies pursued under this conception can be either particular or general. In fact, some focus on specific actors and issues and are more or less policy-oriented; others, though perhaps achieving their ultimate justification in their contribution to policy-making, aim directly at general knowledge relating to classes of actors or kinds of problems.

Though these definitions have merit, they lead to certain difficulties and do not give an entirely satisfying answer to the question of the optimum scope of political science. My definition has been attacked for the focus on struggle. Banfield's wording is less open to question on this score, though the difference is not very great. His focus is also on issues, which implies that controversy, if not struggle, is an essential feature of politics. This, in fact, suggests one of the important questions about both his conception and mine—whether political scientists do or should restrict themselves to issues. It also suggests the question which issues are political and whether all political issues automatically fall within the scope of political science. Perhaps I should add that I am not raising the question whether we as scholars should always address ourselves to precisely the issues faced by practitioners; this would be an absurd rule. The question is whether we should define our field of inquiry in terms of issues, tackling whatever intellectual problems we choose within that field.

One of the obvious implications of a focus on politics defined in terms of issues or controversy is that a very substantial portion of the activities of government are placed beyond the purview of political science. A great deal that goes on in government, and in intergovernmental relations, occurs on the basis of consent. A United States Senator once told me that 90 per cent of the matters that the Senate concerned itself with were nonpolitical; he meant, I am sure, that there was no real controversy over them. The "struggle," if there was one, was to ascertain the relevant facts and perhaps to induce the appropriate persons or agencies to give the problem attention; once this was done, there would likely be consensus on the appropriate action to take. Whether or not 90 is the correct percentage, it is obvious that there is considerable truth in what the Senator said. Congress devotes a significant portion of its time to overseeing, in a nonpolitical way, activities of the executive branch of the government that are themselves generally regarded as nonpolitical. Issues may arise, but many of them relate to questions of fact or judgment rather than to political demands. The same kind of thing happens at the state and local level—probably to an even greater extent. The development and operation of the public educational system are major activities of state and local governments; as the struggle over integration attests, political issues sometimes arise in this field, but still the general record is one of the nonpolitical treatment of educational problems. In the little suburban municipality in which I live, elections are nonpartisan and frequently noncompetitive, and votes on the town council are ordinarily unanimous. Similarly, many aspects of county government are managerial or administrative; politics may determine who holds the offices, but not what they do once they are elected.

Schattschneider points out another difficulty in the notion that political science is the study of politics and that politics centers on issues. To be sure, he takes the view that "the central problem in politics is the management of conflict," but he reminds us that conflict can sometimes be forestalled by preventing issues from arising or by getting them subordinated, obscured, or rendered obsolete. "Some issues are organized into politics while others are organized out."[5] Bachrach and Baratz elaborate on the point: "To the extent that a person or group—consciously or unconsciously—creates or reinforces barriers to the public airing of policy conflicts, that person or group has power." They go on to speak of the possibility that a political actor may have less to do with decision-making than with nondecision-making; that is, he may "limit decision-making to relatively noncontroversial matters, by influencing community values and political procedures and rituals."[6] Of course, action to prevent issues from arising is itself, in a sense, a form of struggle, but the definitions of politics in question are much more likely to suggest open and direct struggle and controversy than subtle strategies that keep issues from coming to the fore.

Apart from the question whether we as political scientists should restrict ourselves to the study of struggle over issues is the question which issues are political and whether all political issues automatically fall within the scope of political science. Suppose, for example, we think of the practice of political scientists concerning research on crime, rioting, revolutions, and war. Most of us, I believe, do not regard it as part of our job to explain crime on the part of private individuals or to try to find out how to deter or to combat it, even though crime is a challenge to law and order and even though the effort to repress it is an important part of governmental activity. In contrast, we do regard it as part of our job to explain war and to try to find out how to deter or combat it. Further, we tend to ignore rioting and to neglect grossly revolutions and civil wars.

[5] E. E. Schattschneider, *The Semisovereign People* (New York: Holt, Rinehart and Winston, 1960), p. 71.
[6] Peter Bachrach and Morton S. Baratz, "Two Faces of Power," *American Political Science Review*, 56 (December, 1962), 949.

What is the rationale of our behavior? One of the rules has just been discussed: that we are commonly not much concerned in the absence of a public issue. And ordinarily there is no public issue about crime on the part of private individuals. The relevant laws are ordinarily not in dispute; and neither are the duties of the police or of the court. So we concede the study of crime to sociologists, lawyers, and others. But the rule concerning the presence or absence of a public issue does not provide a full answer. The question of capital punishment, for example, has sometimes become a minor political issue; but even so, political scientists give it little or no attention, except perhaps in connection with the study of the more general problems relating to the governmental process. My supposition is that we have tacitly added a rule to the one regarding controversy: that where we have conceded a subject to another discipline because it is generally nonpolitical, we tend to maintain the division of labor even when the subject enters the political arena.

No problem arises, of course, about the reason for our concern with war; it is obviously a struggle among actors pursuing conflicting desires on public issues.

Why do we neglect rioting and revolution? The rule that we concern ourselves only with actual or prospective public issues provides part of the answer. Many riots and many mob actions belong in the same category with crime in that no public issue is involved. But this can scarcely be a full explanation. Some riots are clearly the work of actors engaged in struggle on public issues, and all civil wars and revolutions are in this category. I suspect that our neglect of the subject should be treated in part as a lapse. Those who take their cues from the provisions of the Constitution and from our institutional arrangements have not been guided to the subject of riots and revolutions. Moreover, political life in the United States has been so stable for so long that the subject has not come urgently to our attention. Greater national involvement in the life of other countries is changing this situation;

and so, perhaps, is the racial problem that we confront at home.

I should note that problems relating to riots and revolutions can be turned around and stated so as to call for inquiry into the conditions making for the establishment and preservation of stable and effective government. In fact, this is a broader way of stating the problem, for it encompasses studies of such subjects as nation-building, socialization, and political mobilization. As long as American political scientists were focusing their efforts on the United States and western Europe, these subjects got little attention, but developments in the world in the last decade or so have brought significant expansion in the area of our concern.[7]

The question which issues are political, and which ones come within the scope of political science, is somewhat complicated by such adjectives as economic, legal, and military. Obviously, economic, legal, and military issues may also be political, but even so the use of these adjectives suggests that study of them calls for a special expertise that political scientists may not have. We are especially inclined to concede economic issues to economists. We rarely take up questions concerning tax and fiscal policy, except perhaps where budgets and budget-making are directly involved. We provide very little of the knowledge on which the Council of Economic Advisers has to draw. When governments go more and more into the planning of their economies, thus expanding the scope of their activities, we do not expand the scope of political science accordingly. In so far as government and economic life are concerned, the scope of political science is about the same whether the political system studied is that of Spain, the Soviet Union, or the United States. Some political scientists are unhappy about this situation, but few are doing anything to change it. I might note incidentally that several years ago when I was working on the rationale of the American space program, I ran on to a lively controversy on the subject of United States patent policy. The National Aeronautics and Space Administration (NASA) was contracting out research to private firms. When the private firm, using government money, made a patentable discovery, who was or should have been entitled to the patent? Congressional hearings occurred on the question; administrative rulings were made; and legislation was debated. By all the definitions of *political* that I know, the issue was political. But I still wonder whether it fits within the scope of political science and by what overriding criterion it is included or excluded.

Political scientists have not been quite so much inclined to steer clear of a question because it is called legal or judicial, probably because constitutional law has been part of the field from the first. The overlapping of the fields is recognized and accepted. But where the overlapping does or should end is somewhat uncertain. Is jurisprudence, for example, an appropriate part of political science? Is administrative law?

In recent years political scientists—especially some of those concerned with international politics—have been less and less intimidated by the adjective *military*. From our ranks come some of the leading students of such problems as deterrence and limited war. But still there are problems. We have not so far given an unqualified affirmative answer to the question whether all aspects of military strategy come within our field. We have begun to study military strategy when it has a bearing on "political" objectives; but this very way of thinking about the subject implies that some questions of strategy are outside the scope of political science. Perhaps the answer is to be found in our attitude toward crime and criminals. When the goal and the kind of methods to use are both agreed upon, we are willing to leave the subject to others. And in the case of many questions relating to military affairs

---

[7] One of the more recent and novel studies on this subject is Karl von Vorys, *Political Development in Pakistan* (Princeton, N.J.: Princeton University Press, 1965).

and military strategy, this means leaving the subject without an academic home, save in a very few institutions.

Questions arise in the field of human rights that illustrate the problem further. The United States and most other countries have endorsed a Declaration of Human Rights, and the question of concluding one or more international covenants has been debated for many years. Undoubtedly, the general issue of human rights and of national and international action with regard to them comes within the scope of political science. But human rights are said to include the rights of persons of full age to marry and to found a family, the right to equal pay for equal work, the right to periodic holidays with pay, and many other rights. Then do the laws and practices of various countries regarding these rights become suitable subjects for political scientists to inquire into? Should political science journals carry articles on attitudes in various countries concerning the "full age" for marriage and concerning the policies of Liberia relating to forced labor recruitment? Along the same line, consider legislation in the United States concerning marriage and divorce, and concerning birth control. These are areas in which political issues arise, and it would undoubtedly be appropriate for a political scientist to study the kinds of activities in which people engage in order to get the issue resolved as they wish. But would it be appropriate to engage in studies of the age at which marriage occurs, the conditions giving rise to divorce, and the moral and religious questions on birth control?

Analogous problems come up in connection with the study of motives of political behavior and of the psychological effects of political practices and legal rules. Those who try to explain why people vote as they do, or to explain why some people seek political careers, or why governmental leaders follow policies that precipitate or avert war sometimes find psychological or psychiatric data useful. Similarly, in interpreting constitutional provisions bearing on racial discrimination, the Supreme Court has considered the relationship between discrimination and personality development. Does this mean, then, that political scientists should join psychologists and psychiatrists in experiments concerning human motivation and in analyses of the factors that help shape personality?

One of the facts that we are facing here is contained in the extreme statement that everything is related to everything else. We can start out examining what is undoubtedly a political question and undoubtedly a part of political science. But if we pursue the subject very far we may soon find ourselves in the midst of another discipline—or simply in unclaimed territory. And I am not aware of any clear-cut rule for determining whether or at what point we have crossed the outer limits of political science. Even if we say that our central concern is with the struggle over issues, we, in fact, seem to add a rule of convenience or expediency: that we should concede even political issues to scholars in other disciplines—especially to economists—whenever their expertise is especially relevant. Moreover, except where the issue concerns the rules, procedures, and methods of political struggle, we tend to leave it to scholars in other disciplines to supply the data—sociological, economic, psychological—that explain why the issue arose and what consequences follow from any settlement that is reached.

I should, perhaps, add two final notes on the question of the implications for scope of a focus on politics. The first is that the definition of politics with which one starts is obviously of crucial significance. If, instead of using the definitions that I have cited, we were to say, for example, that politics is a struggle for power or a clash of interest groups, somewhat different consequences would ensue. I believe that the difficulties would be greater. The second is that when we talk about politics we pretty regularly mean the politics associated with public government—the kind of government that goes with sovereignty. We may speak of the politics of private groups, but only in exceptional cases is this the kind of politics with which we are concerned.

The Implications of a Focus on Government

If we object to a focus on politics in part because it leaves such a wide range of governmental activity beyond our purview, one possible solution would be to jump the other way and say that our concern is with government. Charles S. Hyneman thinks that, in fact, this is our prime concern. "The central point of attention in American political science . . . is that part of the affairs of the state which centers in government, and that kind or part of government which speaks through law."[8] For short, he speaks of "legal government" as the focus, meaning public as distinct from private government.

Hyneman himself stresses the magnitude of our tasks if we commit ourselves to the idea that legal or public government is the terrain that we should explore.

> Certain controversial subjects of domestic policy in the United States, especially agricultural policy, water resources policy, and a number of matters generally entitled government-labor-business, have come in for increased attention in recent years. But relatively little effort has been made by American political scientists to describe policies adopted in many other highly important areas of public concern, such as health, education, marriage and divorce, crime, and conduct which lies at the edges of legality. . . . As yet there has not appeared a rationale that seems to guide political scientists in the selection of policies to be described.[9]

At other points Hyneman names still other subjects associated with government that political scientists have not explored.

[8]Hyneman, *The Study of Politics*, pp. 26-27.
[9]Hyneman, *The Study of Politics*, p. 39.

Obviously, the statement that we do or should focus on public government eliminates some of the difficulties that attend a focus on issues. At the same time, it does not eliminate them all; for example, it does not eliminate the problem of determining how far into related disciplines political scientists should go in providing the substantive knowledge on which various kinds of decisions should be based. Moreover, it introduces some new difficulties. Though permitting attention to consensual politics, it suggests no limit to the kinds of questions relating to government with which political scientists should be concerned. The United States Weather Bureau, the Food and Drug Administration, the National Institutes of Health, the Tennessee Valley Authority, the Government Printing Office, and the Foreign Service Institute of the Department of State are all parts of public government. If public or legal government is our subject, with what questions concerning these agencies should political scientists deal, and what questions concerning them are outside the optimum scope of our field?

If a focus on legal government opens up too vast a terrain, on the one hand, it traditionally has had connotations that are too narrow, on the other. Traditionally, it has connoted a description of institutional arrangements and of governmental policies and functions, offered with a minimum of attention to theory; too often the descriptions have been formal and legalistic, neglectful of the informal processes and channels through which influence is brought to bear; and too often they have been associated with reformist zeal. Further, those who have accepted government as their focus have tended pretty regularly to neglect international politics, and such attention as they have given to the subject has tended to go to international law and organization. Even today, given the traditional outlook that goes with a focus on government, public law usually means domestic law. The legislative process is usually assumed to be purely domestic. Comparative politics calls for the comparison of domestic political systems, not the comparison of the domestic and the international. The subdivision of political science called political theory concerns itself almost entirely with government within states and with relationships between government and individuals or groups; from the days of Plato and Aristotle on down, political theorists have given very little attention to international politics. Obviously, I think there is too much of a tendency to put international politics off in a compartment by itself, and this leads me to question the use of *government* as the defining term for political science, for the term is associated with the practice of compartmentalization. I might add that those who focus on international politics have not always promoted their own integration into political science. The very fact that they speak of international *relations* creates problems, for many relationships are nonpolitical. I wonder what we would include in the study of domestic politics if we said that the subject encompassed domestic relations.

THE IMPLICATIONS OF A FOCUS ON
THE POLICY PROCESS

Rather than focus on issues or on government, some political scientists choose to focus on the policy process; or they say that political science is or should be a policy science.[10] Like almost all our terms, these are rather vague, and different scholars construe them differently. Reference to the policy process may mean the process by which issues are resolved—the decision-making process. If an issue arises, the object of those focusing on the policy process is to identify the actors and interests, the legal and institutional arrangements, and the conditions, methods, and procedures that relate to the outcome, and to analyze and perhaps assess their relative roles or influence in shaping the outcome. General knowledge of the nature of the process and the factors affecting its functioning is what is sought. The reasons why the issue arose, the relative merits of different solutions, and the consequences of the decision that is made are, I assume, put on or beyond the periphery of the field of political science.

The notion that our focus should be on the policy process has a good deal in common with the notion that it should be on politics, defined as a struggle over issues, and many of the comments already made apply again. In addition, the term *policy process* connotes a narrower range of interest than the term *politics*, pretty much excluding concern for the substance of policy. A political scientist who says that his concern is with the policy process could probably answer my question whether NASA's patent policy is really within our field; he would not concern himself directly with the problem of equity that is involved or with the relationship between patent policy and technological progress; but he would concern himself with the actors engaged in the struggle—including their claims regarding equity and progress—and the methods and procedures by which the struggle is resolved. Similarly, the student of the policy process would be unlikely to study the substance of the foreign policies pursued by a government, or the actual or probable effect of armaments or disarmament on the relationship of states. If he gives any attention at all to international politics, his concern is likely to be with the formulation of policies by governments and with the formulation of policies in international organizations.

The term *policy science* has a broader connotation than the term *policy process*, permitting various kinds of studies—descriptive and prescriptive, if not normative—relating to politics. It does very little, however, to indicate limits. It might, for example, suggest a study of the rules and principles followed by the National Science Foundation (NSF) and the National Institutes of Health (NIH) in making grants.

[10]Hyneman, *The Study of Politics*, pp. 100–108, 165–173.

For that matter, it is not at all clear that the term *policy process* would exclude such matters.

THE IMPLICATIONS OF A FOCUS ON THE POLITICAL SYSTEM

The term *the political system* is so thoroughly identified with David Easton that we should focus on his conceptions and contributions. They are distinctive and impressive, but at the same time they do not resolve all difficulties. It would be too much to expect this of any approach or analytical framework. When Easton says that our concern is with "that system of interactions in any society through which . . . binding or authoritive allocations are made and implemented,"[11] he does a good deal to indicate an appropriate scope for political science. This way of conceptualizing the field reduces the difficulties that attend a focus on politics defined as struggle over issues, for the authoritative allocations of values could be tacit, unattended by current struggle. Easton's definition encompasses the politics of consent as well as the politics of struggle. But it also seems to encompass economic decisions and policies that allocate values authoritatively, and leaves the political scientist uncertain what the division of labor between him and the economist is or should be. More generally, Easton's conceptual scheme does not do much to resolve the question of the extent to which political scientists must also be economists, sociologists, psychologists, and military strategists.

One of the merits of Easton's framework for analysis is the effort to make it applicable to politics at every level. As is already clear, I consider this highly desirable, and I suppose that his scheme could, in fact, be applied to international as well as to domestic politics. At the same time, the fact shows up pervasively in his books that he writes with domestic politics primarily in mind, and the question comes up persistently whether his formulations are as apt or as suitable as they might be for inquiry into the international field. I am fearful that his scheme will, in practice, prove much more readily suitable to the study of domestic politics and that it will thus perpetuate if not accentuate the tendency to treat the two fields separately and differently.

PROPOSALS TO EXPAND THE SCOPE OF POLITICAL SCIENCE

Neither Easton's conceptual scheme nor the definitions cited above call clearly for any change in the present vague but widely accepted attitudes about the scope of political science. Some suggestions along this line have been made, however, most of them relating to the border region between political science and economics. One argument is that as the role of government in economic life expands, the scope of political science should also expand. Another is that we should pay attention to some of the interactions that Easton describes as parapolitical; more specifically, that we should pay attention to private government as well as to public government, that is, to the great corporations which, in American life, make or share in making decisions that have widespread effects.

Michael Reagan makes both of these arguments. With the United States in mind, he speaks of the "merger of the public and private," which manifests itself in several forms: in the Communication Satellite Corporation which is, in part, government-owned; in the insertion in government contracts of terms giving effect to public policy, for example, fair employment practices; in interrelationships between federal regulatory agencies and the industries that they are supposed to regulate; and in the fact of public political pressures (perhaps through congressional hearings) on private business practices.[12]

---

[11]David Easton, *A Framework for Political Analysis* (Englewood Cliffs, N. J.: Prentice-Hall, 1965), p. 50.

[12]Michael D. Reagan, *The Managed Economy* (New York: Oxford University Press, 1963), pp. 190-210.

A number of other developments might also be cited that confuse the distinction between the public and the private. When the United States Mediation Service plays a prominent role in negotiations between management and labor—and when the President of the United States becomes personally involved—are the negotiations private and nonpolitical or public and political? When scientific research is paid for pretty largely by the government, to what extent does the determination of its nature and direction become a matter of public policy? These questions pertain to the United States, but we might note that in some other countries—above all in those that have collectivized their economies—the merger of the private and the public has gone much farther. Reagan's admonition is that "the skills of the political scientist and those of the economist must be fused if adequate analysis of the system and the appropriate prescriptions for public policy are to be developed."

Reagan also points to the widespread effects of decisions made by some of our great private corporations. What General Motors does in the pursuit of its interests clearly affects the public interest. Reagan is even concerned about the sense of social responsibility that some great corporations have developed and about the philanthropy in which they engage, suggesting that we may be heading into a paternalistic system where leading roles are played by benevolent feudal lords. And he urges that political scientists concern themselves with these matters. "The rise of the managed economy reunites politics and economics into a truly political economy."[13] We should be "concerned with popular control over economic decisions affecting the public, whether these are made in the public sector or by the private 'governments' of the corporations." "Public decisions," he argues, are not only those made by government, but those that affect the public; and concern for public decisions, so defined, requires "consideration of the relationship of business management to the people, as well as the relationship of government and the people."[14]

Robert Dahl raises somewhat similar questions. He thinks that it would be an "arid enterprise" to debate the question whether political science should or should not encompass the politics of business firms and of relationships between such firms, but he surveys the state of our knowledge about these questions nevertheless. And he does the same for questions concerning relationships between public government and business.[15] Further, he calls attention to a no man's land that has grown up between political science and economics, and argues that if a theory is to be developed that will help provide answers to questions arising from "the diversity and complexity of contemporary economic orders," economists are likely to have to give more attention to political science, and political scientists are likely to have to give more attention to economics.[16]

Apart from the question of the relationship between political science and economics, other questions have been raised, explicitly or implicitly, about the traditional scope of political science. Oliver Garceau went beyond it when he wrote about *The Political Life of the American Medical Association*. Masters, Salisbury, and Eliot have concerned themselves with *State Politics and the Public Schools*. And in recent years various studies of organizations, organizational behavior, and leadership have proceeded without much regard for a distinction between the political and the parapolitical. Grant McConnell has discussed "The Spirit of Private Government,"[17] focusing on trade unions; but the distinctions that he finds between private and public government

[13] Reagan, *The Managed Economy*, p. 19.
[14] Reagan, *The Managed Economy*, p. 214.
[15] Robert A. Dahl, "Business and Politics: A Critical Appraisal of Political Science," in Robert A. Dahl, Mason Haire, and Paul F. Lazarsfeld (eds.), *Social Science Research on Business: Product and Potential* (New York: Columbia University Press, 1959), p. 5.
[16] Dahl, "Business and Politics: . . ." *Social Science Research on Business: Product and Potential*, pp. 16-17.
[17] *American Political Science Review*, 52 (September, 1958), 754-770.

are so great as to raise the question whether worth-while analogies and comparisons can be made.

CONCLUSION

Obviously, this review does not lead to a clear-cut conclusion either about the actual scope of political science or about its optimum scope. I have dwelt on some of the different conceptions of the nature of the subject that we study, and I think it is clear that each conception carries somewhat different implications for scope. I doubt whether it would be justifiable to describe any of them as either wrong or right. I accept Easton's view that

> no one way of conceptualizing any major area of human behavior will do full justice to all its variety and complexity. Each type of theoretical orientation brings to the surface a different set of problems, provides unique insights and emphases, and thereby makes it possible for alternative and even competing theories to be equally and simultaneously useful.[18]

Further, optimum scope is determined not only by conceptions of the subject of inquiry but also by the kinds of purposes that the scholar pursues and the kinds of methods that he is willing to employ. It would thus be inappropriate for me to make a recommendation concerning optimum scope until questions pertaining to purpose and method have been resolved.

[18] Easton, *A Framework for Political Analysis*, p. 23.

Charles S. Hyneman

Values in Political Inquiry

It will not greatly misrepresent the actual state of affairs to say that two wings of the political science profession in this country wage a continuing battle on the issue of science vs. values. Like so many other controversies, this one can be partly erased by obtaining agreement on use of terms, especially on the meaning to be given the term "value." But the controversy is by no means

Reprinted from *The Study of Politics: The Present State of American Political Science* (Urbana, Ill.: University of Illinois Press, 1959), pp. 174–192, where it appeared as Chapter X, "How Shall We Treat Values?" By permission of the author and the publisher.

wholly semantic. When there is agreement as to what should be called value, there remains a clash between conviction and commitment. The division into opposing camps begins with the very general question of whether we should study values at all, goes on to questions about what kind of study we should make (if we are going to study values) and what are the best methods of making such studies, and comes to its sharpest focus and greatest heat on questions about how to handle personal preference in scholarly literature. We shall examine the debate under three heads. (1) What is the legitimate place of values in political science? (2) What kinds of value study should we make? (3) How should personal preference be handled in scholarly writing?

WHAT IS THE LEGITIMATE PLACE OF VALUES IN POLITICAL SCIENCE?

This is an issue not to be settled in our time. The history of scholarly writing about values makes that an unavoidable conclusion. The most we can hope to do here is locate some main points where battle is joined and venture some judgments about why the combatants refuse to make peace. In such a venture the one who conducts the tour can easily get hurt, no matter how great his care to establish neutrality. . . .

What I have to say settles down to four general observations.

First, failure to agree on the place of values in the study of government and politics is in no small part due to differences in meaning which are put into the word "values." The narrowest meaning for the word makes it applicable only to something thought good for its own sake, to something which wholly supplies its own justification. Matters of this sort were called esthetic values earlier in this essay. Some people call them ultimate or final values.

But the term value is also given a very different application in our conversation and in our literature. In this usage, values are particular to the discourse. In one piece of writing the author examines alternative provisions for election of legislators, seeking to determine their effectiveness for achieving popular control of government. In this item of literature popular control of government is an end and electoral arrangements are means; we also say that in this particular discourse popular control of government is a value and the several provisions for electing legislators are not values. But in a second piece of writing, the author may have fixed as an end the establishment of government whose policies and acts are most widely and most readily complied with, and the purpose of his inquiry may be to determine whether popular control of government is an effective means for achieving this end. What was a value in the first item of literature is thus not a value in the second item. We can easily imagine other particular inquiries in which acceptability of governmental acts to publics is examined as possible means for still other ends, and imagine still other inquiries in which one or another specific arrangement for selecting legislators is viewed as an end and therefore, for that particular discourse, treated as a value.

It is readily seen that talk and writing about the place of values in our scholarly enterprise can become badly confused because of difference in the way "value" is used. The political scientist who says he wholly rejects values as a proper object of study may mean to say only that we ought not be concerned with ultimate, final, esthetic values. Or he may have in mind some earlier point where our inquiry ought to be cut off—any of a number of possible stopping points on the continuum of intermediate or instrumental values ranging from small matters commonly thought of only as means through more inclusive matters which are means in one discourse but ends in another. The same uncertainties exist about the intentions of political scientists who say study of values is a high-priority obligation for the profession. Such a person may intend to say that we should inquire into final, ultimate values. Or he too may have in mind a stopping point earlier on the continuum of instrumental values.

Second, in order to reduce or eliminate misunderstanding which arises out of different uses of the term, it is sometimes proposed that the word value be applied only to ends which are thought to provide their own justification; that things which are acknowledged to contribute to some other end never be called values. This recommendation receives support from some thoughtful individuals who have a strategic objective in mind. They fear that the word value, no matter what the specific referent, will call up in many minds all the implications of esthetic value. To call something a value is to invite the careless thinker to say: "I do not have to justify it. It is enough for you to know that my preferences, my tastes, my sense of right and wrong make it attractive to me. Don't ask me to prove that it is good or bad; values cannot be established by proof." But a thing, no matter how greatly cherished, which is a means to other ends can be justified by proof that it is an effective means to other ends which are also cherished. And no doubt many things which are treated in our literature as ends requiring no justification can be shown by thoughtful analysis to advance or to inhibit realization of other ends which are cherished more highly than this intermediate "value."

In the view of many political scientists, much of the literature produced by other political scientists falls into just this kind of trap. The searching inquiry into what a thing is valuable for, what it contributes to, where it stands in a structure of things valued—such a searching inquiry is avoided because what is actually a means to other ends is treated as if it were an end which provides its own justification. The word value thus becomes a refuge for laziness. A first step to purge our discipline of this refuge, it is argued, is to quit calling things values unless we first convince ourselves that the thing is not a means to other ends; that it does indeed supply its whole justification; that it is desirable for its own sake and not for the sake of something else.

Third, even when there is most careful effort to agree on use of terms, difficulty may arise because of different conceptions of the nature of value. Certainly many political scientists presume that ultimate, final, esthetic values exist and can be identified. Perhaps some who fall in this group are hopeful that the profession might generally agree as to what, for great numbers of people, these ultimate values are. Political scientists who are committed to an idea of ultimate values disagree among themselves as to whether values should be objects of inquiry for political scientists, or if values are inquired into, what kinds of inquiry ought to be made.

Dispute which starts within this sector of the profession is further complicated when issues are joined by those political scientists who reject the notion of ultimate values. I . . . doubt that one can identify any end which provides its whole justification, [it being my] view that anything identifiable acquires its value from the support which it contributes to other things which are valued. I think it likely that a great many American political scientists are in a similar position. If so, they do not always make their position clear. Debate is not likely to advance understanding and alter conviction if it is carried on between parties, neither of whom knows quite what the other is talking about. This appears to be the case much of the time when the argument is about values in political science. Those on one side, taking for granted that everyone believes in ultimate values, fail to say where they stand on this matter; those on the other side doubt that ultimate values exist but do not bother to announce their unusual position.

Fourth, convictions about the proper place of values in political science are greatly affected by methodological commitments. The pro-science wing of the profession thinks political scientists are too much concerned about values; those who have least confidence in science as a way of studying human relationships think political scientists should continue or even increase their attention to values. I think it no exaggeration to say that the controversy about scientific emphasis in our study and the controversy about study of values are essentially one dispute, that science and values are opposite poles in a single area of intellectual conflict. . . .

Our examination, in the next few pages, of controversy over these matters need not be tripped up by the variances in concepts and labelling of concepts discussed above. Some of the statements credited to protagonists in the following paragraphs may have specific application to values which are thought to provide their own justification; other statements, to values acknowledged to be instrumental. This will not generally be the case, however. Regardless of specific referents, the positions and contentions identified in the next paragraphs are, fundamentally, equally applicable to esthetic values and to those instrumental values which, though means to other ends, stand well up the continuum toward esthetic values.

Occasionally one hears a political scientist assert that he would absolutely exclude values from any consideration by political scientists, exclude them wholly and completely from consideration now and at any future time. Presumably he is thinking only of values which provide their own justification and not of instrumental values which are means to identifiable ends. Quite certainly, also, this is a man who thinks our highest duty, if not our only duty, is to make studies of a scientific character. Less extreme positions by those who would depress attention to values and step up the emphasis on scientific inquiry are revealed in statements like the following. "Writing which extols or deplores particular values and urges people to line up for or against particular values ought to be excluded from political science, but we ought to continue writing which identifies what men value, describes value systems, and provides value analysis." Or, "Identification of values, description of value systems, and value analysis are proper activities for political scientists now and in the future, but study directed to these objectives ought to be carried on in closest compliance with scientific method." Or, "Study directed to these objectives is something we should get around to in the future, but such study is premature at this time because we have not yet developed proficiency in scientific study sufficient for effective analysis of such complicated packages of data."

I doubt that any political scientist denies the usefulness to society of literature directed to the three objectives differentiated above. The most any of them will ask for is that political scientists not engage in its production. The same goes for those who think study of values as ends is premature at this time, or think it is all right to go ahead with such study now if we confine our attention to what can be examined scientifically. They do not deny the usefulness to society of literature which extols and deplores, states personal preference, and issues calls to arms; they say only that certain kinds of writing have a legitimate place in political science and others do not. All these critics take their various positions because they think the primary, if not the whole, obligation of political scientists is to carry on studies which eventually settle questions by presentation of proof. Non-scientific inquiry, if widely practiced, diverts from our primary goal too much of the time of the few who are available for scholarly study of political science. Furthermore, extensive and serious preoccupation with non-scientific inquiry nurtures states of mind antagonistic to, if not wholly incompatible with, scientific inquiry. For one who believes these things it is a logical conclusion that political scientists will best serve the society that maintains them if they put aside a kind of writing that has up to now constituted a sizable part of their literature.

This reasoning naturally has little appeal for other political scientists whose methodological commitments are not those of science. Denying that study of legal governments should be limited to what science is good for, they see no reason for withdrawing attention from values, whether instrumental or esthetic. Political science has never been wholly confined or even mainly limited to scientific inquiry, they argue, and it will be a sad day if it ever becomes so restricted. Efforts to state what men value and why they value those things, extolling and deploring, and urging men to take their positions are a central concern of many of the writers whose work we honor as classics of political science. To terminate the stream of literature having this major purpose would be an act of mayhem which political scientists should be the last to propose. It has not been proven that a discipline cannot accommodate both scientific inquiry and extolling and deploring of values; if incompatibility of the two types of inquiry should be proven, it is at least arguable that we should continue inquiry into values and leave scientific studies to others.

What Kinds of Value Study Should We Make?

It is obvious that argument about whether political scientists should study values at all cannot get very far without going into questions of what kind of study should be made. The range of possible approaches to problems of value is too great for anything like a complete examination here. We shall confine our attention at this point to two centers of intellectual conflict: (1) the issue posed by the contention that study which attempts to identify values, describe value systems, or provide value analysis should meet the highest tests of scientific method; and (2) the issue posed by the contention that we place too little emphasis on value analysis. A third center of dispute is reserved for discussion in a separate major division of this chapter—writing in which particular values are extolled or deplored and in which people are urged to take their stand for or against particular values.

1. Scientific Study of Values

Political scientists who are attached to scientific method do not all line up in opposition to study of values. Many who are in the pro-science wing of the profession insist that values ought to be examined and that they can be examined in a scientific way. The determination of what men value, the description of value systems, and value analysis, they argue, are best treated (or only treated well) by scientific study. To find out what men value, you search for evidence. You find out what men attach value to by examining their behavior—what they do and what they say, what they do and what they say put in relation to one another. Dependable, trustworthy descriptions of value systems rest on the same evidential base; you cannot know the total structure of a man's values until you have established his position on particular values and found out as a matter of fact how he accommodates particular values to one another. And you might find out, if you search fully and carefully, that no one has a constant value system; you might find that a man, caught up in one set of circumstances, makes conscious decisions and takes actions which enthrone certain values over others, and later, caught up in a different set of circumstances, makes conscious decisions and takes actions which arrange particular values in a different order of ascendancy and subjection. Finally, value analysis which tries to determine compatibilities and incompatibilities in values (which values support and reinforce one another, which values impair and hinder realization of others) also calls for examination of evidence. You won't know what price you pay for freedom of speech (one value) until you know in what ways and in what degree freedom of speech contributes to anti-Semitic behavior and so impairs toleration of minority groups (another value).

Political scientists who take the stand just set forth believe that dependable, trustworthy statements about values as ends can be supplied only by study which meets the tests of scientific method. They think that too little (some may say that none) of our current writing about values as ends arises out of study that meets these tests. To the extent that current literature of this sort purports to rest on evidence and provide proof, it reveals contentment with a low quality of performance. If it is scientific study, it is sloppy scientific study.

The pleas of defense against this indictment are much the same as those . . . for the defense of our reform literature. *One,* that our writing about values which rests on evidence as to what exists and occurs is better scientific performance than the critics admit. *Two,* that conclusions rest in good part on evidence taken into account by the writer but not disclosed to his reader. And *three* that the importance of the subject to society and its significance for other types of study by political scientists brand as irresponsible and remove from serious consideration a decision to put aside the study of values until the applications of scientific method to investigation of human relationships are more fully developed. People outside the political science profession, including philosophers and students of philosophy, also write inadequately about values, and the special preoccupations and knowledge of political scientists brings something of great usefulness to this body of literature. It is not required that we withhold our contributions until such later day as may find us in mastery of scientific method. It is only required that we be as nearly scientific as we now know how to be in areas of investigation where experience in science provides a guide. That standard we try to meet. If the best of our descriptive writing about values is sloppy science, our best counsel is to try to improve and not to quit.

2. Value Analysis

By value analysis I mean determination of compatibilities and incompatibilities among things valued and determination of how values support and reinforce one another or impair and hinder realization of one another. There is widespread dissatisfaction with our achievement in producing literature of this type. Dispute on this point may align the scientifically oriented against other political scientists, but it also divides political scientists who are alike in their lack of concern with being scientific.

It may be that our final goal in value analysis is nothing more nor less than elaborate description of value systems. That is, it may be that the best way to discover compatibilities and incompatibilities, reinforcements and impairments, is to make searching inquiry into how different values have actually been put together and brought into a balance by different persons or groups, how these structures of associated values are changed, how particular values rise to or fall from ascendancy over others, and so on. If one considers the objective in value analysis to be elaborate description, we may expect him to take the position that political scientists have done very little writing that is entitled to be called value analysis. Indeed, if the premise is granted, I think the conclusion must be accepted. We have produced few items which seriously attempt to describe a value system, and it is doubtful that any attempt that has been made meets a scientist's standard for elaborate description.

Dissatisfaction with our effort at value analysis is by no means confined to political scientists who think it is not sufficiently scientific. Many political scientists who are highly skeptical about the applicability of scientific method to human relationships think we have been too little concerned with value analysis and think that much of what is proffered as value analysis is too narrow in what it comprehends and too wanting in incisiveness to be given that classification. A variety of complaints which I have heard seem to me reducible to the general charge that we engage in one-line reasoning when we ought to engage in multiline reasoning.

What I call one-line reasoning characterizes, so I am told, much of the writing about civil rights which has appeared since World War II. The writer's discourse suggests, if he does not forthrightly declare, that freedom of expression (or privacy, or some other specified value) is so in ascendancy over all other values that there is no need to inquire whether it inhibits realization of other values. The writer thinks it sufficient to state why freedom of expression is good, indicating what other values it supports and reinforces; he affirms its compatibility with certain other values but he does not show an equal concern to find out whether it is wholly or partially incompatible with still other values, hindering their realization.

I presume that everything valued comes at some cost to other things which are also valued, and I have no doubt that all serious writing by political scientists reveals that they recognize this to be the case. The criticism of much of our literature is that it is too close to one-line reasoning; that compatibilities are overemphasized and incompatibilities are underemphasized. If anything said above suggests total failure to identify and put measures on incompatibilities, impairments, and hindrances, those statements can be toned down by the reader to indicate notable lack of adequacy in this respect. I say notable lack of adequacy because that is what I have many times heard said.

The correction for the notable lack of adequacy lies in what I have called multiline reasoning. Multiline reasoning does not preclude focus of attention and treatment which fixes, in a structure of values, the place of the value which is central to the discourse. I understand this is what Aristotle counseled us to do when he recommended a constant attention to *politeia*, a term which some political scientists prefer to translate as regime rather than the more frequent rendition as constitution. . . .

I have never heard or read a defense to the charge that we have put too little emphasis on value analysis. We may waive the rejoinder of the extreme pro-science members of the profession who would say that value analysis should not be attempted at all, or not be attempted at this time, and therefore that we have put too much emphasis on it. Perhaps some of those who count value analysis important would say that we have produced a great deal more of it than the critics seem to be aware of. I suspect that much of what will occur to many as defense of our achievements to date is more appropriate for the defense of personal preference statements than for the defense of what I have identified as value analysis. For that reason I think further consideration of intellectual conflict in this general area can best wait until we have noticed certain issues which arise in respect to preference writing.

How Should Personal Preference Be Handled in Scholarly Writing?

Each item in a body of literature carries with it the special mark of the author. At risk of gross oversimplification I shall say that the author marks his product in three ways, and at risk of gross misuse of a word I shall call each of these an injection of personality. The author injects his personality into his product (a) by assumptions of risk, (b) by presumptions of knowledge, and (c) by incorporation of value preference. We shall be concerned here with the third of these, but because they are often confused with the third, a brief comment must be made on each of the first two.

Assumption of risk is illustrated as follows. I have limited resources for study of a problem. I might pursue study design A or design B but not both. I think A is more likely than B to pay off in findings, and I choose A. The decision is my response to risk. It is an expression of personality, and it has an impact on literature. Further illustration: I have carried my study as far as I can and have found alternative means to an end that appears to be socially desired. The evidence is inconclusive as to whether means A or means B is more certain to achieve the end. All considerations incidental to the two means cancel out. By guessing at unknowns I can put one means ahead of the other, and I do so. This is a response to risk which is an expression of personality and it has an effect on literature. I suspect there is hidden in our literature a lot of response to risk. Choices are made in response to risk, but the discourse at best does not disclose the act of risk; at worst it creates the impression that the writer thought the evidence induced the conclusion which he announced.

As to presumptions of knowledge, no one begins a study with a blank mind. Each student brings to his inquiry a pack of suppositions, beliefs, convictions which he calls knowledge. They are major determinants of his judgments about what to inquire into, what to look for, how to go about the search. Suppositions, beliefs, and convictions which are brought to the study supply tests by which new observations and inferences are evaluated, and provide an intellectual framework into which new accretions of knowledge will be fitted. This professional stock, this intellectual baggage, is peculiar to the individual; no one scholar carries quite the same equipment as any other. It is inevitable that these presumptions of knowledge will indelibly stamp the personality of the author on the product which results from his efforts.

The significance for development of literature of response to risk and presumptions of knowledge—especially the latter—would eminently repay careful investigation, but this is not the place for that inquiry. They are not at the center of intellectual conflict among political scientists at this time. Controversy does abound, however, about the impact made upon our literature by injection of preference which is in response to value position.

Personal preference which is response to the writer's values may be injected into the discourse intentionally or unintentionally, boldly or timorously. It may be fully disclosed to the reader, or it may be scattered about so that the reader can detect its presence but cannot estimate its incidence, or it may be purposely concealed so as to cause the reader to suppose that proof has been provided when in fact evidence was lacking or ran the other way. Preference statements may take a form which makes values the center of attention, extolling and deploring particular values and calling men to line up for or against particular values. Or the writer's preferences may emerge in the discourse as the foundation of judgments about other matters—e.g., as premises in argument about what institutions and practices people will tolerate, or as explanation of why certain conceivable means were ignored in a search for effective means to a specified end.

I think we may presume that all political scientists admit that preference cannot be wholly excluded from scholarly work. The student is a man and he cannot make himself into a different man than he is. His own value commitments, those he is aware of plus those he is not aware of, are a part of his make-up. They help to determine what he gets interested in, what he chooses to look at, what he will see when he looks, and how he will evaluate what he sees. Opinions differ as to how fully we are in bondage to our values, what success we can attain when we try to identify and make allowances for them, what victories we can win when we struggle to suppress them. I suppose that differences in opinions on these matters contribute greatly to intellectual controversy about the way we handle values in discourse, but I have too little knowledge or conviction about this to attempt to relate it to the two issues which I shall discuss. The first issue relates to the aggrandizement of preference; the second to disclosure of preference.

1. *Issue: Aggrandizement of Preference*

The first issue is this: Should political scientists produce literature which is mainly, largely, or even in small degree designed to tell other people what the writer likes or what he thinks is good for society? I think the opposition to such writing was pretty well presented earlier in this discussion of values. It comes mainly if not altogether from political scientists who are devoted to scientific inquiry. They do not contend that such writing is injurious to society in general; they think it is injurious to political science as a discipline. They think the primary obligation of political scientists is to produce a different kind of discourse, and they think that preoccupation with preference statements both diverts men from more useful enterprise and fixes states of mind which inhibit production of more useful literature.

The defense of preference writing was only partially presented in earlier paragraphs. I think the full defense would stand on at least four grounds.

First, a point noted above, preference statements have always been a prominent part of the literature we call political science and their claim to legitimacy is therefore at least as great as the claim of scientific inquiry; there is no more reason for saying that preference writing should be terminated than there is for saying that political scientists should leave scientific investigation to other social study disciplines.

Second, statements of personal preference are helpful to other kinds of inquiry. The utopia is a case. When imagination is set free to ascertain how a particular set of values can be realized in institutions and practices, other students are provided with suggestions as to what may be worth looking into, what unsuspected relationships may lurk in institutions they previously supposed to have been fully explored, what untried means might be added to the list of instrumentalities available for the attainment of a variety of ends. It is not true to say that these aids to scholarship can be as well or better supplied by more objective inquiry. Efforts to restrain or exclude personal preference inevitably become a restraint on imagination. Enthusiasm for the task at hand unleashes powers of observation, of discrimination, of association and summation; enthusiasm brings unsuspected abilities to the aid of scholarship when one makes his boldest statements of what he likes and his most robust refutation or denunciation of what he does not like.

Third, society especially needs in our time the advice which political scientists give in statements of personal preference. Government today pervades virtually all of human affairs; it touches on everything men value. People cannot wait until science provides answers to their questions about how to use government to achieve a good life. They will act, and their acts will be shaped by the advice available to them. Thoughtful political scientists can provide better advice than other people on many things relating to government. This is because they have a more intense preoccupation with legal governments, have mastered a literature relating to legal governments, are intimately associated with people who make the study of legal governments their principal business. If political scientists do not boldly state what they like and dislike, and carefully support their positions by argument, society will fall victim to less well-informed and less thoughtfully established statements of preference.

Fourth, the notion that we can exclude personal preference from writing about legal governments is a delusion anyway. You can write about some matters relating to government with high hopes of excluding personal preference, but these matters are trivial in social significance. When you tackle matters of high significance, including most of the matters political scientists generally are committed to tackle, personal preference will intrude no matter how hard you try to exclude it. It is better to admit it freely to the discourse and honor the admission. The solution of our problems relating to personal preference is not in exclusion of preference statements; it lies in disclosure of what we like and dislike, and advice to the reader as to how he may allow for and correct against the writer's preferences.

2. *Issue: Disclosure of Preference*

We are thus brought to the second issue cited above, disclosure of preference. If we stand on the aphorism that "actions speak louder than words," I think we have to agree that some political scientists believe it is wholly proper for the writer to give his readers no clues to the preferences which affect the discourse, and even that it is proper to present what is really only wishful thinking in language which suggests that it is conclusion forced by evidence. I have not heard or read, and shall not attempt to contrive, a defense for these members of the profession.

We do verbalize differences of opinion as to what constitutes adequate disclosure of preference. I have not been able to identify standards or statements of ideals on which we divide. The dispute is directed to particular cases. We defend or condemn specific writings on the ground that they do or do not make adequate disclosure, or that they do or do not give adequate warning that preference intrudes where it cannot be disclosed. Since the dispute turns on specific cases rather than standards, I will make no effort to identify positions taken or to reproduce arguments pro and con. Some complaints and admonitions I have heard uttered are these.

> He announced a conclusion which was one of several made tenable by the evidence. He should either have given all the alternatives equal status, or have produced evidence that one was more tenable than the others, or have admitted that he adopted one because his values made it more attractive to him.
>
> He detailed the evidence in support of some of his conclusions but let other conclusions stand as fiat. The reader might suppose he had undisclosed evidence which supported the latter conclusions when in fact he had no such evidence. If he guessed his way to some of his conclusions he should

have said so. If he arrived at some of his conclusions because he hoped that is the way things are, he should have said that he was influenced by personal preference.

He made uneven use of evidence, playing up that which supported his conclusions and ignoring or playing down evidence which ran to the contrary. If he didn't know what he was doing, he isn't fit to be in business. If he did it purposely, he should be run out of business.

He prejudiced the reader's ability to evaluate evidence by injecting into what appear to be objective descriptions sweet and sour words, laudatory comments, and crass or subtle aspersions. We assume enough risk when we present one side of an account without making an equal presentation of the other side. It is malpractice when we work on the emotions of the reader to pull his attention away from what we do not wish to present.

He could not possibly have told his readers how his personal preferences affected his discourse. But he could have stated in the preface the nature of the value system he brought to the study. If he could not describe his commitments, he could at least have acknowledged that he was prejudiced. If a man thinks he may be caught with his pants down, he should be the first to say that his pants are down.

Albert Somit and *Joseph Tanenhaus*

The "Behavioral-Traditional" Debate in Political Science

Political scientists have quarreled over many matters in the contemporary period but the most divisive issue by far has been behavioralism. If the controversy it has elicited is any measure, this latest quest for a more scientific politics is easily the paramount development in the discipline's entire intellectual history.[1]

[1] A very incomplete list of the more interesting items in the literature accompanying this controversy is presented at the end of this article.

Reprinted from *The Development of American Political Science: From Burgess to Behavioralism* (Boston: Allyn & Bacon, Inc., 1967), pp. 173–194, where it appeared as Chapter XII, "Political Science as a Learned Discipline: Behavioralism." By permission of the publisher.

Two recent presidential[2] speeches have remarked on the similarities between the post-1945 behavioral movement and the pattern of events which Thomas S. Kuhn has discerned in "scientific revolutions."[3] A "normal science," Kuhn suggests, is characterized by general agreement among its practitioners on the problems which properly concern them and on the concepts and methods whereby these problems are best studied. In his now familiar language, this common set of beliefs constitutes that discipline's "paradigm." Scientific revolutions occur when an existing paradigm gives rise to anomalies (insolvable problems, inexplicable or apparently contradictory findings, etc.) which cannot be handled by the existing conceptual apparatus. Should this happen with some regularity, or should an anomaly occur at a particularly critical juncture, there may emerge a rival definition of concerns,

[2] David B. Truman, "Disillusion and Regeneration: The Quest for a Discipline," *American Political Science Review*, 59 (December, 1965), 865–873 and Gabriel Almond, "Political Theory and Political Science," *American Political Science Review*, 60 (December, 1966), 869–879.
[3] Thomas S. Kuhn, *The Structure of Scientific Revolutions* (Chicago: University of Chicago Press, 1962).

concepts, and techniques. Scientific revolutions can thus be viewed as major scientific advances or as shifts from one paradigm to another.

Whether political science (or any other social science) constitutes a "normal science" in the stricter sense of that term may be a matter of some disagreement. There is also some question whether either traditional political science or behavioralism actually satisfies all the requirements of a "paradigm." Allowing for these objections, the idea of a "scientific revolution" and of a shift of "paradigms" provides a useful framework for the discussion of recent developments within the discipline. From such a vantage point, behavioralism may be treated, if only metaphorically, as an attempt to move political science from a pre-paradigmatic (or literally non-scientific) condition to a paradigmatic stage or, alternatively, as an effort to replace a previously accepted paradigm with one that is more powerful.[4] That the

[4] Without arguing for a close similarity between the natural and the social sciences, Truman inclines to the second of these alternatives. As he sees it, the pre-behavioral discipline was characterized by "six closely related features" of "predominant agreement." These six were:

"(1) an unconcern with political systems as such, including the American system, which amounted in most cases to taking their properties and requirements for granted; (2) an unexamined and mostly implicit conception of political change and development that was blandly optimistic and unreflectively reformist; (3) an almost total neglect of theory in any meaningful sense of the term; (4) a consequent enthusiasm for a conception of "science" that rarely went beyond raw empiricism; (5) a strongly parochial preoccupation with things American that stunted the development of an effective comparative method; and (6) the establishment of a confining commitment to concrete description." Truman, "Disillusion and Regeneration," 866.

We have, of course, traced in some detail the emergence and persistence of these several traits. On the other hand, we have also noted the recurrent attempts, from Burgess through Merriam, to substitute other paradigmatic elements. For this reason, we would be somewhat more dubious than Truman about the predominance of agreement on these commitments.

merits of the undertaking are still being controverted may well be due to the difficulty, indigenous to the social sciences, of demonstrating beyond reasonable doubt the superior explanatory power of the new mode of conceptualization. . . .

In this chapter we will first describe the constellation of beliefs and commitments which collectively constitute the behavioral "paradigm." Next we will summarize the various arguments which have been advanced against behavioralism, or more precisely, against its component doctrines. . . . Finally will come what may well be the critical question: what has been the impact of behavioralism upon the discipline to date?

THE BEHAVIORAL CREED

Ironically, participants in the earlier stages of the behavioral-traditional debate often disagreed as fiercely over the issues to be disputed as over the merits of their respective beliefs. Several factors contributed to this situation. There was, for example, the almost irresistible temptation to attribute to one's opponent an untenable or extreme position and then to demolish what was, in reality, a straw man. More labor and ingenuity sometimes went into this kind of argumentation than in trying to understand what it was that the other person was actually trying to say. The ensuing logic-chopping, hair-splitting, and jesuitry was worthy of an exchange between medieval theologians.

Further contributing to the confusion was the amorphous nature of behavioralism, especially at the outset. During this initial period, as Evron M. Kirkpatrick has written, "the term served as a sort of umbrella, capacious enough to provide a temporary shelter for a heterogeneous group united only by dissatisfaction with traditional political science."[5] Grappling with this same

[5] Evron M. Kirkptarick, "The Impact of the Behavioral Approach on Traditional Political Science," in Austin Ranney (ed.), *Essays on the Behavioral Study of Politics*, (Urbana, Ill.: University of Illinois Press, 1962) p. 11.

problem, another commentator, Robert A. Dahl, concluded that behavioralism was no less a "mood" than a doctrinal commitment.[6] It was even possible, as the literature demonstrates, to quarrel over who was or was not a behavioralist.

The root of the difficulty, unquestionably, is the protean nature of behavioralism. It is less a tightly structured dogma than a congerie of related values and objectives. Those who call themselves behavioralists often differ over component elements of their philosophy, with few accepting the "package" in toto. Similarly, anti-behavioralists tend not to take common exception to all of the behavioralistic tenets but direct their fire at those particular notions which strike them as particularly wrongheaded. A good deal of the argument thus does not concern itself with the merits of behavioralism per se but only with certain of its ideas and aspirations.

Over the past few years the basic outlines of the behavioral position have emerged with increasing clarity. A number of persons—David B. Truman, Robert A. Dahl, David Easton, Heinz Eulau, Evron M. Kirkpatrick, and Mulford Q. Sibley, *inter alia*—have written thoughtful, dispassionate analyses of the movement. While they do not agree on every point, a basic consensus can be discerned. The succeeding paragraphs summarize and describe what are now generally regarded as the major tenets of behavioralism. Before presenting them, however, it is essential to repeat the previous caveat: not even the most committed behavioralist necessarily holds all of these views. Few, however ardent their desire for a truly scientific politics, would be willing to carry every one of these propositions to its logical extreme. Each statement is thus to be read and understood as if it were qualified by such phrases as "to the degree possible," "wherever practicable," and "other things being equal."

With this proviso firmly in mind, the following can be identified as the key behavioralist articles of faith.

1. Political science can ultimately become a science capable of prediction and explanation. The nature of this science, it is generally conceded, will probably be much closer to biology than to physics or chemistry. Given this possibility, the political scientist should engage in an unrelenting search for regularities of political behavior and for the variables associated with them. He should, therefore, eschew purely descriptive studies in favor of the rigorous,[7] analytical treatment essential to the systematic development of political knowledge.

2. Political science should concern itself primarily, if not exclusively, with phenomena which can actually be observed, i.e., with what is done or said. This behavior may be that of individuals and/or of political aggregates. The behavioralist deplores the "institutional" approach because it is impossible properly to study institutional behavior other than as manifest in the actions and words of those who carry out institutional functions.

3. Data should be quantified and "findings" based upon quantifiable data. In the final analysis, the behavioralist argues, only quantification can make possible the discovery and precise statement of relationships and regularities. Associated with this is the aspiration—and occasionally the attempt—to state these relationships as mathematical propositions and to explore their implications by conventional mathematical manipulation.

4. Research should be theory oriented and theory directed. Ideally, inquiry should proceed from carefully developed theoretical formulations which yield, in turn, "operational-izable" hypotheses, that is, hypotheses which can be

---

[6]Robert A. Dahl, "The Behavioral Approach in Political Science: Epitaph for a Monument to a Successful Protest," *American Political Science Review*, 55 (December, 1961), 766–71.

[7]This is one of the most commonly employed terms in the behavioralist vocabulary. For those who are uncertain as to its precise denotation, it means exactly what it says it means—rigorous.

tested against empirical data. Since theory must take into account the nature, scope, and variety of the phenomena under study, the behavioralist speaks of "low-level," "middle-level," and "general" theory. The ultimate objective is the development of "over-arching" generalizations which will accurately describe and interrelate political phenomena in the same fashion, to use a threadbare illustration, that Newton's laws once seemed to account for the physical world.

5. Political science should abjure, in favor of "pure" research, both applied research aimed at providing solutions to specific, immediate social problems and melioratory programmatic ventures.[8] These efforts, as the behavioralist sees it, produce little valid scientific knowledge and represent, instead, an essentially unproductive diversion of energy, resources, and attention.

6. The truth or falsity of values (democracy, equality, freedom, etc.) cannot be established scientifically and are beyond the scope of legitimate inquiry. From this it follows that political scientists should abandon the "great issues" except where behavior springing from or related to these issues can be treated as empirical events (the incidence of a belief in democracy, for example, and the manner in which this belief is reflected in voting behavior would thus be an appropriate subject of study). Needless to say, the contention that political science has no proper concern with moral or ethical questions as such has been one of the most bitterly argued aspects of behavioralism.

7. Political scientists should be more interdisciplinary. Political behavior is only one form of social behavior and the profession would profit tremendously by drawing on the skills, techniques, and concepts of its sister social sciences. Some behavioralists would deny, in fact, that political science constitutes a true discipline in itself.

8. Political science should become more self-conscious and critical about its methodology. Its practitioners should develop a greater familiarity with, and make better use of, such tools as multivariate analysis, sample surveys, mathematical models, and simulation. And, almost needless to say, they should make every effort to be aware of, and to discount, their own "value" preferences in planning, executing, and assessing their research undertakings.

The foregoing represents, we believe, a reasonably complete and accurate catalogue of the intellectual commitments symbolized by the term "behavioralism." While we have grouped them under eight broad headings, these propositions can readily be arranged in some other fashion, with more or fewer categories as desired.

Few, if any, of these ideas are new to political science. There is little in behavioralism that would be completely strange or repugnant to such earlier proponents of a "science of politics" as Merriam, Catlin, and Munro. Or, to go back yet another generation, neither would Lowell, Ford, Macy, and Bentley have found many of these propositions totally novel or unacceptable.

Granting that these ideas can be traced well into the past, there are differences which should not be overlooked. If the basic objectives have not changed radically, the underlying intellectual position is now more systematically developed. The current doctrine is more concerned with formal theory and with fundamental, organizing concepts than was previously the case, although Burgess, Ford, Lowell, Bentley, and Catlin clearly foreshadowed this interest. Lastly, many of the earlier proponents of a scientific politics sought, above all, a political science which could effectively grapple with the practical problems besetting the American democracy. Not for them, certainly, the aseptic aloofness of pure research.

Vindicae Contra Behavioralismos

We have described the leading articles of the behavioralist creed. Equity no less than discretion dictates that we now present the relevant coun-

---

[8]Presumably among the undertakings so proscribed would be the perennially popular projects aimed at democratic citizenship or "better minds for better politics."

tercommitments. Of necessity, only the basic outline of each argument can be indicated. As before, this discussion should be prefaced with the warning that not all anti-behavioralists hold all of these views and that few are inclined to push their arguments to logical extremes. Furthermore, a number of these propositions may well be acceptable to those who regard themselves as behaviorally inclined.

With this in mind, the anti-behavioral brief can be summarized as follows:

1. Political science is not, nor is it ever likely to become, a science in any realistic sense of the term. It cannot become a science for a number of reasons. The phenomena with which political scientists deal do not lend themselves to rigorous study. We cannot treat human behavior, individual or social, with the dispassion needed for scientific knowledge. Neither political science (nor any other social science) is amenable to experimental inquiry. There are too many variables and historical contingencies to permit other than the most general statement of regularities. "Laws" of political behavior cannot be stated for a sentient creature such as man, because he is free to modify his actions in keeping with, or in violation of, such laws once they are made known. Furthermore, though the anti-behavioralist has no objection per se to the use of hypotheses, he argues that rigid adherence to this notion may stifle, rather than advance research. The purely descriptive approach, sometimes the only practicable technique, has a legitimate and an important role to play in inquiry.

2. Overt political behavior tells only part of the story. Different individuals may perform the same act for quite different reasons. To understand what they do, one must go beyond, or behind, observable behavior. Moreover, individuals and groups act within an institutional or a social setting, and a knowledge of that setting is essential to any meaningful explanation of their behavior. The anti-behavioralist holds that the larger part of political life lies beneath the surface of human action and cannot be directly apprehended.

3. Whatever the theoretical merits of quantification, for most practical purposes it is now and will continue to be an unattainable goal. Quantification requires precise concepts and reliable metrics—and political science possesses neither. Significant questions normally cannot be quantified; questions which can are usually trivial in nature. As for mathematics—well, how can one mathematicise that which is both imprecise and immensurable?

4. While it is desirable that research be informed by theory, the behavioralists' aspirations have far outrun their data. It verges on the ridiculous to talk of an "over-arching" general theory when political science still lacks accepted low- and middle-level formulations adequate to the facts at hand. This preoccupation with general theory tends to block less ambitious but in the long run more productive inquiry. At best, it has led to the proliferation of concepts which cannot successfully be operationalized.

5. Applied research and a concern with questions of public policy are, on philosophical and historical grounds, warranted and desirable. American political scientists have a moral obligation to devote some portion of their energies to civic matters, and, just as pure research often yields findings of practical value, so applied and programmatic inquiry may contribute to the better understanding of political and social behavior.

6. Significant political issues invariably involve moral and ethical issues. Political science has historically been, and must continue to be, concerned with questions of right and wrong, even if these cannot be "scientifically" resolved. Were the discipline to turn its back on such matters it would have little justification for continued existence. Going considerably beyond this, one wing of anti-behavioralism denies that values cannot be demonstrated true or false and that political scientists are necessarily condemned to an eternal philosophical relativism.

7. There are many areas where an interdisciplinary approach may be useful but care must be taken to preserve the identity and integrity

of political science. All too often, the anti-behavioralist feels, there has been an indiscriminate borrowing of concepts and techniques which are simply inappropriate for political inquiry.

8. Self-consciousness about methodology can be, and has been, carried too far. Overly critical and unrealistic standards impede rather than advance the pursuit of knowledge. This same obsession has led many behavioralists, it is argued, to exalt technique at the cost of content. Technical, rather than substantive, considerations have been permitted to set the area of inquiry. In any case, many of these technical innovations are still too sophisticated and refined for the raw material with which political scientists must work. As for "scientific objectivity," there is almost universal skepticism among the anti-behavioralists that it is attainable—and considerable doubt that it is inherently desirable.

Just as behavioralists differ among themselves, so do their opponents disagree with each other on a number of matters.[9] One of these deserves specific mention. Some anti-behavioralists are satisfied with political science as it has been practiced in the past and see no cogent reason for drastic change. While they concede that certain aspects of the discipline could be strengthened, they believe that, on the whole, it has been equal to its chosen task. Other anti-behavioralists are less complacent about the state of the discipline. They admit that political science has yet to accumulate a very impressive body of knowledge and may even feel that it has lost ground to the other social sciences. But however these two groups may diverge in their assessment of what has been accomplished to date, they are in accord on a crucial point: behavioralism is not a desirable or viable alternative to the kind of political science it seeks to displace.

The foregoing analysis obviously leaves a major problem unresolved. Many who call themselves behavioralists refuse to embrace the entire octalog of their faith; few of their opponents would reject all eight out of hand. What combination of beliefs, then, held with what relative intensity, makes one a behavioralist? Which cluster of tenets, scorned with how much severity, makes one anti-behavioral?

There is no really satisfactory answer. Of course, we can recast the question by shifting to a more subjective method of classification and have each political scientist fix his own position on the behavioral—anti-behavioral spectrum as he himself defines it. This line of attack is not free of methodological difficulties but it can be operationalized and does have some utility.[10] It points to what seems to be an inescapable conclusion: whether a given political scientist falls into the one or the other category turns, in the final analysis, on his state of mind rather than on readily applicable objective criteria.

The Rise of Behavioralism
Origins and Causes

Orthodoxy has it that the term "behavioral science," subsequently corrupted to "behavioralism,"[11] was coined by a group of quantitatively oriented, "rigorously" inclined social scientists at the University of Chicago. Anxious to secure federal financing for social science research, but apprehensive that some unenlightened "persons confound social science with socialism," they conceived the term "behavioral science." Though "behavior" had been used before, the then most recent example being Herbert Simon's

---

[9]These generalizations rely heavily on the results of the survey reported in Albert Somit and Joseph Tanenhaus, *American Political Science: A Profile of a Discipline* (New York: Atherton Press, 1964).

[10]For all practical purposes this is what we did in our earlier study, *American Political Science*. . . . Although there is no need to summarize those findings here, we might point out that stance on behavioralism (as self-defined) correlated significantly with attitudes toward a variety of other professional issues.

[11]It is interesting to note that the term "behaviorism" was common until the early 1960's, and then gave way to the longer variant.

1947 *Administrative Behavior*, after 1949 "behavioralism" and "behavioral science" came increasingly to connote the kind of social science espoused by the Chicago group.[12]

In political science, behavioralism was unmistakably a lineal descendant of the antecedent "science of politics" movement. Many of its component ideas had been advanced in somewhat cruder form during the 1920's, and were already familiar to the older members of the profession. If behavioralism has a father, paternity belongs to Charles E. Merriam, who "staked out" much of the ground now claimed by it. And if Merriam was the sire, Burgess, Lowell, and Bentley were godfathers to the enterprise.

So much for intellectual genealogy. Now, what was there in the post-1945 climate of opinion that enabled behavioralism to take root so swiftly and to flourish so remarkably? We may be still too close to the event for a definitive explanation but what seem to be the most important predisposing conditions and forces can be tentatively identified. The ensuing list, we should add, is not a rank order.

To begin, there was a widespread dissatisfaction with the "state of the discipline." This stemmed from several sources: the discovery that the talents and skills of political scientists were not highly valued by governmental personnel officers; the disconcerting realization, by those who did spend some time in the public service, of the profound difference between the "accepted wisdom" of the profession and the reality of the governmental process; the inability of traditional political science to account for the rise of fascism, national socialism, and communism, or to explain the continuation of these regimes in power; a growing sensitivity to, and unhappiness with, the basically descriptive nature of the discipline; and a knowledge of apparent advances in other social sciences and a mounting fear that political science was lagging behind its sister professions.

Post-war experiences with technical assistance and economic aid programs contributed to the sense of malaise. The various efforts to export U.S. political and administrative "know-how" forced upon American political scientists a painful awareness that much of their vaunted expertise applied, if at all, only to the type of political and administrative problems encountered in Western, industrialized societies. This discovery provided a powerful impetus to the quest for cross-cultural and trans-national regularities which has characterized one aspect of behavioral inquiry.

The migration of European social scientists to the United States during the 1930's and 1940's also hastened the winds of change. Although few of these scholars were themselves behaviorally inclined, and most in fact hostile to it, they exposed their American hosts to currents of thought (Max Weber, logical positivism, etc.) from which behavioralism was heavily to borrow.

Another factor was the expanding use of public opinion polls and the refinement of survey techniques. These provided instruments for developing vast new bodies of data. Research in this area was greatly facilitated by advances in mathematical statistics and the increased availability of electronic computers to perform what had previously been impossibly tedious computations.

Closely related to the foregoing, and making a good deal of this research possible, was the partiality to behavioralism manifested by those who controlled the allocation of research grants. The Social Science Research Council's Committee on Political Behavior, through which considerable money was channeled, was behaviorally inclined. The foundations in general, and the Ford Foundation (with its Behavioral Science Program) in particular, poured huge sums into behavioral projects. On the federal level, access to public funds was largely limited

[12] James G. Miller, "Toward a General Theory for the Behavioral Sciences," in Leonard D. White (ed.), *State of the Social Sciences* (Chicago: University of Chicago Press, 1956), pp. 29–31.

to the social sciences deemed worthy of the appellation "behavioral sciences."[13] Widespread knowledge of this situation, it is safe to say, did not adversely affect conversions to the faith. Even those who had private reservations about behavioralism sometimes found it possible, when research grants hung in the balance, to render at least lip service to the new creed.

. . . .

The Behavioral Influence

What has been the influence of behavioralism on American political science? The answer is in some ways quite clear. Behavioralism has made the discipline more self-conscious and self-critical. Vast energy has gone into a stocktaking and self-evaluation which, in any case, was long overdue.

Consider also the dramatic changes in vocabulary. An older generation spoke knowingly of checks and balances, *jus soli*, divesting legislation, brokerage function, quota system, bloc voting, resulting powers, proportional representation, pressure group, sovereignty, dual federalism, lobbying, recall and referendum, Posdcorb, quasi-judicial agencies, concurrent majority, legislative court, Taylorism, state of nature, item veto, unit rule, and natural law. From today's younger practitioners there flows trippingly from the tongue such exotic phrases as boundary maintenance, bargaining, cognitive dissonance, community power structure, conflict resolution, conceptual framework, cross-pressures, decision making, dysfunctional, factor analysis, feedback, Fortran, game theory, Guttman scaling, homeostasis, input-output, interaction, model, multiple regression, multivariate analysis, non-parametric, payoff, transaction flow model, role, simulation, political systems analysis, T test, unit record equipment, variance, and, of course, political socialization. It is no longer unusual to find these freshly minted coins of disciplinary discourse dotting the pages of text books written primarily for undergraduates. The vocabulary associated with behavioralism also testifies to the extent that political science has become interdisciplinary, for most of these terms, and the concepts and techniques they symbolize, were borrowed (sometimes rather indiscriminately)[14] from other fields of inquiry.

Another consequence of behavioralism has been a sharply increased attention to research techniques and to analytic theory. Formal courses in methodology, now firmly established in most graduate departments, are now filtering down into the undergraduate curriculum. Further evidence of this interest can be found in the pages of the *Review*. Three two-year periods were singled out and the *Review*'s contents over each of these spans analyzed.[15] The first period, 1946-48, can be regarded as "pre-behavioral" or "very early behavioral"; the second, 1950-52, coincides with the first real blossoming of the movement; and the third, 1963-65, allows a decade for the behavioral influence to be manifest. The results were as follows.

The most compelling change between 1946-48 and 1950-52 was the increased attention to analytic theory, an important, though certainly not an exclusive, concern of behavioralism. This trend continued upward, at a slower rate, between 1950-52 and 1963-65. Almost equally impressive was the increase in the proportion of articles employing more powerful quantitative tech-

---

[13] From political science's viewpoint, perhaps the worst offender in this respect was the National Science Foundation. See Albert Somit and Joseph Tanenhaus, *The Development of American Political Science: From Burgess to Behavioralism* (Boston: Allyn & Bacon, Inc., 1967), p. 154.

[14] On this subject, see the essay by Martin Landau, "On the Uses of Metaphor in Political Analysis," *Social Research*, 28 (1961), 331–53.

[15] Only substantive articles were classified. Communications to the editor and bibliographic essays were excluded. In several instances, very brief "discussion" collections by several contributors were counted as a single item.

Table 1: Impact of Behavioralism as Reflected in the Contents of the American Political Science Review

	1946–8		1950–2		1963–5	
	N	%	N	%	N	%
Low level quantitative techniques	15	10.9	15	11.4	19	18.3
More powerful quant. techniques	1	.7	6	4.5	23	22.1
Discussions of scientific method	6	4.3	5	3.8	2	1.9
Analytic Theory	0	0.0	17	12.9	17	16.3
Other	116	84.1	89	67.4	43	41.3
	138	100.0	132	99.9	104	99.9

niques.[16] While only a modest change occurred between 1946–48 and 1950–52, perhaps because of the time required for the necessary technological retooling, the 22 per cent figure for 1963–65 leaves no doubt as to what transpired. But the data in Table 1 actually understate the case. The articles falling into the "other" and the "low level quantitative techniques" categories in 1963–65 tended to be more analytical and less descriptive than those which appeared in the earlier periods.

Other signs of the behavioral ferment might also be noted. The political science department reputed to have the most distinguished graduate faculty (Yale) is heavily oriented toward behavioralism. The elected officialdom of the Association has increasingly tended to be composed of persons prominent in the behavioral movement. Recent APSA presidents, for example, include Lasswell, Key, Hyneman, Truman, Almond,

[16] A word may be in order about the criteria used in making these classifications. Articles which made use of only percentages and simple counting, no matter how imaginative the analysis, were placed in the "low level quantitative technique" category. To be included in the "more powerful quantitative technique" category, an article had to utilize techniques assuming ordinal or interval measurement, or to employ tests of significance with nominal data. The "analytic theory" category includes articles which sought to articulate or to appraise some type of conceptual scheme such as group, power, systems, elite, etc.

and Dahl. When, in 1963, a random sample of the profession was asked to name the political scientists who have made the most significant contributions to the discipline since 1945, seven of the ten most frequently mentioned (Key, Truman, Dahl, Lasswell, Simon, Almond, and Easton) were of the behavioral persuasion. And, as a last item, the fields of specialization regarded as the most behaviorally oriented were also the fields in which respondents to this survey thought the most significant work was being done.

Unquestionably, behavioralism has had a very substantial impact. But—has the behavioral contribution been sufficiently great to constitute an irrefutable demonstration of the efficacy of its "paradigm"? Or, on the other hand, has its payoff been so meager that a contrary conclusion is justified?

For plentiful and imposing reasons, these questions cannot be answered at the present time. If the single most distinguished graduate department is behavioral in outlook, the five next most distinguished are certainly not. Recent presidents of the Association include several distinguished practitioners (e.g., Cole, Swisher, Redford, and Friedrich) who have not been associated with the behavioral movement. Although behavioralism has left its mark on almost every specialization within the discipline, in only a few (such as community politics, electoral behavior, political socialization, and public opinion) have behavioral theories and techniques proved so obviously superior as to recast drastically an area of inquiry. The *Review*, while more behavioral than it was a few years ago, is not overwhelmingly so. The basic undergraduate course in the discipline remains the traditional offering in American government, and the best-selling textbooks for this course are still predominantly pre-behavioral in conception and substance.

The most direct and most convincing evidence, though, is the discipline's own assessment of behavioralism. A random sample of the profession was asked, in 1963, to respond to two propositions (among others). These propositions, and the responses elicited, were:

1. The really significant problems of political life cannot be successfully attacked by the behavioral approach.

Strongly agree	Agree	Can't say	Disagree	Strongly disagree
14.4%	24.1%	15.8%	31.8%	13.9%

2. Much of the work being done in political behavior is only marginally related to political science.

Strongly agree	Agree	Can't say	Disagree	Strongly disagree
19.0%	21.8%	10.9%	36.0%	12.3%

One need not resort to elegant statistical analysis of these highly correlated sets of responses to conclude that the discipline had not by then developed a consensus on this general issue. Unless the attitude of the profession has changed drastically in the intervening few years, the future of the behavioral movement continues uncertain.

Bibliography

[See footnote 1.]

Arnold Brecht, *Political Theory: The Foundations of Twentieth Century Political Thought* (Princeton: Princeton University Press, 1959).

David E. Butler, *The Study of Political Behaviour* (London: Hutchinson, 1958).

James C. Charlesworth (ed.), *The Limits of Behavioralism in Political Science* (Philadelphia: American Academy of Political and Social Science, 1962).

Robert A. Dahl, *Modern Political Analysis* (Englewood Cliffs, N.J.: Prentice-Hall, Inc., 1963).

David Easton, *A Framework for Political Analysis* (Englewood Cliffs, N.J.: Prentice-Hall, Inc., 1965).

David Easton, *The Political System* (New York: Alfred Knopf, Inc., 1953).

Heinz Eulau, *The Behavioral Persuasion* (New York: Random House, Inc., 1963).

Heinz Eulau, Samuel J. Eldersveld, and Morris Janowitz, *Political Behavior: A Reader in Theory and Research* (Glencoe, Ill.: Free Press, 1956).

Charles S. Hyneman, *The Study of Politics* (Urbana, Ill.: University of Illinois Press, 1959).

Harold D. Lasswell, *The Future of American Politics* (New York: Atherton Press, 1963).

Harold D. Lasswell and Abraham Kaplan, *Power and Society: A Framework for Political Inquiry* (New Haven: Yale University Press, 1950).

Daniel Lerner and Harold D. Lasswell (eds.), *The Policy Sciences* (Stanford, Calif.: Stanford University Press, 1951).

Roy C. Macridis, *The Study of Comparative Government* (New York: Random House, Inc., 1955).

Jean Meynaud, *Introduction à la Science Politique* (Paris, 1959).

Hans J. Morgenthau, *Scientific Man Versus Power Politics* (Chicago: University of Chicago Press, 1946).

Austin Ranney (ed.), *Essays on the Behavioral Study of Politics* (Urbana, Ill.: University of Illinois Press, 1962).

Research Frontiers in Politics and Government (Washington, D.C.: Brookings Institute, 1955).

Herbert J. Storing (ed.), *Essays on the Scientific Study of Politics* (New York: Holt, Rinehart & Winston, 1962).

Leo Strauss, *Natural Right and History* (Chicago: University of Chicago Press, 1953).

David B. Truman, *The Governmental Process* (New York: Alfred Knopf, Inc., 1951).

Vernon Van Dyke, *Political Science: A Philosophical Analysis* (Stanford, Calif.: Stanford University Press, 1960).

Eric Voegelin, *The New Science of Politics: An Introductory Essay* (Chicago: University of Chicago Press, 1952).

Dwight Waldo, *Political Science in the United States of America, A Trend Report* (Paris, 1956).

T. D. Weldon, *The Vocabulary of Politics* (Harmondsworth, England, 1953).

Roland Young (ed.), *Approaches to the Study of Politics* (Evanston, Ill.: Northwestern University Press, 1958).

Robert A. Dahl, "The Behavioral Approach in Political Science: Epitaph for a Monument to a Successful Protest," *American Political Science Review*, 55 (Dec., 1961), pp. 763–772.

III THE NATURE OF BEHAVIORAL POLITICAL INQUIRY

Inquiry can be conducted in a variety of ways, but regardless of the means selected, certain guidelines should be followed. During all stages of the research process, the investigator must be guided by the logic, rationale, and principles governing that type of inquiry. When political scientists began to explore behaviorally oriented research, they sought to learn more about this kind of inquiry in order to use it most intelligently and rewardingly. This has meant a growing emphasis in political science upon training in the methods and techniques of research.

Although behavioral forms of inquiry differ in several respects from other forms of inquiry, it should be emphasized again that behavioralism does not involve a set of rigid rules or specific procedures. Rather, it is a mood or disposition toward research. A major aspect of this mood is the desire to extend scientific forms of inquiry from the natural to the social sciences. It is assumed that great expansion of knowledge is possible through "scientific discovery," a phrase descriptive of the general research procedures and methods of proof frequently used in behavioral inquiry. Moreover, scientific inquiry is a creative effort; inspiration and insight are as important to the scientist as to the artist. In the first reading Arnold Brecht concentrates on the general operations identified with "scientific" forms of inquiry. Although he discusses research strategies useful for achieving valid and accurate conclusions, Brecht clearly argues that these steps are *not* binding on the individual researcher. Hopefully, Brecht's explanation of scientific operations and the scientific mood will clarify many popular misconceptions regarding this kind of inquiry.

Certain general steps are followed in conducting scientific inquiry. Fred N. Kerlinger points out that before any research can be planned or any data gathered, the investigator must have a clear idea about the problem he intends to investigate. The purposes behind such a statement of the problem are twofold: first, a lucid conception of the problem prevents the investigator from being diverted to tangential problems and permits him to concentrate on studying what really concerns him. In short, it prevents wasted effort. Second, such a statement usually indicates what is manageable and what is out of proportion to research effort. Investigators may find that without a statement of the dimensions of their research, they try to cover more than is realistically possible.

Kerlinger continues by indicating that once the problem has been conceptualized the investigator must generate statements purporting to explain the problem, which can be subjected to verification. Such conjectural statements, in the language of inquiry, are hypotheses. The generation of hypotheses, as we have said before, is a critical stage in the research process because the suspected relationships among factors set forth in these hypotheses are the relationships the researcher seeks to confirm.

When formulating the problem and generating hypotheses the researcher may use abstract ideas, which by themselves are not always useful in research. Inquiry using such ideas as "peace," "happiness," "love," and "hate" would probably soon become mired in the problem of clarity and precision because of individualized notions about the meaning and measurement of these ideas. An investigator can only observe their manifestations, not the ideas themselves. Because divorcing the use of concepts from inquiry is not realistic, some alternate strategy must be implemented.

Carlo Lastrucci discusses the problems of conceptual definition and application. The guidelines he sets out are intended to assist the researcher in translating vague and sometimes ambiguous notions into clear and precise formulations useful for inquiry. Concepts must be translated into forms that allow for observation of that to which they refer. Sometimes this translation comes as a function of the definitional process; otherwise additional manipulation is necessary. Paul F. Lazarsfeld considers the nature of these other nondefinitional processes, especially the development of indices to measure concepts. The goal of concept definition and index formation is to develop manageable and meaningful hypotheses by using clear, concise, and specific expressions.

Most political scientists would probably agree that the purpose of their investigations is to gain understanding of the political world. This understanding includes the ability to describe, explain, predict, or establish causal relationships among variables. As we previously indicated, such goals involve constructing general statements called theories. Behavioral theories in turn can be thought of as describing, explaining, perhaps predicting, and possibly even making causal statements about events. In discussing explanation, Warren Weaver points out that an adequate explanation of some phenomenon creates an intellectually "comfortable" feeling about it. Explanation thus promotes an understanding of why one observed what one did. To the scientist, knowledge about why some phenomenon occurred as it did is an important step toward predicting and controlling the future, and predicting future events has been a matter of great interest to all social science disciplines. Irwin D. Bross clarifies the notion of prediction by setting forth alternate procedures for achieving this goal. These include predicting the future as a replication of the past, an increment on the past, a cycle of the past, or as an association among factors. Related to this notion of prediction is that of causation. In treating this subject, long avoided by many political scientists, R.M. MacIver points out that a consideration of causation cannot be circumvented regardless of the investigator's predispositions. The notion of causality is inherent in all investigations and in all language used when discussing that investigation. Thus, claims by political scientists that they are not seeking to investigate causal relationships do not exempt them from the causal associations implied in their findings.

Description, explanation, prediction, and causation are all integral aspects of theory building. Useful theories may combine all of these aspects into one unified statement about the nature of reality. In his selection, Eugene Meehan clarifies the role of theory in research.[1]

In a further elaboration of theory's role in inquiry, Robert Merton explores the functions of theory that are beyond the frequently discussed ones of testing and verifying hypotheses. The excerpt included here concentrates on the importance of theory as a means of defining knowledge and for guiding research. Theory, according to this view, is not only a consequence of research but also a source for suggesting additional avenues for inquiry.

---

[1] See also his more recent book, *Explanation in Social Science: A System Paradigm* (Homewood, Ill.: Dorsey Press, 1968), especially Chap. 2.

Arnold Brecht

Scientific Operations and Scientific Justification

The term "scientific method" can be used in more than one sense. In its broadest meaning it sets off any method that is considered "scientific" from any other that is considered "nonscientific," without indicating what makes it such. When used this way the term designates a problem—the problem of what is scientific—rather than an answer to it.

In its more specific sense, however, the term refers to a definite type of method. Hereinafter, when so used, the words "Scientific Method" will be capitalized. There still remain two alternative possibilities. Scientific Method may be considered the only method that can claim to be scientific, all others being called unscientific; or it may be singled out only because it is well defined and scientific beyond doubt, without the claim that it is necessarily the only scientific method.

. . . .

To begin a book on the foundations of scientific political theory in the twentieth century with a theoretical analysis of Scientific Method is justified not only on the ground that this method has dominated scientific work; it is advisable also for several other reasons. It provides us with a clearly defined model of scientific operations, and

Reprinted from *Political Theory: The Foundations of Twentieth-Century Political Thought* (Princeton, N.J.: Princeton University Press, 1959), pp. 27-30 and 113-115, where it appeared under the title, "Theory of Scientific Method: Facts and Logic." By permission of the author and the publisher.

with unambiguous terms of reference for any basic remarks, . . ., apt to throw them automatically into sharp relief as being either in line with or in opposition to this model. It supplies an appropriate framework within which fundamental philosophical problems of science can be discussed, each in a setting that indicates the relevance of such examinations. And it offers the welcome opportunity for clarifying the meaning of Scientific Method . . . regarding points where clarity is lacking or opinions are not yet unified. But nothing said . . . should be considered prejudicial to the question whether there are other, better, or at least supplementary, scientific methods.

. . . .

In every inquiry—and that means inquiry within the social as well as the natural sciences—Scientific Method concentrates on the following "scientific actions," "scientific operations," or "steps of scientific procedure:"

1. *Observation* of what can be observed, and tentative acceptance or nonacceptance of the observation as sufficiently exact.
2. *Description* of what has been observed, and tentative acceptance or nonacceptance of the description as correct and adequate.
3. *Measurement* of what can be measured; this being merely a particular type of observation and description, but one sufficiently distinct and important to merit separate listing.
4. *Acceptance* or nonacceptance (tentative) as *facts* or *reality* of the results of observation, description, and measurement.
5. *Inductive generalization* (tentative) of accepted individual facts (No. 4), offered as a "factual hypothesis."
6. *Explanation* (tentative) of accepted individual facts (No. 4), or of inductively reached factual generalizations (No. 5), in terms of relations, especially causal relations, offered as a "theoretical hypothesis."
7. *Logical deductive reasoning* from inductively

reached factual generalizations (No. 5) or hypothetical explanations (No. 6), so as to make explicit what is implied in them regarding other possible observations (No. 1), or regarding previously accepted facts (No. 4), factual generalizations (No. 5), and hypothetical explanations (No. 6).

8. *Testing* by further observations (Nos. 1-4) the tentative acceptance of observations, reports, and measurements as properly made (Nos. 1-3), and of their results as facts (No. 4), or tentative expectations as warranted (No. 7).

9. *Correcting* the tentative acceptance of observations, etc., and of their results (Nos. 1-4), of inductive generalizations (No. 5) and hypothetical explanations (No. 6), whenever they are incompatible with other accepted observations, generalizations, or explanations; or correcting the previously accepted contributions.

10. *Predicting* events or conditions to be expected as a consequence of past, present, or future events or conditions, or of any possible constellation of such, in order either

(a) to test factual or theoretical hypotheses (Nos. 5 and 6), this being identical with steps 7 and 8; or

(b) to supply a scientific contribution to the practical process of choosing between several possible alternatives of action.

11. *Nonacceptance* (elimination from acceptable propositions) of all statements not obtained or confirmed in the manner here described, especially of "a-priori" propositions, except when "immanent in Scientific Method" or offered merely as "tentative assumptions" or "working hypotheses". . . .

Before we discuss these scientific actions, operations, or steps in detail,[1] several points should be made perfectly clear. In the first place,

[1]F. S. C. Northrop, *The Logic of the Sciences and the Humanities* (New York: Macmillan Company, 1947), p. 18, uses the term "scientific method" for each individual step (observation, description, forming a hypothesis, etc.). Then, of course, "There is no one scientific method" (pp. ix, 19, 107). I follow here the prevalent usage, epitomizing its meaning by capitalization. . . .

our enumeration is not meant to express the postulate that Scientific Method proceed exactly in the given order. On the contrary, actual procedure will generally begin with a tentative working hypothesis ventured upon not infrequently on the basis of an as yet most cursory knowledge of facts and used as a trial balloon to guide more systematic research. In the absence of conclusive data, the inquiry may even start out from a purely factual assumption.

We must go even further and recognize that in order to engage in scientific work at all, as distinct from other activities, the investigator must always begin by forming in his own mind some tentative *ideas* about (1) the objective of his inquiry, i.e. the question for which the answer is sought, (2) the relevance of this question for human knowledge in general, as distinct from a merely private interest of the investigator, and (3) the relevance of the scientific actions the investigator is about to take for his purpose of finding the answer.[2] . . .

Secondly, it should be noted from the beginning that the various steps here enumerated are not always neatly separable in actual scientific working practice. Frequently the forming of a hypothesis and its testing by observation, the engaging in inductive and deductive reasoning, in generalization and specification, follow one another so quickly that they seem to constitute a whole rather than a sequence of actions. How-

[2]Similarly, John Dewey, *Logic, the Theory of Inquiry* (New York: H. Holt and Co., 1938, *passim*) held that inquiry begins, neither with facts nor with hypotheses, but with the recognition of a "problematic situation," and F. S. C. Northrop, *The Logic of the Sciences and Humanities*, pp. 18, 28, that "analysis of the problem" is a second important step, equally preceding the forming of a hypothesis. Both views are in line with those expressed above. However, finding that a situation is scientifically problematic, and engaging in its analysis, are two steps that in turn must be preceded by the investigator's forming some idea or "working hypothesis" that the situation *may* conceal a significant problem and that its analysis may be relevant for human knowledge in general. "Analysis" is merely one of various steps in the complex process of either forming definite hypotheses for the solution of a "problematic situation" or discovering that, after all, it is not problematic.

ever, analytically the various steps can and must be distinguished.

. . . .

We now shall take up the discussion, thus far postponed, of the exclusive or nonexclusive character of Scientific Method. All scientists, both in the natural and in the social sciences, agree that Scientific Method *is* scientific, i.e., that contributions made according to its principles are contributions to science. Objections are raised solely against claims that it *alone* is scientific and that results reached through other methods, unless checked in line with Scientific Method, apart from being "nonscientific-in-terms-of-Scientific-Method," are nonscientific also in every other sense, so as not to add to science in any reasonable sense of the term. These objections are tantamount to denying that the elements of Scientific Method are *necessary* elements of every scientific procedure.

The leading adherents of Scientific Method have indeed raised the claim to exclusiveness for it. Felix Kaufmann, for example, calls certain basic rules (see below, note 5) "invariable"; they are common to all sciences *qua* sciences and, therefore, "a priori for science because science is defined in terms of them," he said.[3] This is a good illustration of what we have called an "immanent methodological a priori."

The decisive question, of course, is not whether science *is* defined in these terms, but whether it *must* be so defined and not otherwise, since it is our purpose here to present a *theory* of Scientific Method and not just a description of it. Consequently, it is not sufficient for us to insist on the a-priori character of the rules of procedure. The question remains, Why must these rules be so? Why can they not be otherwise? Or is there "no ultimate justification of these rules," as Kaufmann said, adding that "we cannot go beyond them in discriminating between correct and incorrect scientific decisions"?[4]

If there were no justification, how then could insistence on Scientific Method be defended in view of the manifold attacks that have been leveled against its claim to exclusiveness? Yet there is one distinguishing merit of Scientific Method, among others that are less relevant in the present context. It supplies knowledge that can be transmitted from person to person *qua* knowledge (here called "intersubjectively transmissible knowledge," or briefly "transmissible knowledge"). To transmit knowledge *qua* knowledge is more than merely to communicate the assertion that I have such knowledge, or even to add how I came by it. This latter type of communication is always possible, of course, where the usual means of communication are at our disposal; with their aid we can relate to others anything that we consider part of our own subjective knowledge and the fact that we do so consider it and why, limited only by the shortcomings of language and the receptiveness of the hearer or reader. But mere communication does not transmit our subjective knowledge to others *qua* knowledge. To do that requires more, and Scientific Method supplies this more regardless of individual beliefs. Indeed, it supplies a type of knowledge that can be transmitted from *any* person who has such knowledge to *any* other person who does not have it but who can grasp the meaning of the symbols (words, signs) used in communication and perform the operations, if any, described in these communications. It submits observational testimony, refined as far as feasible by testimony on the conditions under which the observations were made, in order to enable ourselves and others the better to appraise the exactness of the observations made and the plausibility of the conclusions derived therefrom. Yet . . . Scientific Method does not insist that any other scientist must "accept" as unchal-

---

[3] Felix Kaufmann, *Methodology of the Social Sciences* (New York: Oxford University Press, 1944), pp. 45, 47.

[4] Kaufmann, *Methodology of the Social Sciences*, p. 230, No. 4.

lengeable either the exactness of the observations or the validity of the conclusions, beyond the purely logical implications of meaning. Such acceptance is ultimately left to the judgment of the receiver of the communication. What is intersubjectively transmitted *qua* knowledge is the evidence, not the conclusions therefrom. Scientific Method, quite consistently, is zealous to exclude evidence that cannot be transmitted.

This justification of the claim to exclusiveness seems to me to provide a good ground as far as it goes. Couched in terms of logical reasoning it would read, "*If* we decide to call scientific only those methods which supply intersubjectively transmissible knowledge (as defined above), *then* only such and such methods are scientific." But placing ourselves on this ground we must not ignore the fact that there has been at all times in the world a considerable amount of subjective knowledge not intersubjectively transmissible *qua* knowledge yet still firmly held, and frequently on excellent grounds, such as childhood remembrances no longer shared by any other person, or the knowledge of a wrongly condemned person that he has not committed the crime, or reminiscences by a statesman of important political conversations he had with others none of whom is still alive or willing to tell the truth about it. There may be such correct yet nontransmissible knowledge also in metaphysical questions, if there be metaphysical agents that reveal knowledge to us, or to some of us. But, although all this must, I think, be conceded, we cannot lump these types of knowledge indiscriminately with knowledge that is intersubjectively transmissible. Should someone want to use the term "science" for both types of knowledge . . . then there would be an obvious need for distinguishing them in some other manner, giving them either different compound names—such as "science in the sense of intersubjectively-transmissible-knowledge" and "science in the sense of not-intersubjectively-transmissible-knowledge"; or different symbols, such as S_1 and S_2, respectively. To these two categories a third might be added, "science in the sense of speculations-not-considered-knowledge," or S_3.

The problem, then, of the exclusive character of Scientific Method boils down to that of whether there is "intersubjectively transmissible knowledge" (S_1) outside of Scientific Method. . . . [S]ome proposals will be made that lead to broadening the conception of transmissible knowledge by including the category of "universal and inescapable" elements in human thinking and feeling to the extent that their existence can be empirically ascertained. By and large, however, with due reservation for these proposals, I consider the claim of Scientific Method to exclusiveness in category S_1 justified.

All rules of procedure enumerated at the beginning . . . are essential for science as a supplier of intersubjectively transmissible knowledge, except that No. 10 b (prediction) is optional, of course, and that some of the others may be omitted when, *but only when*, the strength of the grounds provided by the other steps suffices for intersubjective transmissibility.[5]

[5]Felix Kaufmann, (*Methodology of the Social Sciences*, pp. 39 ff.) singled out seven fundamental rules as "invariable." These are, briefly summarized: (1) Distinction between "accepted" and "non-accepted" propositions; (2) transfer from one to the other class not to be arbitrary; (3) such transfer to be based only on grounds accepted at the time of the decision on the transfer; (4) no decision irrevocable; (5) no incompatible propositions to be tolerated within the "accepted body of knowledge"; (6) no proposition that falls within the subject-matter of a science to be a priori excluded from acceptance; and (7) results of observation to play a key role as grounds for acceptance.
The last four are generally recognized by all adherents of Scientific Method. As to the first three, concerning acceptance, I have expressed my criticism. . . . Kaufmann himself is compelled to admit that some rules can be "modified and changed"; these are obviously those which govern acceptance, not the other principles. On the other hand, his seven invariable rules omit any special reference to tests and to logical reasoning, criteria that are generally held to be essential for empirical science, but seem to be drowned in Kaufmann's broad requirement of submission to the rules governing acceptance. Limitation of his list to its last four items would, therefore, not be satisfactory either. This explains why I have considered it necessary to refer to all the eleven steps enumerated in our initial survey, subject to the modifications mentioned in the text above.

Fred N. Kerlinger

Problem Formulation and Hypothesis Generation

If research were limited simply to gathering so-called facts, scientific knowledge could not advance. Many people think that science is basically a fact-gathering activity. It is not. As Cohen says:

> There is . . . no genuine progress in scientific insight through the Baconian method of accumulating empirical facts without hypotheses or anticipation of nature. Without some guiding idea we do not know what facts to gather. Without something to prove, we cannot determine what is relevant and what is irrelevant.[1]

The scientifically uninformed person often has the idea that the scientist is a highly objective individual who gathers data without preconceived ideas. Poincaré long ago pointed out how wrong this idea is. He said:

> It is often said that experiments should be made without preconceived ideas. That is impossible. Not only would it make every experiment fruitless, but even if we wished to do so, it could not be done.[2]

[1] M. Cohen, *A Preface to Logic* (New York: Meridian Publishing Company, 1956), p. 148.
[2] H. Poincaré, *Science and Hypothesis* (New York: Dover Press, 1952), p. 143.

Reprinted from Chapter 2 of *Foundations of Behavioral Research* by Fred N. Kerlinger. Copyright © 1964 by Holt, Rinehart & Winston, Inc. Reprinted by permission of Holt, Rinehart & Winston, Inc.

PROBLEMS

It is not always possible for a researcher to formulate his problem simply, clearly, and completely. He may often have only a rather general, diffuse, even confused notion of the problem. This is in the nature of the complexity of scientific research. It may even take an investigator years of exploration, thought, and research before he can clearly say what questions he has been seeking answers to. Nevertheless, adequate statement of the research problem is one of the most important parts of research. That it may be difficult or impossible to state a research problem satisfactorily at this time should not allow us to lose sight of the ultimate desirability and necessity of doing so. Nor should the difficulty be used as a rationalization to avoid stating the problem.

Bearing this difficulty in mind, a fundamental principle can be stated: If one wants to solve a problem, one must generally know what the problem is. It can be said that a large part of the solution of a problem lies in knowing what it is one is trying to do. Another part of the solution lies in knowing what a problem is and especially what a scientific problem is.

What is a good problem statement? Although research problems differ a great deal, and although there is no one "right" way to state a research problem, certain characteristics of problems and problem statements can be learned and used to good advantage. To start, let us take one or two examples of published research problems and study their characteristics. First, take the problem of the study by Hurlock. . . . What are the effects on pupil performance of different types of incentives?[3] Note that the problem is stated in question form. The simplest way is here the best way. Also note that the problem states a relation between variables, in this case between the variables incentives and pupil performance (achievement).

[3] E. Hurlock, "An Evaluation of Certain Incentives Used in Schoolwork," *Journal of Educational Psychology*, 16 (1925), 145–59.

A *problem*, then, is an interrogative sentence of statement that asks: What relation exists between two or more variables? The answer to this question is what is being sought in the research. If the problem is a scientific one, it will almost always contain two or more variables. In the Hurlock example, the problem statement relates incentive to pupil performance. Another problem, by Page, is: Do teacher comments cause improvement in student performance?[4] One variable is teacher comments (or reinforcement), and the other variable is student performance. The relational part of the question is expressed by the word "cause." Still another problem, by Harlow, is more complex: Under what conditions does learning how to learn transfer to new situations?[5] One variable is "learning how to learn" (or set); the other variable is transfer (of learning).

Criteria of Problems and Problem Statements

There are three criteria of good problems and problem statements. One, the problem should express a relation betsween two or more variables. It asks, in effect, questions like: Is A related to B? How are A and B related to C? Is A related to B under conditions C and D? There are exceptions to this dictum, but they are rare. They occur mostly in taxonomic or methodological research. . . .

Two, the problem should be stated clearly and unambiguously in question form. Instead of saying, for instance, "The problem is . . .," or "The purpose of this study is . . .," ask a question. Questions have the virtue of posing problems directly. The purpose of a study is not necessarily the same as the problem of a study. The purpose of the Hurlock study, for instance, was to throw light on the use of incentives in school situations. The problem was the question about the relation between incentives and performance. Again, the simplest way is the best way: ask a question.

The third criterion is often difficult to satisfy. It demands that the problem and the problem statement should be such as to *imply* possibilities of empirical testing. A problem that does not contain implications for testing its stated relation or relations is not a scientific problem. This means not only that an actual relation is stated, but also that the variables of the relation can somehow be measured. Many interesting and important questions are not scientific questions simply because they are not amenable to testing. Certain philosophic and theological questions, while perhaps important to the individuals who consider them, cannot be tested empirically and are thus of no interest to the scientist as a scientist. The epistemological question, "How do we know?," is such a question. A medieval theological classic is "How many angels can dance on the head of a pin?" Education has many interesting but nonscientific questions, such as, "What effect is the changing ethos of American education having on American children?" "Does democratic education improve the learning of youngsters?" "Are group processes good for children?" These questions can be called metaphysical in the sense that they are, at least as stated, beyond empirical testing possibilities. The key difficulties are that some of them are not relations, and most of their constructs are very difficult or impossible to so define that they can be measured.

HYPOTHESES

A *hypothesis* is a conjectural statement of the relation between two or more variables. Hypotheses are always in declarative sentence form, and they relate, either generally or specifically, variables to variables. There are two criteria for "good" hypotheses and hypothesis statements. They are the same as two of those for problems and problem statements. One, hypotheses are statements about the relations between variables. Two, hypotheses carry clear implications for

[4]E. Page, "Teacher Comments and Student Performance: A Seventy-Four Classroom Experiment in School Motivation," *Journal of Educational Psychology*, 49 (1958), 173–81.
[5]H. Harlow, "The Formation of Learning Sets," *Psychological Review* 56 (1949), 51–65.

testing the stated relations. These criteria mean, then, that hypothesis statements contain two or more variables that are measurable or potentially measurable and that they specify how the variables are related. A statement that lacks either or both these characteristics is no hypothesis in the scientific sense of the word.[6]

Let us take three hypotheses from the literature and apply the two criteria to them. First, consider a very simple hypothesis: Group study contributes to higher grade achievement. We have here a relation stated between one variable, group study, and another variable, grade achievement. Since measurement of the variables is readily conceivable, implications for testing the hypothesis, too, are readily conceivable. The criteria are satisfied. A second hypothesis is different because it states the relation in the so-called null form: Practice in a mental function has no effect on the future learning of that mental function. Note that the relation is stated directly and clearly: one variable, practice in a mental function, is related to another variable, future learning, by the words "has no effect on." On the criterion of potential testability, however, we meet with difficulty. We are faced with the problem of so defining "mental function" and "future learning" that they are measurable. If we can solve this problem satisfactorily, then we definitely have a hypothesis. Indeed, we have a famous one—but one that has usually not been stated as a hypothesis but as a fact by many educators of the past and the present.

The third hypothesis represents a very numerous and important class. Here the relation is indirect, concealed, as it were. It customarily comes in the form of a statement that Groups *A* and *B* will differ on some characteristic. For example, Middle-class children more often than lower-class children will avoid finger painting tasks. Note that this statement is one step removed from the actual hypothesis which might be stated: Finger painting behavior is in part a function of social class. If the latter statement were the hypothesis stated, then the first statement might be called a subhypothesis, or a specific prediction based on the original hypothesis.

Let us consider another hypothesis of this type but still one more step removed: Individuals having the same or similar occupational role will hold similar attitudes toward a cognitive object significantly related to the occupational role. ("Cognitive objects" are any concrete or abstract things perceived and "known" by individuals. Tables, houses, people, groups, the government, and education are examples of cognitive objects.) The relation in this case, of course, is between occupational role and attitudes (toward a cognitive object related to the role, for example, role of educator and attitudes toward education). In order to test this hypothesis, it would be necessary to have a least two groups, each representing a different occupational role, and then to compare the attitudes of the groups. For instance, we might take a group of teachers and compare their attitudes toward education to those of, say, a group of businessmen. Thus the hypothesis, as stated, is really a "difference" hypothesis. Still, it, too, could be reduced to the general relational form with which we started: Attitudes toward cognitive objects significantly related to occupational roles are in part a function of the behavior and expectations associated with the roles.

The Importance of Problems and Hypotheses

There is little doubt that hypotheses are important and indispensable tools of scientific research. There are three main reasons for this belief. One, they are, so to speak, the working instruments of theory. Hypotheses can be deduced from theory and from other hypotheses. If, for instance,

---

[6]There are legitimate hypotheses that, at least on the surface, lack the relation criterion. For instance, in factor-analytic investigations, . . . we might have some such problem statement as: What are the factors underlying social attitudes? An hypothesis such as this might be used: There are two underlying factors behind social attitudes: (I) liberalism and (II) conservatism. . . . [Here], however, only relational statements will be considered.

we are working on a theory of aggression, we are presumably looking for causes and effects of aggressive behavior. We might have observed cases of aggressive behavior occurring after frustrating circumstances. The theory, then, might include the proposition: Frustration produces aggression. From this broad hypothesis we may deduce more specific hypotheses, such as: To prevent children from reaching goals they find desirable (frustration) will result in their fighting with each other (aggression); if children are deprived of parental love (frustration), they will react in part with aggressive behavior.

The second reason is that hypotheses can be tested and shown to be probably true or probably false. Isolated facts are not tested, as we said before; only relations are tested. Since hypotheses are relational propositions, this is probably the main reason they are used in scientific inquiry. They are, in essence, predictions of the form, "If A, then B," which we set up to test the relation between A and B. We let the facts have a chance to establish the probable truth or falsity of the hypothesis.

Three, hypotheses are powerful tools for the advancement of knowledge because they enable man to get outside himself. Though constructed by man, hypotheses exist, can be tested, and can be shown to be probably correct or incorrect apart from man's values and opinions. This is so important that we venture to say that there would be no science in any complete sense without hypotheses.

Just as important as hypotheses are the problems behind the hypotheses. As Dewey has well pointed out, research usually starts with a problem, with a problematic situation. Dewey says that there is first an indeterminate situation in which ideas are vague, doubts are raised, and the thinker is perplexed.[7] He further points out that the problem is not enunciated, indeed cannot be enunciated, until one has experienced such an indeterminate situation.

[7] J. Dewey, *Logic: The Theory of Inquiry* (New York: Holt, Rinehart & Winston, Inc., 1938), pp. 105–107.

The indeterminacy, however, must ultimately be removed. Though it is true, as stated earlier, that a researcher may often have only a general and diffuse notion of his problem, sooner or later he has to have a fairly clear idea of what the problem is. Otherwise he can hardly get very far in solving it. Though this statement seems self-evident, one of the most difficult things to do, apparently, is to state one's research problem clearly and completely. In other words, you must know what you are trying to find out. And when you finally know this, the problem is a long way toward solution.

VIRTUES OF PROBLEMS AND HYPOTHESES

Problems and hypotheses, then, have powerful virtues in common. One, they direct investigation. The relations expressed in the hypotheses tell the investigator what to do, in effect. Two, problems and hypotheses, because they are ordinarily generalized relational statements, enable the researcher to deduce specific empirical manifestations implied by the problems and hypotheses. We say: "If Hypothesis 1 is true, then perhaps Hypothesis 2 is also true and 3 not true." Then we test Hypotheses 2 and 3. If Hypothesis 2 is true and Hypothesis 3 not true, as predicted, Hypothesis 1 is confirmed.

The third point is closely related to the second and refers to a difference between problems and hypotheses. Hypotheses, if properly stated, can be tested. While a particular hypothesis may be too broad to be directly tested, if it is a "good" hypothesis, then, as indicated under the second point, above, other testable hypotheses can be deduced from it. The point is that facts or variables are not tested as such. The relations stated by the hypotheses are tested. Another point is that a problem really cannot be scientifically solved if it is not reduced to hypothesis form because a problem is a question, usually of a broad nature, and is, in and of itself, not directly testable. One does not test the question: Does anxiety affect achievement? One tests one or more hypotheses implied by this question, for example: "Test anxiety reduces achievement test scores," or "Anxiety-provoking test situations will depress achievement test scores."

The fourth point is that problems and hypotheses advance scientific knowledge by helping the investigator to confirm or disconfirm theory. Suppose a psychological investigator gives a number of subjects three or four tests, among which is a test of anxiety and an arithmetic test. Routinely computing the intercorrelations between the three or four tests, he finds that the correlation between anxiety and arithmetic is negative. He concludes, therefore, that the greater the anxiety the lower the arithmetic score. But it is quite conceivable that the relation is fortuitous or even spurious. If, however, he had hypothesized the relation in advance on the basis of theory, the investigator could have greater confidence in the results. The investigator who does not hypothesize a relation in advance, in short, does not give the facts a chance to prove or disprove anything.[8]

This use of the hypothesis is similar to playing a game of chance. The rules of the game are set up in advance, and bets are made in advance. One cannot change the rules after an outcome, nor can one change one's bets after making them. That would not be "fair." One cannot throw the dice first and then bet. Similarly, if one gathers data first and then selects a datum and comes to a conclusion on the basis of the datum, one has violated the rules of the scientific game.

The reason is that the game is not "fair." And it is not fair because the investigator can easily capitalize on, say, two significant relations out of five tested. What happens to the other three? They are usually forgotten. But in a fair game every throw of the dice is counted in the sense that one either wins or does not win on the basis of the outcome of each throw. The main point, perhaps, is that the purpose of hypotheses is to direct inquiry. As Darwin pointed out long ago, all observations have to be for or against some view if they are to be of any use.

The last point to be made about hypotheses has already been made, but it needs formal statement, even repetition. Hypotheses incorporate the theory, or part of it, in testable or near-testable form. Earlier an example of reinforcement theory was given in which testable hypotheses were deduced from the general problem. The importance of recognizing this function of hypotheses may be shown by going through the back door and using a theory that is very difficult, or perhaps impossible, to test. Freud's theory of anxiety includes the construct of repression. Now, by repression Freud meant the forcing of unacceptable ideas deeply into the unconscious. In order to test the Freudian theory of anxiety it would be necessary to deduce relations suggested by the theory. These deductions would of course have to include the repression notion which includes the construct of the unconscious. Hypotheses can be formulated using these constructs, and, in order to test the theory, they would have to be so formulated. But testing them is another, more difficult matter because of the extreme difficulty of so defining terms like "repression" and "unconscious" that they can be measured. Up to the present, no one has succeeded in defining these two constructs without seriously departing from the original Freudian meaning and usage. Hypotheses, then, are important bridges between theory and empirical inquiry.

PROBLEMS, VALUES, AND DEFINITIONS

To clarify further the nature of problems and hypotheses, two or three common errors will now

---

[8] The words "prove" and "disprove" are not to be taken here in their usual literal sense. It should be remembered that a hypothesis is never really proved or disproved. To be more accurate we should probably say something like: The weight of evidence is on the side of the hypothesis, or the weight of the evidence casts doubt on the hypothesis. Braithwaite says: "Thus the empirical evidence of its instances never proves the hypothesis: in suitable cases we may say that it *establishes* the hypothesis, meaning by this that the evidence makes it reasonable to accept the hypothesis; but it never *proves* the hypothesis in the sense that the hypothesis is a logical consequence of the evidence." R. Braithwaite, *Scientific Explanation* (Cambridge: Cambridge University Press, 1955), p. 14.

be discussed. First, scientific problems are not moral and ethical questions. What is the best way to teach fourth-grade children? Are punitive disciplinary measures bad? Is an authoritarian school system bad for the children's personal and social development? To ask these questions is to ask value and judgmental questions that science cannot answer. Many so-called hypotheses are not hypotheses at all. For instance, The small-group method of teaching is better than the lecture method. This is a value statement; it is an article of faith and not a hypothesis. If it were possible to state a relation between the variables, and if it were possible to define the variables so as to permit testing the relation, then we might have a hypothesis. But there is no way to test value questions scientifically.

A quick and relatively easy way to detect value questions and statements is to look for words like "should," "ought," "better than" (instead of "greater than"), and similar words that indicate cultural or personal judgments or preferences. Value statements, however, are tricky. While a "should" statement is obviously a value statement, certain other kinds of statements are not so obvious. Take the statement: Authoritarian methods of teaching lead to poor learning. Here there is a relation. But the statement fails as a scientific hypothesis because it uses two value expressions or words, "authoritarian methods of teaching" and "poor learning," neither of which can be defined for measurement purposes without deleting the words "authoritarian" and "poor." The word "poor" is obviously a value word: it expresses a value judgment. Again, science does not pass value judgments. Scientists may do so, but science never does. To attain scientific respectability, the expression "poor learning" would have to be deleted and some expression substituted like "decreased problem solving behavior," which implies measurement possibilities but no value judgment. The expression "authoritarian methods of teaching" is perhaps almost hopeless, at least at present, although its definition is conceivable if very difficult. The trouble is that the mere use of the word "authoritarian" expresses a value judgment, at least in this case. As used today, it says, in effect, that such methods are "bad." Another difficulty is that at present we do not know what "authoritarian methods of teaching" means. Most often it seems to mean the personal teaching biases of the person using this method.[9]

Other types of statements that are not hypotheses or are poor hypotheses are frequently formulated. One type, fortunately infrequent, is the definition. Consider, for instance, "The core curriculum is an enrichening experience." Another type, unfortunately frequent, is what might be called the vague generalization. Examples are: People react to special social resources; The existing curriculum thoroughly prepares students for successful teaching; Scholastic achievement is the major consideration in predicting the success of doctoral candidates; Listening ability can be increased in the third grade; Reading skills can be identified in the second grade; Arithmetic skills lend themselves to teaching.[10] Comment hardly seems necessary.

Another common defect of problem statements often occurs in doctoral theses: the listing of methodological points or "problems" as subproblems. These methodological points have two characteristics that make them easy to detect: (1) they are not substantive problems that spring from the basic problem; and (2) they clearly relate to techniques or methods of sampling, measuring, or analyzing. They are usually not in question form, but rather contain the words "test," "determine," "measure," and the like. "To determine the reliability of the instruments used in this research," "To test the significance of the differences between the means," and "To assign pupils at random to the experimental

[9] An almost classic case in education of the use of the word "authoritarian" is the statement sometimes heard among educators: The lecture method is authoritarian. This seems to mean that the speaker does not like the lecture method and he is telling us that it is bad. Similarly, one of the most effective ways to criticize a teacher is to say that he is authoritarian or rigid.

[10] All of these statements were prepared by doctoral students in the writer's classes in research design and methodology.

Generality and Specificity of Problems and Hypotheses

One of the difficulties that the research worker usually encounters and that almost all students working on a thesis find bothersome is the generality and specificity of problems and hypotheses. If the problem is too general, it is usually too vague and cannot be tested. Thus, it is scientifically useless, though it may be interesting to read. Problems and hypotheses that are too general and too vague are common in the social sciences and education. For example, Creativity is a function of the self-actualization of the individual. Another is: Democratic education enhances social learning and citizenship. Still another is: Authoritarianism in the classroom inhibits the creative imagination of children. These are interesting problems. But, in their present form, they are worse than useless scientifically, because they cannot be tested and because they give one the spurious assurance that they are hypotheses that can "some day" be tested.

Terms like "creativity," "self-actualization," "democracy," "authoritarianism," and the like have, at the present time at least, no adequate empirical referents.[11] Now, it is quite true that we can define "creativity," say, in a limited way by specifying one or two creativity tests. This may be a legitimate procedure. Still, in so doing, we run the risk of getting far away from the original term and its meaning. This is particularly true when we speak of artistic creativity. We are often willing to accept the risk in order to be able to investigate important problems, of course. Yet terms like "democracy" are almost hopeless to define, since the problem of measurement is very difficult. Even when we accomplish it we often find we have destroyed the original meaning of the term.

The other extreme is too great specificity. Every student has heard that it is necessary to narrow problems down to workable size. This is true. But, unfortunately, we can also narrow the problem out of existence. In general, the more specific the problem or hypothesis the clearer are its testing implications. But triviality may be the price we pay. While the researcher cannot handle problems that are too broad because they tend to be too vague for adequate research operations, in his zeal to cut the problems down to workable size or to find a workable problem, he may, as indicated above, cut the life out of it. He may make it trivial or inconsequential. A thesis, for instance, on the simple relation between speed of reading and size of type, while important and maybe even interesting, is too thin for doctoral study. Too great specificity is perhaps a worse danger than too great generality. At any rate, some kind of compromise must always be made between generality and specificity. The ability effectively to make such compromises is a function partly of experience and partly of much critical study of research problems.

● Concluding Remarks—The Special Power of Hypotheses

One will sometimes hear that hypotheses are unnecessary in research, that they unnecessarily restrict the investigative imagination, that the job of science and scientific investigation is to find out things and not to labor the obvious, that hypotheses are obsolete, and the like. Such statements are quite misleading. They misconstrue the purpose of hypotheses.

It can almost be said that the hypothesis is the most powerful tool man has invented to achieve

---

[11] Although many studies of authoritarianism have been done with considerable success, it is doubtful that we know what authoritarianism in the classroom means. For instance, an action of a teacher that is authoritarian in one classroom might not be authoritarian in another classroom. The alleged democratic behavior exhibited by one teacher might even be called authoritarian if exhibited by another teacher. Such elasticity is not the stuff of science.

dependable knowledge. Man observes a phenomenon. He speculates on possible causes. Naturally, his culture has a stock of answers to account for most phenomena, many correct, many incorrect, many a mixture of fact and superstition, many pure superstition and mythology. It is the scientist's business to doubt most explanations of the phenomena of his field. His doubts are systematic. He insists upon subjecting explanations of phenomena to controlled empirical test. In order to do this, he must so formulate explanations that they are amenable to controlled empirical test. He formulates the explanations in the form of theories and hypotheses. In fact, the explanations *are* hypotheses. The scientist simply disciplines the business by writing systematic and testable hypotheses. If a causal explanation cannot be formulated in the form of a testable hypothesis, then it can be considered to be a metaphysical explanation and thus not amenable to scientific investigation. As such, it is dismissed by the scientist as being of no interest.

The power of hypotheses goes further than this, however. A hypothesis is a prediction. It says that if x occurs, y will also occur. That is, y is predicted from x. If, then, x is made to occur (vary), and it is observed that y also occurs (varies concomitantly), then the hypothesis is confirmed. This is more powerful evidence than simply observing, without prediction, the covarying of x and y. It is more powerful in the betting-game sense discussed earlier. The scientist makes a bet that x leads to y. If, in an experiment, x does lead to y, then he collects his money. He has won the bet. He cannot just enter the game at any point and pick a perhaps fortuitous common occurrence of x and y. Games are not played this way (at least in our culture). He must play according to the rules, and the rules in science are made to minimize error and man's fallibility. Hypotheses are part of the rules of the game.

Even when hypotheses are not confirmed, they have power. Even when y does not covary with x, knowledge is advanced. Negative findings are sometimes as important as positive ones, since they cut down the total universe of ignorance and sometimes point up fruitful further hypotheses and lines of investigation. *But the scientist cannot tell positive from negative evidence unless he uses hypotheses.* It is possible to conduct research without hypotheses, of course, particularly in exploratory investigations. But it is hard to conceive modern science in all its rigorous and disciplined fertility without the guiding power of hypotheses.

Carlo Lastrucci

Concepts in Empirical Research

A. Concept Definition

. . . Since science strives to achieve accuracy, every field of scientific endeavor develops a continuously refined set of concepts which, to the initiated, mean the same thing at all times under stated conditions. Thus it is imperative at the outset of any research effort to define clearly every *concept* (i.e., an idea, or a generalized idea of a class of objects), or *construct* (i.e., an idea expressing an orderly arrangement of concepts into a single whole) that will be employed. (The term "construct" is often employed to refer to abstract or purely synthetic formulations having no counterpart in observable reality—e.g., "force," "symbiosis," "status," "value." Treating such abstractions as though they do exist in tangible form—e.g., visualizing the "status hierarchy" as a real pyramid of roles, or the earth's core as a magnet inducing gravity—is termed "reification." To reify is a common error made by those who think primitively.)

. . . The problem of accurate definition is of fundamental importance in science. One of the basic rules of definition is that a definition can be neither true nor false—i.e., it is not a factual proposition. A definition is simply an explicit declarative statement or resolution; it is a contention or an agreement that a given term will refer to a specific object. One may question the intelligibility or the usefulness of a given definition, but he cannot logically test its truth; for its "truth" is established by declaration—i.e., it is what the definer says it is.

. . . In order to promote clarity and precision, several basic rules of definition are consistently followed by all competent scientists in accordance with accepted principles of logic. Stated briefly they are, first, that a definition must denote the unique or distinctive qualities of that which is being defined. The term employed must be the symbolic equivalent of the thing it stands for; it must be applicable to every instance of that thing and to nothing else. In other words, it must be inclusive of all things denoted by it and yet exclusive of all things not denoted by it. This quality of precision and distinctiveness is of particular significance when building a *taxonomic* system (i.e., a system of classification of objects wherein each class bears some logical relationship to every other class); for the usefulness of any taxonomy is directly related to the precision of its various classes of objects. It would be of little value, for example to classify certain objects as "foods" if such things could not be further defined more precisely and exclusively in terms of specific qualities (e.g., their use by different species, their caloric content, their commercial value, their nutritive value or their scarcity). The development of a comprehensive, precise and functional set of classifications is a significant index to the degree of maturity of any particular science.

. . . The second rule of definition states that a definition must not be circular—i.e., it must not contain within itself either directly or indirectly any part of the thing being defined. The error of *tautology* (i.e., of defining a thing by itself—e.g., "A man is a person having masculine qualities"), is a feature of vague exposition such as is commonly found in small, cheap dictionaries. The difficulty here, however, lies in the fact that very few terms have true equivalents in any language; so the problem of tautology is sometimes difficult to resolve. To say, for example, that a boat is "A small, open vessel or water craft,"

Reprinted from *The Scientific Approach: Basic Principles of Scientific Method* (Cambridge, Mass.: Schenkman Publishing Company, Inc., 1963), pp. 78–86, where it appeared under the title, "The Role of Concepts in Scientific Approaches." By permission of the author and the publisher.

may seem clear enough until one checks the definition of the term "vessel," only to find that it refers to "A craft for travelling on water," while a "craft" refers to "a boat, ship or aircraft." This second rule, therefore, refers mainly to gross or obvious errors of tautology such as might be the case when one defines a "crowd" as "a group lacking organization." A contemporary example of a popular tautology is the case of defining the aged in our society as "senior citizens." Later, under a discussion of common errors of conceptualization, this type of error will be analyzed in more detail.

. . . The third rule of definition states that a definition should not be stated negatively when it can be stated positively. This rule, however, is not as binding as are the others; for in some instances the exclusiveness feature of a good definition demands that it be made clear what a term does not stand for.

. . . .

. . . Furthermore, some concepts are essentially negative in character. A drunk, for example, is essentially a person who is not sober, and a loser is simply one who is not a winner. Whenever possible, however, clarity is enhanced by the employment of positive rather than of negative denotations. In cases of doubtful interpretations, both the positive and negative features of a term may be utilized to insure clarity.

. . . The fourth rule of definition states that a definition should be expressed in clear and unequivocal terms, not in obscure or figurative terminology. The problem here, however, is often that of common agreement and understanding of terms. What might appear as clear and unequivocal to one person might appear vague, amorphous or obscure to another. As a case in point the term "disease" might prove illustrative. To a physiologist the term may have a very precise meaning; but the same term in the hands of a social reformer (who might speak of a "diseased society") might well be vague, amorphous and highly general or ambiguous. Scientists themselves, for example, often employ terms which may appear vague or obscure to the uninformed person, but which within their group are quite clearly and meaningfully perceived. (Examples of such terms might be "meson track," "cartilaginous tissue," "temperament," "status index.")

. . . While on this subject of definition it perhaps would be useful to consider also a basic error often made in the employment of concepts. The error, essentially, is that of assuming that changing or manipulating a word (symbol) changes the thing for which it stands; or, in other words, that things can be changed simply by changing their names. The converse of this error is that of assuming that the same name necessarily implies the same meaning, or that things can be made the same simply by giving them the same names. Examples of this general class of errors in definition, conceptualization or symbolism come easily to mind. Though it may be quite true that "A rose by any other name would smell as sweet," it is highly dubious that a belligerent military program is any the less offensively oriented simply because it is called a "defense program," or that deceit in business can be denied simply by calling it "shrewdness." Since there are so many types of communicative errors possible in any discourse, and since understanding and avoiding such errors is fundamental to scientific competence, an exposition of some major conceptual errors is presented later in this chapter.

. . . Whenever possible, concepts and constructs should be defined either (a) objectively or (b) operationally—i.e., they should be defined (a) in terms of empirically verifiable and standardized referents (such as rulers, thermometers, scales, etc.), which leave little room for dispute among competent observers; or they should be defined (b) in terms of specific operations, behaviors, processes or effects which likewise leave little room for serious dispute.

. . . Thus terms such as "beauty," "good," "rich," "intelligent," "hot," "wide," "dark," etc., are useless for research purposes until they are transposed into objective or operational referents. "Beauty," for example, might be defined objectively as "A score between seven and ten on the Smith Scale of Beauty Ranking," or it might be defined operationally as "Evidenced by winning a contract to pose for perfume advertisements." "Good," for example, might be defined operationally as "A good student is one who is never cited for disciplinary infractions." "Rich," of course, might easily be defined objectively in terms of gross income, or savings, or property owned. "Intelligent" might be defined objectively as an I.Q. score above a certain designated point; or it might be defined operationally as "Solving the maze within thirty seconds." "Hot," of course, can be translated into certain objective units of Centigrade temperature, or in terms of volume of heat by "BTU's" (British Thermal Units); just as "wide" can be defined objectively in terms of inches, yards or meters; while "dark" might be defined objectively as "Any condition less than ten candle-power of measured light."

. . . A particular problem of definition arises in the employment of measuring instruments. The standard practice is to define an instrument in terms of two qualities: validity and reliability. An instrument (a scale, ruler, balance, meter, questionnaire, attitude test, etc.) is said to be *valid* when it measures that which it is purported to measure. It is said to be *reliable* when it gives consistent results under comparable conditions. These two features are not necessarily related. The validity of an instrument is generally established by consensual definition—for example, the agreement that a valid measure of foot candles (of light) or of viscosity (of a liquid) shall be that as determined by such a body as the American Standards Association or by the Society of Automotive Engineers. If the definition is either ambiguous or not agreed upon, however, it is impossible to determine the validity of an instrument. In the social sciences, for example, the lack of a clear agreement on the essential objective attributes of the term "social class" would be a case in point—making, in this case, the development of a valid measure of social class impossible.

. . . A related aspect of the problem of determining the validity of an instrument arises in those cases where the logical implications of a definition can be seriously questioned. A case in point again can be drawn from the behavioral sciences in the matter of defining, say, intelligence. To argue that a test validly measures intelligence simply because one has declared that the items *are* measures of intelligence—of a particular sort, at least—is to raise the question of the requirements of a good definition. Unless it has been established that the items tested are the unique and inherently distinctive qualities of what is normally connoted by the term "intelligence," it is dubious that one is actually measuring what is logically implied in the connotation of the term. This same point could be illustrated in the case of such terms as "adjustment," "interaction," "status," "institution." In short, an instrument is said to be valid when it measures those qualities or attributes clearly and objectively defined according to a logically defensible empirical connotation.

. . . The relative merits of denotative, connotative and operational definitions, as employed in science, comprise a lengthy topic of debate. The major arguments of that debate, however, can be summarized as follows: A *denotative* definition has a nonverbal referent—i.e., it "points to" the object (often by use of an illustration). A *connotative* definition, on the other hand, implies or describes—usually by listing the attributes or features of the object—that which the term names (i.e., the referent or object). An *operational* "definition," as discussed earlier, describes or prescribes the steps or procedures required to carry out the idea in question. Each of these types of definition has its particular strengths and weaknesses when employed in science.

. . . The strength of a denotative "definition" is its essentially nonverbal character. Any normal person can perceive a "horse," a "motor," a "house," or a "microscope," when these objects are indicated by various denotative definitions. Suppose, however, that not all horses, motors, houses or microscopes look, act, or feel alike? How is one then to perceive their common qualities? Herein lies the weakness of a denotative definition. Unless a class of objects has a common, easily perceivable set of attributes—which many classes of phenomena do not—it becomes quite difficult to designate such classes by a simple denotative term. Particularly in the employment of constructs—which by definition are synthetic abstractions having no physical or behavioral referents—a denotative definition is practically useless.

. . . The strength of a connotative definition—the type most often employed in dictionaries—lies in its synthesis of the inherent and unique qualities of an object in terms already understood. A "horse," for example, may be defined connotatively as a "four-legged animal about four to six feet tall, having a long bushy tail," etc. Furthermore, abstractions (e.g., "honor," "beauty," "integration," "parabolic") can only be defined connotatively. The weakness of such definitions, however, should be quite obvious. Suppose a person does not understand the referents themselves? How, for example, would one connotatively explain to a primitive person the essential meaning of such terms as "expansion coefficient," "osmosis," "trauma," or "adjustment" if the person did not already possess some related or synonymic referents to which such terms allude? Defining a concept in terms of other, already understood concepts assumes that one understands the "other" terms. As anyone well knows from the experience of consulting a dictionary, a connotative definition is often nothing more than an exercise in increasing confusion.

. . . To reiterate, then, it is often advisable and sometimes even necessary in science to define concepts both denotatively and connotatively. In such cases the combined definition practically becomes *operational*—i.e., the concept (but not the construct) is defined in terms of the procedural steps involved (such as "cooking," "heating," "measuring," "testing"). . . . In spite of all its obvious virtues, however, the essential weakness of an operational definition should be mentioned, to wit: there is always the possibility that various users will not agree to define an operational concept in a similar manner. Clay, a bench, an apple pie, or even a specific painting may all be produced by varying operations; and one set of procedures to create, interpret or construct such objects may be just as legitimate (and even useful) as another. In short, an operational definition often lacks the feature of exclusiveness—a feature basic to all good (i.e., precise) definitions employed in science. Other weaknesses of operational definitions—though they interest the logician of science—need not concern us at this point.

. . . There are many occasions in science when a concept or construct cannot very well be "defined" in either empirical or operational terms. This occurs in the case of theoretical definitions serving as explications of large classes of phenomena. A general theory of energy, for example, or of mutation, or of social change, or of "homeostasis" (i.e., the tendency of an organism to seek a balance between its tensions and its environment), may refer to abstractions which exist only conceptually but not empirically—and hence not operationally. True, the abstract concept in such cases may be further defined and hence related to specific definitional attributes which can then be related to empirical qualities; but the original definition remains substantially a theoretical abstraction. After all, no one can see, hear, touch, smell or taste "force," "energy," "adjustment," "leadership" or "morale." He can only experience the empirical consequences of such abstract notions when such consequences have been designated by definition.

B. UNITS OF STUDY

. . . Clear definitions have their first applicability when a study delineates its specific phenomena of interest. In some instances the phenomena may be gross classes (e.g., the world's human population, all living bacteria, all forms of heat, the whole planetary system). In most cases, however, a study is concerned only with specific sub-classes of phenomena; hence it must designate such sub-classes in terms of relatively precise units of reference. Such units are the specific features of the phenomena that interest the researcher (e.g., the "income" of a population, the "mobility" of a social group, the "tensile strength" of a metal, or the "viability" of an organism). In order to increase accuracy and precision, therefore, a satisfactory unit of scientific analysis should possess at least five clarifying attributes: appropriateness, clarity, measurability, comparability and reproducibility. The following discussion will indicate the role of these attributes.

. . . The first requirement of an accurate unit of study is its *appropriateness*; that is, the unit selected must focus attention upon the essential object of study. Thus an analysis of income differences would need to prescribe whether "income" shall mean the gross salary and earnings, or the take-home pay after taxes have been deducted, or the take-home pay after taxes and contributions and union dues have been deducted, etc. A study of comparative birth rates would need to specify who is being compared: the total population of two cities, or all the females of two cities, or all the married females of two cities, or all the marriageable females of two cities, etc. In general, both a clearly defined question and experience in the particular field of study (besides a clearly formulated hypothesis) are necessary before the appropriateness of the units can be determined. But in the first as well as in the final analysis, the appropriateness of a unit will be determined by its role in the total study design—i.e., by whether or not it fulfills the needs of the study and accurately conveys what is intended.

. . . The second requirement of an accurate unit of study is *clarity*. Essentially this is a problem of precise and unambiguous definition. In speaking of "colds," for example, can the unit be precisely defined so that it means the same thing to all students of the subject? The same can be asked of such units as "crime" (All types? Just those called to the attention of the police? Both major crimes and misdemeanors? etc.), or "wars" (Only those officially declared? What about so-called insurrections or revolutions? What about so-called "crime-wars"? And what is a "cold war"?), or even "religiosity" (Going to church? If so, how often? Organized into sects and denominations only? Any differences between believers and doers? etc.). This problem of clarity is particularly acute when a study attempts to employ such subjective, relative or abstract units as "peace," "good government," "economic recession," "morality," etc. In the last analysis clarity is a matter of the degree of specificity, on the one hand, but also of the character of the concept on the other. Some objects inherently permit clear and discrete definition (e.g., cast iron, teeth, pregnancy, death), while others are inherently obtuse or indiscrete (e.g., windy, will-power, illness, reverence). Wherever possible, therefore, clarity should be viewed not as an absolute necessity without which a study could not proceed but, rather, as an ideal goal constantly to be striven for even if never completely achieved.

. . . The third requirement of an accurate unit of study is its *measurability*. Essentially this means that one should strive constantly to devise units which permit quantification and therefore mathematical manipulation. Admittedly this may not always be possible; but since mathematics is the most precise, logically consistent, universal and standardized language of science, mathematical measurement is the optimal tool of all scientific endeavor. In any event a unit of study is improved to the extent that it can be defined in measurable terms.

. . . The fourth requirement of an accurate unit of study is its *comparability*. This means essentially that the units to be studied and compared should be of a like order. (This again is a definitional, hence taxonomic problem.) Divorces, for example, are hardly comparable (as any student of the subject knows), because of the differences in divorce laws among various states and countries. Nor are such general phenomena as crime, drug addiction, infant mortality, unemployment or migration comparable. In all these cases, although the units are the same, the ways of determining them vary widely. Thus it is necessary that the researcher demonstrate at the outset the comparability of his units. If he is to gather examples of religious behavior, or of drunkenness, or of wife beating, or of poverty, etc., he must be sure that his units refer to the same phenomena wherever and however derived.

. . . The fifth requirement of an accurate unit of study is its *reproducibility*. Since science is concerned with general and not with unique phenomena, any study which employs units that cannot be reproduced defies verification. The study of history quite clearly illustrates this particular deficiency of an accurate unit of study. Since most historical events are presumably unique, they cannot be reproduced; so it becomes impossible to restructure them in a manner permitting restudy or verification. Once again, however, it might be well to point out that this requirement, like that of clarity, is neglected more often than is apparently necessary—i.e., it is highly questionable that many presumably unique phenomena actually are unique. (A highly publicized study of human sex behavior, for example, contended that even such intimate—and therefore presumably individualistic or unique—behaviors as sexual relations occur with much greater consistency and similarity than is commonly assumed to be the case.) Final determination of reproducibility or uniqueness is oftentimes a matter of opinion rather than of established fact; but the competent scientist deals only with demonstrably reproducible phenomena and employs them in such manners or designs that both the phenomena and the design can be reproduced by other investigators interested in verifying his conclusions. A major difference between the so-called natural sciences (i.e., physics, astronomy, chemistry and biology) and the social sciences is this very attribute of reproducibility. "Replication" studies are rare in the social sciences though very common in the natural sciences.

Paul F. Lazarsfeld

The Translation of Concepts into Indices[1]

No science deals with its objects of study in their full concreteness. It selects certain of their properties and attempts to establish relations among them. The finding of such laws is the ultimate goal of all scientific inquiries. But in the social sciences the singling out of relevant properties is in itself a major problem. No standard terminology has yet been developed for this task. The properties are sometimes called aspects or attributes, and often the term "variable" is borrowed from mathematics as the most general category. The attribution of properties is interchangeably called description, classification, or measurement.

When social scientists use the term "measurement," it is in a much broader sense than the natural scientists do. For instance, if we are able to say that one department in a company has higher morale than another, we would be very pleased with ourselves and we would say that we had performed a "measurement." We would not worry that we cannot say that it is twice as high or only 20 per cent higher. This does not mean that we make no efforts to arrive at

[1] Publication No. A-276 of the Bureau of Applied Social Research, Columbia University. I am indebted to Prof. Allen Barton and Mr. Herbert Menzel for their contributions to this review.

Reprinted from *Daedalus*, Vol. 87, No. 4 (Fall, 1958), "On Evidence and Transference," 99–130, where it appeared under the title, "Evidence and Inference in Social Research." By permission of the author and *Daedalus*, Journal of the American Academy of Arts and Sciences, Boston, Mass.

measurements in the traditional sense, with a precise metric. Some success has been achieved, but these efforts are only beginning, and they represent merely a small part of measurement activities in the broader sense.

Keeping in mind this generalized idea of measurement, let us see how social scientists establish devices by which to characterize the objects of empirical investigations. There appears to be a typical process which recurs regularly when we establish "variables" for measuring complex social objects. This process by which concepts are translated into empirical indices has four steps: an initial imagery of the concept, the specification of dimensions, the selection of observable indicators, and the combination of indicators into indices.

1. Imagery

The flow of thought and analysis and work which ends up with a measuring instrument usually begins with something which might be called imagery. Out of the analyst's immersion in all the detail of a theoretical problem, he creates a rather vague image or construct. The creative act may begin with the perception of many disparate phenomena as having some underlying characteristic in common. Or the investigator may have observed certain regularities and is trying to account for them. In any case, the concept, when first created, is some vaguely conceived entity that makes the observed relations meaningful.

Suppose we want to study industrial firms. We naturally want to measure the management of the firm. What do we mean by management and managers? Is every foreman a manager? Somewhere the notion of management was started, within a man's writing or a man's experience. Someone noticed that, under the same conditions, sometimes a factory is well run and sometimes it is not well run. Something was being done to make men and materials more productive. This "something" was called management, and ever since students of industrial organization have tried to make this notion more concrete and precise.

The same process happens in other fields. By now the development of intelligence tests has become a large industry. But the beginning of the idea of intelligence was that, if you look at little boys, some strike you as being alert and interesting and others as dull and uninteresting. This kind of general impression starts the wheels rolling for a measurement problem.

2. Concept Specification

The next step is to take this original imagery and divide it into components. The concept is specified by an elaborate discussion of the phenomena out of which it emerged. We develop "aspects," "components," "dimensions," or similar specifications. They are sometimes derived logically from the over-all concept, or one aspect is deduced from another, or empirically observed correlations between them are reported. The concept is shown to consist of a complex combination of phenomena, rather than a simple and directly observable item.

Suppose you want to know if a production team is efficient. You have a beginning notion of efficiency. Somebody comes and says, "What do you really mean? Who are more efficient—those who work quickly and make a lot of mistakes, so that you have many rejections, or those who work slowly but make very few rejects?" You might answer, depending on the product, "Come to think of it, I really mean those who work slowly and make few mistakes." But do you want them to work so slowly that there are no rejects in ten years? That would not be good either. In the end you divide the notion of efficiency into components such as speed, good product, careful handling of the machines—and suddenly you have what measurement theory calls a set of dimensions.

The development of dimensions can go quite far. One university in California has made a study under a Navy contract of an airplane factory, aimed at determining what is really efficient management on the lowest level. The notion of efficient management was divided into nineteen components, some of which were: absence of dissensions in the group, good communication downward, not too much compulsion, consistency of command, the size of command, and so on.

This can probably be overdone. I have rarely seen a concept that needed nineteen dimensions. But as a general principle, every concept we use in the social sciences is so complex that breaking it down into dimensions is absolutely essential in order to translate it into any kind of operation or measurement.

3. Selection of indicators

After we have decided on these dimensions, there comes the third step: finding indicators for the dimensions. Here we run into a number of problems. First of all, how does one "think up" indicators? The problem is an old one.

William James has written in *The Meaning of Truth*:

> . . . Suppose, e.g., that we say a man is prudent. Concretely, that means that he takes out insurance, hedges in betting, looks before he leaps. . . . As a constant habit in him, a permanent tone of character, it is convenient to call him prudent in abstraction from any one of his acts. . . . There are peculiarities in his psychophysical system that make him act prudently. . . .

Here James proceeds from an image to a series of indicators suggested directly by common experience. Today we would be rather more specific about the relation of these indicators to the underlying quality. We would not expect a prudent man always to hedge in betting, or to take out insurance on all possible risks; instead we would talk about the probability that he will perform any specific act as compared with a less prudent individual. And we would know that the indicators might vary considerably, depending on the social setting of the individual. Among students in a Protestant denominational college, for instance, we might find little betting and rare occasions for taking out insurance. Still a measure of prudence could be devised which was relevant to the setting. We might use as indicators whether a student always makes a note before he lends a book, whether he never leaves his dormitory room unlocked, etc.

The fact that each indicator has not an absolute but only a probability relation to our underlying concept requires us to consider a great many possible indicators. The case of intelligence tests furnishes an example. First, intelligence is divided into dimensions of manual intelligence, verbal intelligence, and so on. But even then there is not just one indicator by which imaginativeness can be measured. We must use many indicators to get at it.

There is hardly any observation which has not at one time or another been used as an indicator of something we want to measure. We use a man's salary as one of the indicators of his ability; but we do not rely on it exclusively, or we would have to consider most businessmen more able than even top-ranking university professors. We take the number of patients a doctor has cured as another indicator of ability in that setting; but we know that a good surgeon is more likely to lose a patient than is a good dermatologist. We take the number of books in a public library as an indicator of the cultural level of the community; but we know that quality of books matters as much as quantity.

When a battery of indicators is being drawn up, one difficult problem is to decide where to stop. Which indicators are considered "part of" the concept, and which are considered independent of or external to it? If we start listing indicators of the "integration" of a community, is the crime rate a part of the conception of integration, or is it an external factor which we might try to predict from our measure of integration? Here again, as with the problem of projective indices, knowing the laws which relate indicators to one another is of great importance. Even if we exclude crime rates from our image of an "integrated" city, they might be so highly correlated, as a matter of empirical generalization, that we could use them as a measure of integration in situations where we could not get data on the indicators which we "really" want to call integration. To do this, of course, we must first have "validating studies" where we correlate crime rate with the other indicators of integration and establish that it is generally closely related. We should also know whether there are other factors besides integration influencing crime rate which might confuse our measurements if we used it alone to measure integration, so that we can check on these other factors, or add enough other indicators so as to cancel out their influence.

4. FORMATION OF INDICES

The fourth step is to put Humpty Dumpty together again. After the efficiency of a team or intelligence of a boy has been divided into six dimensions, and ten indicators have been selected for each dimension, we have to put them all together, because we cannot operate with all those dimensions and indicators separately.

For some situations we have to make one over-all index out of them. If I have six students and only one fellowship to give, then I must make an over-all rating of the six. To do this I must in some way combine all the information I have about each student into an index. At another time we may be more interested in how each of several dimensions is related to outside variables. But, even so, we must find a way of combining the indicators, since by their nature the indicators are many, and their relations to outside variables are usually both weaker and more unstable than the underlying characteristic which we would like to measure.

To put it in more formal language, each individual indicator has only a probability relation to what we really want to know. A man might maintain his basic position, but by chance shift on an individual indicator; or he might change his basic position, but by chance remain stable on a specific indicator. But if we have many such indicators in an index, it is highly unlikely that a large number of them will all change in one direction, if the man we are studying has in fact not changed his basic position.

To put the matter in another way, we need a lot of probings if we want to know what a man can really do or where he really stands. This, however, creates great difficulties in the fourth step of the measurement sequence which we described above. If we have many indicators and not all of them move in the same direction, how do we put them together in one index? Only recently have we raised the question: can you really develop a theory to put a variety of indicators together? The subject is a large one, and it is impossible to go into details here. The aim always is to study how these indicators are interrelated with each other, and to derive from these interrelations some general mathematical ideas of what one might call the power of one indicator, as compared with another, to contribute to the specific measurement one wants to make.

In the formation of indices of broad social and psychological concepts, we typically select a relatively small number of items from a large number of possible ones suggested by the concept and its attendant imagery. It is one of the notable features of such indices that their correlation with outside variables will usually be about the same, regardless of the specific "sampling" of items which goes into them from the broader group associated with the concept. . . .

Warren Weaver

Explanation and Scientific Investigation[1]

The explanations of science are often regarded as so complete, so precise, so irrefutable, and so ultimate in nature that other types of interpretation of experience are rudely crowded out. Numbers, charts, equations expressed in abstract variables—these are by some supposed

[1]The author is vice president of the Alfred P. Sloan Foundation, New York. This article is adapted from the 4th Daniel Coit Gilman lecture, given at the John Hopkins University School of Medicine, Baltimore, Md., on 24 September 1963.

Reprinted from *Science*, Vol. 143, No. 3612 (March 20, 1964), 1297–1300, where it appeared under the title "Scientific Investigation." By permission of the author and the publisher, *Science*. Copyright 1964 by the American Association for the Advancement of Science.

to constitute the final, absolute, and complete explanation of all phenomena. And as science successfully moves from its major conquest of the physical world to equally promising attacks on the biological world, including all mental and emotional phenomena—indeed as science moves towards an analysis of the nature of life itself—this assumption that science "explains everything" becomes more and more formidable.

Thus it seems useful to examine the character of scientific explanation. Possessing not even the vocabulary of philosophy, I propose to phrase my comments in very simple language—what my friend Fred Mosteller calls "kitchen words." What is lost in the appearance of scholarship will, I hope, be at least somewhat compensated by clarity.

Consider, then, the person who is confronted by something which he does not understand. He goes through some process of talking, or listening, or reading, or thinking, or experimentation, or perhaps of all of these. It may take 10 minutes, or it may take years. Subsequent to that process, he says, "Well, at least I have made a start in a good direction, for now I understand better than I did."

Goals of Inquiry: Explanation

What has happened to that person in the interval between his complaint that he does not understand and his later feeling that he now does partially or even fully understand? What, in other words, is the nature of this strange process we call explanation?

Perhaps it will be well, at the outset, to note that a person usually considers a statement as having been explained if, after the explanation, he feels intellectually comfortable about it. I am sure that this criterion is too vague to be approved by the philosophers; but I am also sure that it expresses something that is widely understandable and acceptable. The average person applies this criterion to the explanation of a machine, or a process, or any natural phenomenon. The scientist is influenced by similar considerations. After a good explanation he is likely, because of his special interest, not only to feel intellectually more comfortable but also to experience a very active satisfaction or a very real esthetic pleasure. And *scientific* explanations characteristically have two further very important aspects.

First, the scientist who "understands" a phenomenon is almost always in a position to *predict*. He can say with confidence, "under such and such conditions, such and such will happen." If he has rather full understanding, he may be able to add, "If you change the conditions in such and such a way, then the results will be altered in such and such a way."

The second aspect flows directly out of the last preceding remark. For if the scientist knows how to change the result by altering the attendant circumstances, then he is well along toward accomplishing *control*. And the control of natural phenomena, to bring them more effectively to the service of men, is obviously one of the major aims of science. Indeed a scientist is very likely to say that he "understands" a phenomenon if he can predict it and can control it. If he can express this in mathematical equations, and if he can thus relate the phenomenon in question to a wide range of other phenomena, then he is likely to consider the explanation satisfactory and complete.

THE TWO TYPES

With these preliminaries behind us, we can now begin a direct discussion of the nature of explanation. I suggest that there are two main types of explanation, very different in character, and useful in different circumstances.

The first, the more familiar, the older, and by far the more popular, consists of explaining something by *restating or describing the unfamiliar in terms of the familiar.*

This is the way the dictionary explains the meaning of a word. You may not understand the word *euphroe*; but when Webster tells you that it is a little block of wood with a hole in each end, used to cinch a tent rope, then the meaning of the word has been explained.

This is what happens when a person, completely mystified by the idea of electromagnetic waves spreading out from a radio station, is told: "You have, of course, seen the circular ripples expand on the surface of a still pond when you drop a pebble. Notice that the ripples get weaker as they get further and further from the center. (Do you live so far from your station that the signals are weak?) You have doubtless noticed that, just behind a rock which sticks up above the surface, there is calm water with no ripples; but a few feet beyond the rock the ripple patterns from either side join up and show little or no residual effect of the rock. (If you live just 'behind' a big steel-frame building you will have trouble, won't you, in getting the radio signals; but a half mile beyond, the building does not shield or interfere.) And remember that the water ripples are two-dimensional waves on the surface of the pond; but the radio waves are three-dimensional, going out in spherical form like the successive spherical shells of a magic onion which keeps growing larger and larger."

Quite apart from the obvious incompleteness of these remarks, this is a reasonable example of at least the early stages of an "explanation" of the sort that many persons find satisfying. Indeed, examples not very much more sophisticated than this have played exceedingly important roles in the development of scientific theories, particularly in physics.

We should note, parenthetically, that every poet should feel friendly toward this type of scientific explanation; for it is essentially equivalent to the similes and metaphors that, explicit or implied, are the very essence of poetry.

Indeed, the matter goes deeper than this. As Bronowski has said . . . "The scientist or the artist takes two facts or experiences which are separate; he finds in them a likeness which had not been seen before; and he creates a unity by showing the likeness."[2]

When one thinks a little it is promptly clear that this first type of explanation (which points out that the unfamiliar is in certain respects like the familiar), when considered simply *as* an explanation, is almost completely illusory. It restates in terms which are very familiar, to be sure, but which, upon really honest examination, are just exactly as little "understood" as the unfamiliar concept, procedure, or phenomenon for which explanation is sought. A person may be very familiar with expanding ripples on a pond without actually having any clear idea at all as to *how* these ripples propagate, interfere, or attenuate. Long familiarity has dulled penetrating curiosity, and one just unthinkingly accepts the familiar as understood.

It does not, however, at all follow that this type of explanation is silly or useless. For it is a rather remarkable fact that there is a tremendous amount of non-obvious isomorphism in the logical structures of natural phenomena, especially when all the phenomena in question are broadly macroscopic in scale—say, involving space dimensions ranging from those of optical microscopy up to planetary and very possibly to galactic dimensions. That is, there are very numerous pairs, A and B, of "things" in the physical world which in important respects behave similarly; so that it often constitutes a useful, illuminating, and suggestive explanation, when meeting a strange pair of related variables, to be reliably told that this unfamiliar pair is "like" a familiar pair. This turns out to be mentally satisfying; and it often suggests the application, to the new and initially strange pair, of a lot of procedures which have previously been found useful in the case of familiar pairs. Presently, of course, the new pair becomes a familiar pair.

This recognition of similarity of behavior is one of the major ways in which science moves forward in its great task of bringing its type of order and beauty out of confusion. Bronowski, in his superb little book *Science and Human Values*, says, "All science is the search for unity in hidden likenesses.[3] . . . The scientist looks for order in the appearances of nature by exploring such likenesses.[4] . . . The discoveries of science, the works of art, are explanation—more, are explosions, of a hidden likeness. The discoverer or the artist presents in them two aspects of nature and fuses them into one. This is the act of creation, in which an original thought is born, and it is the same act in original science and original art."[5] . . .

There are serious limitations to this first type of explanation, but before commenting on them, I will say a little about the second type of explanation.

The second type of explanation has no concern whatever with familiarity. On the contrary, it characteristically describes a phenomenon or a statement in terms which are almost indefinitely *less* familiar, or in any event more basic and more abstract, then the phenomenon or statement being explained.

A good example is furnished by mathematics. Suppose one is confronted by a theorem which is complicated, subtle, and wholly unfamiliar. The first stage of explanation (useful in the case of a person well trained in the mathematical field in question) consists of *proving* this theorem by showing that it logically follows from various theorems previously proved. At this stage the second type of explanation shares, to a mild degree, the essential feature of the first type. For

---

[2] J. Bronowski, *Science and Human Values* (New York: Harper and Row, 1959), p. 35.
[3] *Ibid.*, p. 23.
[4] *Ibid.*, p. 24.
[5] *Ibid.*, p. 30.

the trained mathematician will in fact be familiar with the previously known theorems which he uses in proving the new theorem.

The essential difference, however, is that the mathematician, while he enjoys the fact that he is familiar with the older theorems, does not in the least base his just-gained confidence in the new equation upon that element of familiarity. For it is not at all familiarity which makes him trust and understand the older theorems. His trust and understanding rest solidly on the fact that these older theorems have, in their turn, been logically deduced from a still older, still more primitive, set of theorems.

One must not underestimate the power and excitement of this step-by-step procedure of proof. His friend John Aubrey described the reaction of Thomas Hobbes, who, about 1630 when he was 40 years old, accidentally looked at a copy of Euclid's *Elements* open at the pages containing the proof of the famous theorem of Pythagoras.

> "By G- -," sayd he (He would now and then sweare, by way of emphasis) "By G- -, sayd he, this is impossible!" So he read the demonstration of it, which referred him back to such a proposition; which proposition he also read. *Et sic deinceps,* that at last he was demonstratively convinced of that trueth. This made him in love with geometry.

Now this procedure of pushing the explanation down, step by step, to lower and lower, more and more primitive levels of explanation obviously requires examination. For where does this descent stop?

As an historical fact, there have been three answers to this question. The first and least satisfactory answer comes from those individuals who, so to speak, descend one or two or three steps and then get so bored or so confused that they are content to give up. For them this descending set of steps ends in a fog.

The second kind of answer is that which appealed to Euclid. One descends, step by step, until one reaches a "Bottom Step," on which are found statements (axioms) which supposedly are so obviously true, so clearly necessary, and so patently clear that all reasonable men are supposed to agree to these statements and to accept them without any further examination.

This kind of answer seemed satisfactory to many persons over hundreds of years. But, as every schoolboy now knows, this kind of answer is no good. For it turned out that the axioms on this supposedly bottom step simply were *not* obviously true, nor were they *necessary*. Euclid considered it unthinkable to question the statement "through a point not on a straight line it is possible to draw one and only one straight line parallel to the given line." But we now know that a person can perfectly well assume that through this external point one can draw *more* than one straight line parallel to the given line. The result of this second assumption is not logical chaos or contradiction but an alternative geometry, rich, consistent, beautiful, and useful.

So it is not tolerable to let these descending stair steps, on which are found successively more and more simple explanations, terminate in a fog, nor is it tolerable to let them terminate in a universally accepted bottom step. What then can be done? It was a triumph of 19th-century mathematics to see a respectable alternative. It is to descend down to a step which is *not* labeled "Unique Bottom Step" but which *is* labeled "This Is As Far As We Go." On this step one does not find axioms—statements which are supposed to be necessary and obviously acceptable to all. One finds postulates—statements which, for the purposes in hand, are simply assumed to be true. Two different mathematicians can perfectly properly, even when working in the same field, assume two quite different sets of postulates. If you do not like a given set of assumptions, there is no compulsion. You can just decide not to play and can take your doll rags and go home. Or you may decide to accept the set of postulates, this not at all meaning that you "believe" them but simply meaning that you adopt them, and that you will now, starting there, apply logical procedures and lift yourself, step by step, to ever higher levels of complicated and sophisticated deductions.

The "explanation" of a statement on a high-level step is now clearly to be obtained by tracing the relationship between that statement and statements on the next lower step, the second lower step, the third lower step . . ., until one reaches the step labeled "This Is As Far As We Go." On this step one finds nothing "obvious," nothing "true." One finds only statements which have a footnote, "It seems interesting to assume this set of remarks." And it seems to me that this type of explanation is precisely the process of descending to the step on which we agree to stop.

We may again parenthetically note that poets should find nothing alien or objectionable here, but rather ought to be enthusiastic about the element of ultimate mysticism that exists at the bottom of this type of scientific explanation.

Interrelationship

In terms of the stair-step metaphor we have been using, we can now describe the interrelationship between the two types of explanation. In the first type, being located for the moment on a step which contains strange and not-understood things, one looks horizontally about him and observes a neighboring set of steps. This second set—or at least a few steps of it—are friendly and familiar. One has been on them many times. And one notices (or is told) that the strange elements on the step one is now occupying are "like" elements over on a familiar step (presumably at about the same level) of the other set.

It is a curious fact that this observation is very comforting, whether or not one has ever visited more than one or two steps of the second set.

In terms of this metaphor, one can usefully refer to this first type as "horizontal explanation." The procedure does not move vertically downward to deeper levels of simplicity or abstraction but moves horizontally over to more familiarity. And although this is an interesting, pleasant, and clearly useful procedure, it seems clear that it does not, in any sophisticated sense, and certainly not in any ultimate sense, constitute "explanation" at all.

The second type is "vertical explanation." And it seems fair to say that it is deep and logical, but that, again, it certainly is not ultimate.

At this point it is useful to return to some remarks I made earlier, when referring to the nonobvious isomorphism of natural phenomena—namely, remarks concerning the importance of the scale of the events. For it seems at least generally true that horizontal explanation is useful when the scale of the events being explained is roughly the same as the scale of the more familiar events used in the explanation. The electrical oscillations in circuits are usefully discussed in terms of the oscillations of a taut string; and the sizes, masses, periodic times, and so on, in the two cases are, very roughly at least, of the same large-scale order of magnitude. This statement about spatial and temporal size is an exceedingly rough one. It can be expressed (and very probably this is not accidental) by saying that the distances and times and masses involved must not be too extremely small (or large) as compared with the dimensions and mass of a man, and as compared with the times (years, days, hours, minutes, and seconds) which enter directly into human experience. The horizontal-explanation method has worked surprisingly well on a planetary, or even larger, scale; and surprisingly well down to molecular dimensions. But when one tries to push this method down to atomic dimensions, and certainly when one tries to push it to nuclear dimensions, then the method of horizontal explanation (at least so I believe) collapses entirely.

For example, you will read in articles on modern physics that the density within the nucleus of an atom is of the order of ten thousand million tons per cubic inch. I refuse to gasp and say, "Isn't that amazing!" For it seems to me that such a statement is simply meaningless, and the collapse of meaning has resulted from trying to use man-sized language, and horizontal explanation, where they are totally inapplicable.

The situation at the cosmic scale does not seem so clear; but perhaps attempts to answer such questions as, "Is the universe expanding?" or "What is the age of the universe?" all experience difficulties which result from the inappropriateness of using "human-size" concepts for such problems. . . .

The yearning for "physical explanation" (which as far as I can see always means horizontal explanation) is an urgent one, which extends to all levels of sophistication in science. It is clear that Einstein never gave up the idea that physical interpretation of the unitary events of physics was both possible and desirable. There is a long list of earnest and able individuals who have been puzzled by "action at a distance," and who have sought some other model with macroscopic properties which would help them escape the, for them, intolerable fact that action at a distance is not "understandable" (although, curiously, action *not* at a distance presents equally grave difficulties). All of these persons have, in my judgment, not faced up to the nature of explanation. Vertical explanation has not been satisfying to them; and their concern has been with cases to which horizontal explanation is not applicable.

I want to emphasize an aspect which the two types of explanation have in common. It is an aspect, moreover, which the scientist values very highly indeed, for both practical and esthetic reasons. Namely, either type of explanation addresses itself to an element of our experience and gives meaning to it, gives new significance and richness to it, suggests new usefulness for it, in short *explains* it, by placing it in a broader context. Horizontal explanation does this vividly, but narrowly. An electromagnetic wave is put into the context of more familiar mechanical or hydrodynamical waves on strings or ponds. Vertical explanation probes ever so much deeper into the isomorphism of phenomena and puts the case under study within the total context of all the possible phenomena which conform to all the relationships deducible from their common origin—namely, the postulates on the bottom step. The electromagnetic wave thus is placed within the broad context of all possible types of solutions of certain very general types of differential equations. All the practical and esthetic values which result from this recognition of relatedness constitute, I think, the important essence of explanation.

Irwin D. Bross

Predicting Future Events

. . . How can man foresee the future? For thousands of years man has searched for the answer—the secret that would endow its possesor with riches, fame, and possibly happiness.

Reprinted with permission of The Macmillan Company from *Design for Decision* by Irwin D. Bross. Copyright 1953 by The Macmillan Company.

Many seers have come forth with the proclamation that the secret was theirs, that their eyes could see beyond today and into tomorrow. Some have generously offered to share their secret, for a small fee, with their fellow men. These claims have never stood up when judged by the pragmatic principle.

Yet man has never abandoned his efforts to read the future—too much depends on it. He has, however, become more modest in his demands. Since he cannot peek into the future, he will settle for shrewd guesses. The more modest question: How can man predict the future? has at least a partial answer. The answer: By studying the past.

Even in the mysterious and erratic world in

which we live, there are some threads of continuity. There is chaos and confusion all about, but also some system and stability. Our progress in the real world is like driving along a road that is shrouded in a heavy fog; there are no sharp, clear details, but only vague outlines. By looking very hard through the swirling, random fog shadows we can distinguish enough of the more permanent road shadows to enable us to go ahead successfully if we go slowly and use caution.

Similarly, the first step toward prediction is the search for stable characteristics—those characteristics which persist over a period of time. In fact the simplest procedure for prediction is a method often called *Persistence Prediction*.

Persistence Prediction means nothing more than the prediction that there will be no change. If one wishes to predict the weather tomorrow by this method, one simply describes the weather today. Sometimes this device works out surprisingly well.

In weather forecasting, for example, Persistence Prediction is hard to beat. The modern meteorologist uses the data from hundreds of weather stations, combined with a complicated air-mass theory, in order to arrive at weather forecasts. But in one hundred predictions the scientific weatherman will (on the average) be right in only about ten more cases than a weatherman who used Persistence Prediction. This is not because modern methods are bad, but because persistence methods are good (they give the correct prediction about three quarters of the time so that there is not too much room for improvement).

A fan who used Persistence Prediction would do just as well as the sports experts in predicting the outcome of the National and American League pennant races. This is especially true in the American League over the last few years! Some sports writers seem to have noticed this fact and follow the policy of sticking with the champion.

The method has its limitations, of course. It only works in relatively stable or slowly changing situations. It is often of little practical value because, as in the stock market, the money is to be made by predicting *changes*. Nevertheless, Persistence Prediction is the basis of a number of successful predicting systems including those used by insurance companies. Insurance life tables rely on the fact that death rates, while not actually constant, change rather slowly.

A second scheme for forecasting is *Trajectory Prediction*. This scheme assumes that, although there is change, the extent of change is stable. If noon temperatures were recorded on successive days as 75, 76, and 77 degrees then the Trajectory Prediction for the next day would be 78 degrees. In making this prediction we have assumed that the rise of one degree per day will continue. This method may give fairly good predictions for the next time-interval, but it can also lead to ridiculous long-range forecasts. If we used the assumption of a one-degree rise per day to predict the temperature a year ahead, we would obviously be in hot water.

Trajectory methods are used in artillery fire control, some weather forecasting, short-range stock market prediction, and in estimating the size of human populations. The word "trend" is often used instead of trajectory.

Cyclic Prediction is based on the principle that history repeats itself. The method had some notable early successes: the first effective long-range predictions made by man employed this device to foretell eclipses and other astronomical events.

In Cyclic Prediction, it is assumed that cycles or patterns of events are stable. The method has been used in predicting the return of comets, the occurrence of sunspots, insect plagues, high and low agricultural yield, weather, stock prices, and even (by Spengler) the course of our civilization.

The early successes of Cyclic Prediction stirred great hopes in the breast of man that here, at last, was the long sought-for secret of prophesy. Even today some investigators, notably in the stock market, are still striving—but without much luck—to realize this ancient promise.

Astrology, a perversion of the method, still survives on the strength of this old prestige.

In going from Persistence Prediction to Cyclic Prediction, there is a utilization of more and more data. The former needs only the most recent occurrence of the event, while in the latter the available historical information is used—in fact, the standard alibi for the failure of Cyclic Prediction is that the record does not go back far enough.

Associative Prediction differs from the foregoing in that it uses the data from one type of event to predict a second type. Conditioned response is an example of Associative Prediction. Pavlov made dogs salivate by ringing a bell. To accomplish this, he rang a bell just before feeding and repeated the pattern over a period of time. The association of the two different types of events, ringing of the bell and feeding, is very similar to a causal event chain. In both cases the stable element that is the basis of prediction is the stability of a *relationship* between two events.

In politics, economics, and everyday life, Associative Prediction is the favorite method. Commodity market speculators feel that they must stay abreast of national and international events in order to judge the movements of the market. Even the general public is aware of the violent gyrations of prices immediately after war scares or peace scares.

A relationship between events is often expressed by the word "cause." People say that a large national debt "causes" inflation, that overproduction "causes" unemployment, that armament races "cause" war. In everyday life overeating "causes" indigestion, nasty remarks "cause" hard feelings, and extravagance "causes" ruin.

In all of these examples one type of event, the cause, generally precedes the second type of event, the effect. From the point of view of Pavlov's dogs the bell "caused" the feeding. As long as the word "cause" is used in this sense, it serves a useful descriptive purpose.

If we stick to the simple meaning of the word "cause," our quest for causes will not go off on wild goose chases. All that we really want to do is to identify the bell that comes before feeding.

This is not always easy to do because Associative Prediction greatly enlarges the area that must be searched for clues. If we want to predict the price of a stock, we cannot focus our attention solely on the previous history of the stock; we may have to examine events of many different kinds. Our "bell" may be an event that takes place ten thousand miles away—a political speech by a foreign leader or the report of some new scientific discovery. Of the many events that we might study, only a few will have any discernible association with the events we wish to forecast.

A serious weakness of Associative Prediction is that unless a great deal of care is exercised in the selection of the "bell," the whole process may degenerate into nonsense. Fortune-tellers use Associative Prediction, their "bells" being such events as the fall of cards or the configurations of tea leaves. They have never demonstrated that the events they use in exposing the future are *relevant* to the events predicted.

Analogue Prediction sets up a correspondence between two sets of events. One of the sets is simple, or at least familiar, and consequently predictions can be made for this set of events. The analogues of these predictions are then made for the second set.

If modern nations with atom bombs are analogous to small boys playing with sticks of dynamite (as has often been suggested) then the fate of the nations can be predicted by analogy with the fate of the boys.

Analogy is one of the most potent gimmicks in an author's arsenal, especially if he is dealing with strange or difficult topics. Not only will an apt analogy make a reader feel that he understands what the author is saying, but it also may convince the reader that the author knows what he is talking about. However, verbal analogies have a dangerous tendency to blow up in the user's face, especially when they are carried too far. Because of the great overuse of analogy, argument by analogy is no longer in good standing with logicians.

Nevertheless, when properly used, analogy may be a powerful tool for prediction. This is especially true if a *mathematical* analogy (or model) can be constructed. By mathematical arguments, the performance of this model can be predicted. Events in the real world may then be forecast by analogy.

The use of scale-model airplanes in wind tunnels to predict the performance of full-sized aircraft and the use of experimental animals to test drugs destined for human consumption are two examples of Analogue Prediction in the field of science.

This list of techniques for prediction is not intended to be exhaustive. Many special techniques have been developed to meet the many different prediction problems that arise.

There is one method of prediction that deserves mention here because it (and it alone) is 100 per cent successful. This is the technique of *Hindsight Prediction*, the prediction of an event after it has already occurred. Radio commentators, newspaper columnists, economic authorities, and politicians use this I-told-you-so method with excellent results. All that is necessary to apply the method are the ability to make ambiguous (or even contradictory) remarks and a talent for selective amnesia. Examples of Hindsight Prediction abound in the writings of historians and philosophers.

R. M. MacIver

Causation in Social Research

Primacy of the Concept

Back of all our conscious activity, whether we think things or do things, lies some concept of causation. Some philosophers, like David Hume, have denied that we have any valid ground for the attribution of cause and effect; but there has been no one who did not at every turn act and think, live and breathe, as though that ground existed.

Reprinted from *Social Causation* (New York: Harper & Row, Publishers, 1964), pp. 5–10 and 73–77, where it appeared under the title, "Causality." Copyright by Ginn & Company, 1942. By permission of the author and the publishers.

It was all very well, and no doubt very salutary, for David Hume to challenge us to discover anything more in nature or in experience than *one object following another*,[1] but in putting that challenge on paper did he not expect that it would have some *effect*, at least on the thoughts of others? When we speak of cause and effect we certainly do not *mean* "one object followed by another." The light of mid-day is followed by the darkness of night; it is a sequence perhaps more invariable than any other, but we do not think of the light as the cause of the darkness. The noise of a train is heard before the train appears in sight, but we do not think of the noise as the cause of the appearance of the train. The fall of a stone in water is followed by a splash, and we do think of the stone as the cause of the splash. The concept of sequence, even invariable sequence, is one thing; that of cause is definitely another. We are not here concerned with the validity, but only with the universality, of the application of the

[1] David Hume, *An Essay on the Human Understanding*, Sec. 7.

concept. Whenever we set about any task we assume causation. Whenever we use an active verb we postulate a cause, and whenever we use a passive one we postulate an effect. Whenever we attribute continuity or process, we imply causation. In all doing and suffering we experience, or at least believe we experience, cause and effect. Hume took the stand that we do not *perceive* causation, that we have no sense "impression" of it, and therefore no right to impute it. The obvious answer is that, if we employ the term as he did, we have no sense "impression" of anything that is a relationship. We have no sense "impression" even of succession, but only a succession of sense "impressions."[2] Hume's one object following another becomes instead one sensation, or one mental image, habitually associated with another. We are transported into a shadow-land of impotent "impressions," and there is no way out any more into the objective realm. The "phenomena" are in our "minds," not in nature. And our minds, too, become a succession of moments of awareness, save that even the succession becomes a figment of the mind figment that thinks it. Dissolve all relationships into Hume's sense "impressions," and we dissolve our world. Deny the concept of cause, and every other concept—succession, change, continuity, time itself—vanishes into thin air. Let us consider, for example, the concept of time. Time has meaning for us as the precondition not merely of change, but of change as caused. We cannot conceive of change except as caused change. Otherwise it would be the idle flux of meaningless appearances, at most the moving picture on a screen, which can be run backward as well as forward, which can be slowed or accelerated, which can be repeated again and again—provided that, though there is no causation in change itself, there is beyond it some cause that idly plays with the idle sequence. Change obviously implies three things, that which changes, that which is constant relative to that which changes, and the span of time in which the change takes place. But if change were uncaused, then it would have no necessary direction in time, and time itself would become directionless and be no more, since the one-way signposts of change are all effaced.

The world we experience is a world of continuity and change. We perceive change and therefore continuity. We perceive continuity and therefore change. It is a world in which the concept of causation reigns over our experience. Our life is a process signalized by events. Succession is transition, a one-way road from the past to the future. The road has no breaks in it. The discontinuous signposts, the events of experience, merely reveal the continuity of passage. Things happen *to* us and *in* us continually. We react to the happenings, and the totality of happenings and reactions—our strivings, emotional tensions, controls, comprehensions, anticipations—is our experience. Our experience has the finality that belongs not to mere change, but to irrevocable change. As an ancient Greek poet said, even God cannot make undone the things that have been done. Hence our experience must always assume the character of a causal nexus. There is no escape from the web of cause and effect.

Whether we could derive any conception of causation from the mere observation of external change, may be doubtful. Possibly from that alone we could get no further than the concept of succession, that of "one object followed by another," to which Hume reduced our knowledge of cause and effect. But speculation on this head is vain, since the capacity to observe external change already implies the causative activity of the observer. To live . . . is itself a special causal activity, and to be aware of anything is also to be aware of this causal activity of ourselves. In being aware there is at least implicit the concept of causation. But the causal nexus we attribute to the changes of the external world differs in one important respect from the causal nexus we discover when we are ourselves the

---

[2]See, for example, G. H. Mead, *The Philosophy of the Act* (Chicago: University of Chicago Press, 1938), pp. 646–51.

authors or sources of change. In the latter case we often foresee, in some part, the changes we bring about; we undertake the activity in order to bring about the foreseen change. Such activity introduces a factor of causation that physical nature nowhere reveals to us. Besides other differences, there is here a different relation of time and change. In the physical universe each instant, so far as our science goes, determines that which succeeds it. Given the moment, there is present all that the next moment requires for its coming to birth. But for teleological activity, wherever or in whatever degree we find it, the moment is not enough. Here the causal process has another quality, that of *duration*. The image of what is yet to be informs the process of its becoming. We do not merely observe in the present the signs of the direction of approaching change. The teleological impulse, previsioning the change, is itself dynamic. It exists prior to the change and persists through the process of its accomplishment. It may lie low and work in the twilight of consciousness or it may emerge into conscious purpose. In the latter form it constitutes the distinctive quality of causation within the sociopsychological realm. In the teleological process the effect, not yet as actuality but as prevision and end of action, operates to bring itself about. The effect, dimly adumbrated or clearly conceived, exists as projection in advance of the physical process from which, as actuality, it emerges. This reversed relationship is the essential factor in the control that the living thing, or the social group, exercises within its environment, when it is no longer content merely to adapt itself to that environment but has undertaken the greater adventure of adapting also that environment to itself. When it reaches this stage, the living being exists not only in the ever-enduring present; it exists, as it were, *in the time dimension*, embracing at each moment the future and the past with the present. So the concept of time and the concept of causation are essentially interdependent.

There are those who, acknowledging the paramount role of the concept of causation in human experience, nevertheless conclude that it has no scientific warrant when applied to the outer world of physical nature. "Technical and mathematical language . . . is surely, if slowly, replacing expressions of causal relations with mathematical functions or equations, which are neutral to all anthropomorphic hypotheses."[3] . . .

We experience change everywhere, and wherever we experience change we summon the concept of causation. For human life change is a one-way sequence. It is here, in the irreversible character of change, that we find the essence of causality, something that no "mathematical functions or equations" can ever represent, or even suggest. Moreover, the concept of causation embraces not only that which changes but also that in terms of which it changes. Change is always relative to something else that does not similarly change. The earth moves in relation to the sun, the train moves in relation to the earth, the passenger moves in relation to the train, and his hand moves in relation to his organism. Nothing changes, nothing passes away, except as against that which by contrast endures. A change, then, is a difference occurring within a relatively determinate system. Change and the unchanging are correlative, and as we cannot think of the one without the other we cannot think of either except in the light of the principle of causation.

Everything is in process of change, except conceivably the infinitesimal units of "matter" or "energy" of which all things consist. Everything that is now is different from what it was: its form, place, properties, relations, functions, undergo change incessantly. The elements alone —or rather most of them—seem indifferent to the processes they endure, for though changed by changing temperature, pressure, and exposure to the action of other elements, they alone can return the road they travelled, back to their former state again. But for all constructed things,

[3] M. R. Cohen, *Reason and Nature* (New York, 1931), pp. 224-25.

for all things that assume character and function in space and time, for all the works of nature and the little works of man, for all things that expend energy, for all that live, there is no return to a former state. For them all the process of change is irreversible.

This primary datum gives determination, law, meaning, finality, within the universe. Were the process reversible, like a mere mathematical function, there would be no significant history. Life would lose its order, its quality, its sting—and so would death. Change would be an irrelevant incident, an episode in chaos, affecting the whole as little as the waves affect the surface of the sea.

. . . .

Facile Imputation

Numerous books and articles offer causal explanations of social phenomena. Some deal with social movements—the rise of fascism, the ebb and flow of democracy, the growth of some new cult, economic or religious. Some deal with social trends—urbanization, the tempo of technological exploitation, the changing phases of the "economic cycle," the decline of the birthrate. Some deal with sociopathological problems, explaining the relative frequency of crime, delinquency, pauperism, divorce, political corruption, and so forth. The incessant impact of social change in its myriad aspects is the perpetual challenge to which these responses are made. Not only our scientific interest but also the imperative demands of public policy here give weight and urgency to the investigation of causes. . . .

For reasons that will appear more fully in the sequel, the methodological investigation of the causes of social phenomena is beset by peculiar difficulties. Any social change we seek to explain is meshed in a tangled web of its inclusive history. It is dependent on conditions arising within every order of reality—physical, biological, psychological, and social. The factors we invoke embarrassingly combine the universal elements of physical causation and the human elements of teleological causation. The phenomena themselves are often hard to demarcate. What shall we include in a social movement? What is socialism or fascism or democracy? What is common to the different instances to which we may agree to apply the name? Each of the movements or systems we study presents a different aspect to different observers, according to their focus of interest, their experience, and their temperament. These systems, however impressed on the face of things, are not themselves "things" in the sense of being detachable or demarcated items of the objective world. As operative social systems they are the partial objectification of human conceptions, of interest and beliefs, of purposes and dreams. They have no clear boundaries marking where one ends and another begins. They have ever-changing configurations. All the time each conjuncture is undergoing subtle modifications. There is unbalance and mobility, the constant emergence or injection of new elements. What we call the movement or the trend is an aspect of an endlessly variant flux. We cannot isolate it for experimental study. We cannot find two identical instances of it, or even two instances with no relevant difference between them. Nor can we find two instances that differ only by the presence or absence of a single clearly designated factor.

In the face of these perplexities it is not surprising that much investigation into the causes of social phenomena is of a haphazard hit-or-miss character, or that various devices are employed to skirt round the difficulties of a direct attack on the problem. For example, it is not unusual to proceed on certain assumptions that simplify the task of imputation but render the conclusions thereby attained dubiously applicable to the actual world. To this order belongs the economist's conception of an equilibrium that is being forever disturbed and forever reasserting itself. To the same order belongs the recourse to the formula, "other things being equal," even where all our experience tells us that they are not so. Another device is to limit the search for causes to factors falling within the same category as the phenomenon to be explained, as when we try to account for unemployment or a decline in the volume of business solely with reference to the internal functioning of the economic mechanism, regarding political, cultural, and social changes as merely incidental disturbances, "random factors," and so forth. Such departmentalism has been widely prevalent in the social sciences. The search for causes cannot be so confined. We can describe and classify social phenomena according to the categories of tradition or of academic division. Our interest may very properly be focussed within the range of any of these categories. In fact without these categories we could scarcely make any advance towards articulate knowledge.

But when we tackle the central issue of causation we must ignore all frontiers. For the quest leads us to the common substratum of social phenomena. . . .

We could take in turn each of the social sciences, or branches of social science, and show that the degree of their inability to grapple with the issue of causation is the main obstacle to their advance. For unless we can discern the causal nexus of things we do not know the way they belong together or the way they are set apart, we do not know the nearer and the more inclusive systems they constitute, we do not know their behavior or their properties or the routes they follow in their changing relationships. We are limited to description and to insecure classification.

Eugene J. Meehan

Empirical Theory: Explanation in the Social Sciences

Theories are explanatory devices, and something more. In an explanation, isolated observable phenomena, loosely defined, are brought together and related systematically. As a first step, general statements of various sorts classify observables according to their properties, and a classification system is the simplest form of single-step explanation. Particular events are explained by "bringing them under" the general

Reprinted with permission from Eugene J. Meehan, *The Theory and Method of Political Analysis* (Homewood, Ill.: The Dorsey Press, 1965), pp. 128–34 and 145–50.

statements in the classification system, deductively or in some other way. The explanatory hierarchy is always open at the top, for it is always reasonable to demand an explanation of the general statements used in explanation, and then to demand a further explanation of the structure used to explain the generalizations. At any given time, some level of explanation exhausts our knowledge, it lies beyond further explanation, but only in practice and never in principle.

A theory is a generalization, or set of generalizations, that explains general statements, or explains other theories. Common usage, even in philosophy of science, is not exactly clear about the distinction, largely because certain applications of terminology were historically determined, but the general principle is safe. The answer to a request for an explanation of a general statement or law is always a theory. The relation between theory and observables is always indirect; generalizations mediate between theory and direct evidence. Theories explain general statements by relating them to one another, much as general statements relate individual phenomena, deductively or in some other way. A theory supplies the larger conceptual scheme in which generalizations find a place and thereby are explained. And like a good generalization, a strong theory does more than simply relate what has already been observed. It "spills over" into areas as yet unexplored, suggesting relationships not yet observed and generalizations not yet asserted. A theory is more than a means of explaining; it is also a guide to research, a means of substantiating other theories, an aid to discovery.

On this view, a theory is simply an instrument for ordering and arranging general statements that man creates for his own purposes and not, in some way, a map or picture of "reality." This is the "instrumentalist" conception of theory, and though it is sometimes disputed by philosophers of the "realist" school, the dispute is best left to philosophy of science. The significant point for the political scientist is that there are no good grounds for asserting that one type of theory is "more scientific" than another if theories are instruments. Man may use any and all theories that explain and, hopefully, assist him to discover new knowledge and integrate what is already known. Nor is a construction to be despised because it performs only some, not all, of the functions of theory. A theory may explain but not predict, for example, and if that is a weakness, it is not a reason for discarding the theory. All that can ever be decided is whether or not a given theory is useful in a particular inquiry, and here the subject-matter specialist is king. If the theory explains, if it suggests new relationships and generalizations, if it opens new lines of inquiry, or if it "satisfies" the person familiar with the data—if it does one or more of these things—then it is a "good" theory and worth using. The grounds on which theories are evaluated are somewhat vague. When a theory relates generalizations deductively, validation is relatively simple, but not entirely cut and dried; when theories are partial and weak, as is likely to be the case in political science on most occasions, evaluation is an act of judgment, not the application of formal rules of validation. In all cases, the value of a theory will depend upon its "usefulness" to the inquiry in which it is employed.

Theories and Generalizations

Theories explain or relate generalizations. This is a useful minimal criterion for a theory. In practice, however, some general statements, like the "Laws of Nature," are so broad and powerful that they actually explain other generalizations, though they are commonly referred to as "laws." There is no conflict in principle here; terminology, particularly when it originates in the distant past, is not standardized. Furthermore, a theory may contain only one general statement, though more often it contains several. So long as the structure is able to explain generalizations, we will consider it to be a theory, disregarding current usage. The distinction between law and theory is not quite so clear as we might wish, even in physical science, but a brief comparison of the two types of constructions will help define the properties of each and thus set them apart reasonably well.

General statements are related to observables directly and immediately; theories relate to observables only indirectly—through general statements. This is a useful means of separating the two types of constructs. A generalization classifies observable phenomena (loosely defined to avoid confusion with scientific constructions like atoms or genes which are not, strictly speaking, observables) according to their properties. Each general statement is an independent proposition, established by its own body of evidence; it can be considered an inductive generalization from a body of observed data. The meaning of a generalization is independent of any theory in which it may appear and its validity depends primarily upon its relationship to singular facts. This is not the case with a theory.

Theories are abstract and symbolic constructions, and not descriptions of data or inductive generalizations from observations. Theories relate general statements by appealing to underlying similarities, not to observable properties.

Theories supply a wider framework in which generalizations appear as special cases of broader principles. The link between theory and generalization is conceptual and not empirical. Some of the nonlogical terms in a theory will refer to observables, but others will refer to nothing that is directly observable. These "theoretical terms" are extremely important, for they give a theory its generalizing power and provide the linkage between theory and generalizations. "Gravity," for example, is a theoretical term, as are terms like "instinct," or "motive," used to explain human behavior. The relationship between theoretical concepts and observables is always a little vague. Of course, some inferential relationship must exist, for theories are meant to explain observed phenomena—they are not merely logical constructions. But the exact meaning of theoretical terms is defined implicitly by the terms of the theory, and not explicitly and precisely in terms of observables. A theory must include some statements that relate theoretical terms to observable data, but not all of the theoretical terms are so related, and even when the relationship is specified, it is often loose and imprecise.

The loose fit between theoretical terms and observable data is worth a great deal of emphasis. Theories often employ terms that are idealizations of observed phenomena (frictionless motion or absolute zero, for example); no physicist really expects his experimental data to fit theories employing such terms precisely. But even in those cases where the terms are not idealizations, the results of observation will not fit exactly. For that reason, theories cannot be proved or disproved in absolute terms. There is always some element of judgment involved in the evaluation of a theory, in the decision to accept or reject it. A theory is not simply an attempt to generalize or describe the real world in terms of observables. It is a conceptual structure—using abstract, symbolic terms—in which empirical generalizations find a place through their relationship to these symbols. The observable referrents of the symbols are imprecise. The symbol is "more general" than the observables it implies in a theory. This is the reason why theories are powerful explanatory tools. The looseness of fit, in other words, is an asset, not a liability.

The Structure of Theories

An empirical generalization is always a single statement. A theory *may* be a single statement but more often it comprises a number of general statements linked together in various complex ways. In the first case, the theory *is* the general statement: in the second, the theory is the sum of the general statements and the set of relationships that bind them. Often, the set of relations really define a theory; for example, in factor theories, where none of the generalizations employed are sufficiently broad to include other generalizations, the theory consists of the *selection* of general statements and the rules of interaction governing their application.

We will classify theories into two broad types, depending upon the kind of relationship that obtains between the various elements of the structure. If the relationship is deductive, the theory forms a hierarchical construction and the general statements in the theory (theorems) are logical derivatives of a few basic axioms or postulates. We will refer to theories of this type as hierarchical or *deductive* theories. If the general statements in the theory are not related deductively but held together by some other factor like relevance to a common class of phenomena, we will refer to them as *concatenated* theories, following Abraham Kaplan. Kaplan's terminology is preferred to Quentin Gibson's use of "factor" theories because it is broader and therefore more useful in political science. Social theories are at present all of the concatenated type, and even the limited form of factor theory that Gibson favors is often more than political science can produce. In philosophy of science, a common definition of a theory is "a deductively related group of general statements," and discussions of theory center around this conception. But that is far too strict for political science, for it requires the use of universal general statements and they are not found in abundance in politics. There seems no good reason why concatenated theories cannot perform various useful and legitimate functions in the study of politics, and if the restricted form of deductive theory is easier to use, and more powerful, that is not a good reason for eliminating concatenated theories from consideration. Indeed, I shall suggest, in another part of the chapter, the use of concatenated theories or "quasi-theories" so loose and weak that they can hardly be called "theories" at all, though they seem likely to prove useful in the discipline.[1]

The division of theories into deductive and concatenated types underlines the importance of the kinds of general statements that are available in a discipline. Without empirical laws, or universal general statements, deductive theories are impossible. If political science were limited to deductive theory there could be no political theory, for most of the general statements that political science disposes are tendency statements and they cannot be combined deductively. Political theories, if we are to have them, must be connected nondeductively—an awkward statement implying only that the relationships among the general statements in a theory are not formal or logical. Concatenated theories will not be able to perform all of the functions that deductive theories perform, but that need not impugn their status for not everything in man's universe is logical. The weakness of political theory is not fatal; it only makes discovery a more difficult task, and it opens up a wider area for individual judgment and disagreement. But so long as political scientists are aware of the nature and source of the weakness and evaluate their work accordingly, this can do no harm. The chief danger in the use of weak theories arises out of a failure to appreciate their weakness, to demand too much of them.

Unfortunately, philosophers and methodologists have paid very little attention to nondeductive theories, and the social scientists themselves have not exploited them systematically. What follows is, therefore, somewhat speculative and incomplete. Nevertheless, a beginning must be made, and exposure to criticism seems the best way to increase the quality of the formulation. And, in any case, the best strategy open to political science is clearly to exploit the possibilities of such theories as we are able to produce at the present time, continuing the search for a deductive base that can augment the weaker forms now in use.

.   .   .   .

What Theory Is Not

It will help to begin by eliminating from consideration certain kinds of intellectual constructions which, if they are often mistaken for theories, in fact perform no explanatory functions. It would be a hopeless task to seek a complete enumeration of such structures, but the major sources of confusion are fairly readily identified. In fact, they often include the term "theory" in their nomenclature, as is the case with "field theory" and "power theory."

Series of Definitions

No series of nominal definitions is a theory.[2] Although some of the nonlogical terms in a theory may be nominally defined, some at least must be tied to observables if the theory is to explain. A logical structure, or a sequence of vaguely related definitions that has no connection to observable evidence, cannot explain anything;

---

[1] The discussion of theory in Quentin Gibson, *The Logic of Social Enquiry* (London: Routledge and Kegan Paul, 1960), chap. xiii, and Abraham Kaplan, *The Conduct of Inquiry: Methodology for Behavioral Science* (San Francisco, Calif.: Chandler Publishing Co., 1964), chap. viii is excellent. It is interesting to note that neither R. B. Braithwaite nor Ernest Nagel deal with nondeductive theories in their respective works on scientific explanation.

[2] See Robert Bierstedt, "Real and Nominal Definitions," in L. Gross (ed.), *Symposium on Sociological Theory* (New York: Harper & Row, Publishers, 1959).

the whole of pure mathematics falls into this category. "Game theory," for example, is a well-developed mathematical structure but it is in no sense a social theory or a scientific theory; it is only a formal logical system. Of course, formal structures can be transformed into theories, they can be applied to observable data, if, and only if, the axioms of the system can be linked to observable data through a set of rules of correspondence. Plane geometry, for example, is a formal structure, but it can be applied by surveyors, on the earth's surface at least, because conditions on earth fit the axioms of the system. Similarly, if a social situation can be found in which the axioms of the mathematical theory of games are fulfilled, then the implications of that situation can be explored through the use of the mathematical structure. In political science, this has thus far proved almost impossible. In any event, when formal theories are applied, they cease to be nominally defined and are linked directly to observable data.

List of Factors

A theory is not simply a list of factors that may influence a given phenomenon, no matter how complete and exhaustive the list may be. That would be roughly equivalent to telling an author that everything he might conceivably have to say could be found in one book—a dictionary. Now factor theories do, as we shall see, involve the enumeration of factors involved in a particular phenomenon, but the list is always selective; no factor can be included in the theory unless it is relevant and operative, and no factor should be omitted when it is influential. A mere list of potential factors explains nothing. Indeed, a list that is purportedly exhaustive may well retard theoretical development by prejudging the issue. In any case, there can be no grounds for making this assertion. If a list of factors is derived by logical inference from nominal definitions, the result is a formal system . . . and subject to the same limits. If the list has been adduced from empirical observation there can be no grounds for asserting that it is exhaustive. In no case should a list of factors be given status that might inhibit the search for new points of view.

Approaches

A theory is not merely an "approach" to a discipline or topic—a suggested framework for investigation. What is sometimes called "power theory," for example, is no more than a statement that, in principle, generalizations about political behavior ought to be stated in terms of the power-seeking propensities of individuals. "Systems theory," similarly, implies no more than the view that society is best investigated within a framework provided by a "system," nominally defined. Neither of these conceptions is a theory, and it is puzzling to try and think out the reasons why men would be willing to argue in favor of the primacy of one or another of these approaches. A theory could, certainly, be constructed within this framework; theories can function with any kind of generalization, whatever the terms they employ. But theories are definite and particular; they are attempts to explain real phenomena. Power theories and systems theories are research strategies, or statements about the content that political generalizations ought to include. The assertion that they are the best possible research strategies is logically and practically dubious. Systems theory, to take one example, is open to very serious methodological criticism.[3] The concept of a "system" is highly ambiguous, as is the concept of function which is needed to clarify the operation of the system. It is very easy to confuse "system" with "association" or to take one as model for the other. The precise bound-

[3]See particularly, C. G. Hempel, "The Logic of Functional Analysis," in L. Gross (ed.), *Symposium on Sociological Theory* (New York: Harper & Row, Publishers, 1959), and Ernest Nagel, *Structure of Science* (New York: Harcourt, Brace & World, 1961), pp. 520–55. The criticisms in both works seem to me to be devastating and unanswerable.

aries of any system are empirically difficult to define. And a meaningful system, like a human society, is so fearfully complex that the task of describing its "state," a necessary feature of systems analysis, seems unlikely to be fulfilled. When we add the problem of showing how a given phenomena contributes to the maintenance of the system, which is the basic strategy of explanation employed in systems theory, there seem to be formidable limits and restrictions on the use of the approach.

"Field theory" offers another illustration of the use of the term "theory" to identify what is essentially only a method—a strategy for analyzing social relations. Kurt Lewin, one of the leading proponents of field theory, defined the status of the structure very accurately:

> If one proceeds in physics from a special law of theory . . . to more general theories . . . one does *not* finally come to field theory. . . . In other words, field theory can hardly be called a theory in the usual sense. Field theory is probably best characterized as a method: namely, a method of analyzing causal relations and building scientific constructs.[4]

Again, a theory can certainly be constructed that will explain general statements phrased in "field" terms, and such theories are quite popular in physical science. But the generic conception of field theory has no explanatory value whatever.

Models

Finally, a theory should be differentiated clearly from a model. A model is not an explanatory instrument, in the sense that theory explains generalizations. Models are an aid to understanding, and they may contribute indirectly to explanation, but they do not have a part in the process. A model, very generally, is an analogy, an isomorphic construction that is similar in some, but not all, respects to the theory or phenomenon for which it is a model. To use a model is to use an analogy, and it is subject to all of the limits imposed on analogous reasoning. In no case is the model causally related to the theory or object, and in all cases the model is a simplification, a partial isomorph.[5]

Models are very useful tools, of course, and a model of a complex theory or phenomenon can facilitate comprehension and discussion enormously. What are called "postulational" models, formal structures in which the postulates are "educated guesses," can aid the exploration of observable phenomena and relationships. Max Weber's "Ideal types," which are essentially models and not theories, do this very well, as do many of the models used in economics. Such "quasi-theories" have an important role to play in political science, as I hope to show presently, but they are not properly entitled to be called theories.

The uses of models in engineering, medicine, biology, science, and elsewhere are so obvious and manifold that they hardly need further elaboration here. But the results obtained from the study of models are not knowledge, and strictly speaking a model is not an explanatory tool. Models can perform some of the functions of theories, properly employed. But the very properties that make models useful also make their use dangerous. It is very easy to confuse model and theory, to assume that a property of a model is a property of a theory or object. It is also easy to forget that models are always partial and incomplete, that some variables have been eliminated, that some relationships have been dropped, that structure has been simplified. No amount of study of formal models can produce knowledge unless the axioms of the model are linked to observables by rules; in that case, the model

---

[4]Kurt Lewin, *Field Theory in Social Science* (New York: Harper, 1951), p. 45. Italics in the original.

[5]Kaplan, *The Conduct of Inquiry*, chap. vii, contains an excellent discussion of the role of models and their limitations.

becomes a theory. Finally, as Kaplan rightly points out, extensive reliance upon models can lead to overemphasis on rigor and precision, hence to unrealistic and unattainable criteria of inquiry. In the world of observable facts, rigor must always be sacrificed to human capacity and precision is limited by the tools of inquiry. Rigor may be bread and butter to the logician but it can be an extremely expensive luxury for the field worker in social science.

Robert K. Merton

Research and Theory Building[1]

THE THEORETIC FUNCTIONS OF RESEARCH

With a few conspicuous exceptions, recent sociological discussions have assigned but one major function to empirical research: "testing" or "verification" of hypotheses. The model for the proper way of performing this function is as familiar as it is clear. The investigator begins with a hunch or hypothesis, from this he draws various inferences and these, in turn, are subjected to empirical test which confirms or refutes the hypothesis.[2] But this is a logical model, and so fails, of course, to describe much of what actually occurs in fruitful investigation. It presents a set of logical norms, not a description of the research experience. And, as logicians are well aware, in purifying the experience, the logical model may also distort it. Like other such models, it abstracts from the temporal sequence of events. It exaggerates the creative role of explicit theory just as it minimizes the creative role of observation. For research is not merely logic tempered with observation. It has its psychological as well as its logical dimensions, although one would scarcely suspect this from the logically rigorous sequence in which research is usually reported.[3] It is both the psychological and logical pressures of research upon social theory which we seek to trace.

It is my central thesis that empirical research goes far beyond the passive role of verifying and

[1] Paper read before the annual meeting of the American Sociological Society, Cleveland, Ohio, March 1-3, 1946. This may be identified as Publication No. A-89 of the Bureau of Applied Social Research, Columbia University. Manuscript received April 19, 1948.

Reprinted from *American Sociological Review*, Vol. 13, No. 5 (October, 1948), 505-15, where it appeared under the title, "The Bearing of Empirical Research Upon the Development of Social Theory." By permission of the author and the publisher, the American Sociological Association.

[2] See, for example, the procedural review of Stouffer's "Theory of Intervening Opportunities" by G. A. Lundberg: "What are Sociological Problems?" *American Sociological Review*, 6 (1941), 357-59.

[3] See R. K. Merton, "Science, Population and Society," *The Scientific Monthly*, 44 (1937), pp. 170-71; the opposite discussion by Jean Piaget, *Judgment and Reasoning in the Child* (London, 1929), Chaps. V, IX, and the comment by William H. George, *The Scientist in Action* (London, 1936), p. 153. "A piece of research does not progress in the way it is 'written up' for publication."

testing theory: it does more than confirm or refute hypotheses. Research plays an active role: it performs at least four major functions which help shape the development of theory. It *initiates*, it *reformulates*, it *deflects* and *clarifies* theory.[4]

1. *The Serendipity Pattern*

(The unanticipated, anomalous and strategic datum exerts a pressure for initiating theory.)

Under certain conditions, a research finding gives rise to social theory. In a previous paper, this was all too briefly expressed as follows: "Fruitful empirical research not only tests theoretically derived hypotheses; it also originates new hypotheses. This might be termed the 'serendipity' component of research, *i.e.*, the discovery, by chance or sagacity, of valid results which were not sought for."[5]

The serendipity pattern refers to the fairly common experience of observing an *unanticipated, anomalous and strategic* datum which becomes the occasion for developing a new theory or for extending an existing theory. Each of these elements of the pattern can be readily described. The datum is, first of all, unanticipated. A research directed toward the test of one hypothesis yields a fortuitous by-product, an unexpected observation which bears upon theories not in question when the research was begun.

Secondly, the observation is anomalous, surprising,[6] either because it seems inconsistent with prevailing theory or with other established facts. In either case, the seeming inconsistency provokes curiosity; it stimulates the investigator to "make sense of the datum," to fit it into a broader frame of knowledge. He explores further. He makes fresh observations. He draws inferences from the observations, inferences depending largely, of course, upon his general theoretic orientation. The more he is steeped in the data, the greater the likelihood that he will hit upon a fruitful direction of inquiry. In the fortunate circumstance that his new hunch proves justified, the anomalous datum leads ultimately to a new or extended theory. The curiosity stimulated by the anomalous datum is temporarily appeased.

And thirdly, in noting that the unexpected fact must be "strategic," *i.e.*, that it must permit of implications which bear upon generalized theory, we are, of course, referring rather to what the observer brings to the datum than to the datum itself. For it obviously requires a theoretically sensitized observer to detect the universal in the particular. After all, men had for centuries noticed such "trivial" occurrences as slips of the tongue, slips of the pen, typographical errors, and lapses of memory, but it required the theoretic sensitivity of a Freud to see these as strategic data through which he could extend his theory of repression and symptomatic acts.

The serendipity pattern, then, involves the unanticipated, anomalous and strategic datum which exerts pressure upon the investigator for a new direction of inquiry which extends theory. Instances of serendipity have occurred in many disciplines.

. . . .

2. *The Recasting of Theory*

(New data exert pressure for the elaboration of a conceptual scheme.)

But it is not only through the anomalous fact that empirical research invites the extension of theory. It does so also through the repeated

---

[4]The fourth function, clarification, [was] elaborated in [the same issue] by Paul F. Lazarsfeld [Ed].

[5]R. K. Merton, "Sociological Theory," *American Journal of Sociology*, 50 (1945), 469n. Interestingly enough, the same outlandish term "serendipity" which has had little currency since it was coined by Horace Walpole in 1754 has also been used to refer to this component of research by the physiologist Walter B. Cannon. See his *The Way of an Investigator* (New York: W. W. Norton, 1945), Chap. VI, in which he sets forth numerous instances of serendipity in several fields of science.

[6]Charles Sanders Pierce had long before noticed the strategic role of the "surprising fact" in his account of what he called "abduction," that is, the initiation and entertaining of a hypothesis as a step in inference. See his *Collected Papers VI* (Cambridge, Mass.: Harvard University Press, 1931–35), pp. 522–28.

observation of hitherto neglected facts. When an existing conceptual scheme commonly applied to a given subject-matter does not adequately take these facts into account, research presses insistently for its reformulation. It leads to the introduction of variables which have not been systematically included in the scheme of analysis. Here, be it noted, it is not that the data are anomalous or unexpected or incompatible with existing theory; it is merely that they have not been considered pertinent. Whereas the serendipity pattern centers in an apparent inconsistency which presses for resolution, the reformulation pattern centers in the hitherto neglected but relevant fact which presses for an extension of the conceptual scheme.

. . . .

3. The Re-Focussing of Theoretic Interest

(New methods of empirical research exert pressure for new foci of theoretic interest.)

To this point we have considered the impact of research upon the development of particular theories. But empirical research also affects more general trends in the development of theory. This occurs chiefly through the invention of research procedures which tend to shift the foci of theoretic interest to the growing points of research.

The reasons for this are on the whole evident. After all, sound theory thrives only on a rich diet of pertinent facts and newly invented procedures help provide the ingredients of this diet. The new, and often previously unavailable, data stimulate fresh hypotheses. Moreover, theorists find that their hypotheses can be put to immediate test in those spheres where appropriate research techniques have been designed. It is no longer necessary for them to wait upon data as they happen to turn up—researches directed to the verification of hypotheses can be instituted at once. The flow of relevant data thus increases the tempo of advance in certain spheres of theory whereas in others, theory stagnates for want of adequate observations. Attention shifts accordingly.

In noting that new centers of theoretic interest have followed upon the invention or research procedures, we do not imply that these alone played a decisive role.[7] The growing interest in the theory of propaganda as an instrument of social control, for example, is in large part a response to the changing historical situation, with its conflict of major ideological systems; new technologies of mass communication which have opened up new avenues for propaganda; and the rich research treasuries provided by business and government interested in this new weapon of war, both declared and undeclared. But this shift is also a by-product of accumulated facts made available through such newly developed, and confessedly crude, procedures as content-analysis, the panel technique and the focused interview.

. . . .

4. The Clarification of Concepts

(Empirical research exerts pressure for clear concepts.)

A good part of the work called "theorizing" is taken up with the clarification of concepts—and rightly so. It is in this matter of clearly defined concepts that social science research is not infrequently defective. Research activated by a major interest in methodology may be centered on the *design* of establishing causal relations without due regard for analyzing the variables involved in the inquiry. This methodological empiricism, as the design of inquiry without correlative concern with the clarification of substantive variables may be called, characterizes a large part of current research. Thus, in a series of effectively designed experiments, Chapin finds that "the rehousing of slum families in a public housing project results

---

[7]It is perhaps needless to add that these procedures, instruments and apparatus are in turn dependent upon prior theory. But this does not alter their stimulating effect upon the further development of theory. *Cf.* Merton, "Sociological Theory," 463n.

in improvement of the living conditions and the social life of these families."[8] Or through controlled experiments, psychologists search out the effects of foster home placement upon children's performances in intelligence tests.[9] Or, again through experimental inquiry, researchers seek to determine whether a propaganda film has achieved its purpose of improving attitudes toward the British. These several cases, and they are representative of a large amount of research which has advanced social science method, have in common the fact that the empirical variables are not analyzed in terms of their conceptual elements.[10] As Rebecca West, with her characteristic lucidity, put this general problem of methodological empiricism, one might "know that A and B and C were linked by certain causal connexions, but he would never apprehend with any exactitude the nature of A or B or C." In consequence, these researches further the procedures of inquiry, but their findings do not enter into the repository of cumulative social science theory.

But in general, the clarification of concepts, commonly considered a province peculiar to the theorist, is a frequent result of empirical research. Research sensitive to its own needs cannot avoid this pressure for conceptual clarification. *For a basic requirement of research is that the concepts, the variables, be defined with sufficient clarity to enable the research to proceed*, a requirement easily and unwittingly not met in the kind of discursive exposition which is often miscalled "sociological theory."

The clarification of concepts ordinarily enters into empirical research in the shape of establishing *indices* of the variables under consideration. In non-research speculations, it is possible to talk loosely about "morale" or "social cohesion" without any clear conceptions of what is entailed by these terms, but they *must* be clarified if the researcher is to go about his business of systematically observing instances of low and high morale, of social cohesion or cleavage. If he is not to be blocked at the outset, he must devise indices which are observable, fairly precise and meticulously clear. The entire movement of thought which was christened "operationalism" is only one conspicuous case of the researcher demanding that concepts be defined clearly enough for him to go to work.

. . . .

What often appears as a tendency in research for quantification (through the development of scales) can thus be seen as a special case of attempting to clarify concepts sufficiently to permit the conduct of empirical investigation. The development of valid and observable indices becomes central to the use of concepts in the prosecution of research.

. . . .

There remain, then, a few concluding remarks. My discussion has been devoted exclusively to four impacts of research upon the development of social theory: the initiation, reformulation, refocusing and clarification of theory. Doubtless there are others. Doubtless, too, the emphasis of this paper lends itself to misunderstanding. It may be inferred that some invidious distinction has been drawn at the expense of theory and the theorist. That has not been my intention. I have

---

[8] F. S. Chapin, "The Effects of Slum Clearance and Rehousing on Family and Community Relationships in Minneapolis," *American Journal of Sociology*, 43 (1938), 744–63.

[9] R. R. Sears, "Child Psychology," in Wayne Dennis (ed.), *Current Trends in Psychology* (Pittsburgh: University of Pittsburgh Press, 1947), pp. 55–56. Sears's comments on this type of research state the general problem admirably.

[10] However crude they may be, procedures such as the focused interview are expressly designed as aids for detecting possibly relevant variables in an initially undifferentiated situation. See R. K. Merton and P. L. Kendall, "The Focused Interview," *American Journal of Sociology*, 51 (1946), pp. 541–557.

suggested only that an explicitly formulated theory does not invariably precede empirical inquiry, that as a matter of plain fact the theorist is not inevitably the lamp lighting the way to new observations. The sequence is often reversed. Nor is it enough to say that research and theory must be married if sociology is to bear legitimate fruit. They must not only exchange solemn vows—they must know how to carry on from there. Their reciprocal roles must be clearly defined. This paper is a brief essay toward that definition.

IV METHODS AND TECHNIQUES FOR BEHAVIORAL INQUIRY

The methods for gathering and evaluating data must be guided by the logic and principles discussed in Part III. In the final analysis, the significance of research findings depends in large measure upon the rigor and precision of the research tools employed. The emphasis in behavioralism is upon empiricism—perception of phenomena through sense experience and evaluation of this experience through impersonal techniques—and therefore certain research techniques are more appropriate than others. For example, data collection through direct observation is preferable to data collection through intuition. Although some use of empirical techniques has occurred throughout the history of political science, support for vigorous application of these techniques is comparatively new. Information about political events has always been gathered through observation and verbal reports, but only recently have these techniques been systematically and stringently applied.

One of the first steps in behavioral research is to identify the data that will be used as evidence. The nature of the problem under study and the hypotheses generated in order to examine this problem will indicate which data must be collected. The types of information needed will in many cases also suggest which techniques are appropriate. One problem of which the researcher must be continually aware is the impact of the researcher himself upon the data, especially the influence of his perceptions, prejudices, and values.

A number of empirically based data collection techniques are available to the behavioral political scientist. Systematic and objective observation is the most fundamental. During observation the investigator uses his own senses, or mechanical extensions of them, to collect information and thus does not need secondary sources such as informants or recall. Raymond L. Gold considers the nature of observation as a data collection technique with special reference to the role of the investigator. Observation in the physical sciences poses different methodological problems, because interaction between the investigator and the research phenomenon is less likely to produce biased results; however, in social research, several alternate interaction roles are available to the investigator. At one extreme the social researcher is totally involved in the ongoing situation; at the other he is a spectator watching the proceedings from the sidelines. The types of research problems under investigation and the kinds of information required determine the role the investigator takes.

Observation, however, is inadequate or inappropriate for getting certain types of information. Frequently, observations of human behavior do not tell why people behave the way they do. In other words, explanatory factors for human behavior are not necessarily revealed in the behavior itself. In seeking these factors one might employ techniques for probing beyond actions themselves. Information concerning the motivation of behavior may be obtained by a wide variety of interview techniques, ranging from psychoanalysis to mass public opinion surveys. William J. Goode and Paul K. Hatt discuss the utility of the interview, and they conclude that interviewing can yield reliable information regarding behavioral factors that cannot be obtained by systematic and objective observation. As a technique the interview has been useful for political scientists in gathering information about political actors as individuals, as members of groups, and as members of the larger body politic.

With the refinement of interviewing techniques and methods of data analysis, the use of large-scale interviewing has come into its own as a major research technique. As Herbert McClosky points out, this use of interviewing, labeled survey research, has provided major breakthroughs in the study of public opinion. Today survey research is used extensively in areas ranging from marketing of consumer products to campaigning for public office. In terms of the discipline, this has meant that political scientists are now capable of gauging the opinions of large numbers of people on topics relevant to the study of politics, whereas before they could only guess about the nature of public opinion. The survey research technique thus has been responsible for much of the expansion of our knowledge about how people view politics.

As R. M. MacIver noted in Part III, one goal of inquiry is the investigation of causal relationships. Data collected through observation and interviewing frequently enable the researcher to identify relationships among factors. This type of information does not mean, however, that cause-effect conclusions can be formulated. The essential difference is that causal information permits the investigator to conclude that some combination of factors is necessary or sufficient or both, to produce certain consequences. Usually causal relationships in both the natural and the social sciences are investigated through experimentation. This data collection technique, discussed by Barry Anderson, can be implemented only when the researcher is able to manipulate causal factors and when he can control the environment surrounding these factors. Such manipulation and control facilitate the researcher's ability to determine which specific set of factors produce which specific effects. The social sciences are clearly limited in their ability to use experimentation, because researchers are rarely capable of manipulating causal variables and controlling the environment.

One form of experimentation that has become popular in the social sciences, especially for studying small group interaction and decision-making situations, is the technique of simulation. With simulation the researcher achieves the ability to manipulate and control by artificially constructing the phenomenon to be investigated. In the selection included here, Richard E. Dawson compares simulation to the creation of a model. With this model, the researcher is able to manipulate factors in order to determine their effects on the outcome of the situation. One form of simulation can be seen in political "games," which were developed so that people might try their hand at playing different strategies in order to reach some goal. By manipulating the situation and controlling the persons playing the game, the researcher is in a position to conduct an experiment.[1] A more abstract form of simulation has been devised that uses computers rather than people. With this scheme the computer accomplishes the manipulation of factors and determines the outcome through trial and error.

After the necessary information is gathered, data are manipulated and scrutinized so that they may be used to advance understanding of the subject under investigation. This process, data analysis, does not change the data but alters their form to facilitate interpretation. Such further investigation is necessary because the meaning of the data is not a part or function of the data themselves. What the data mean is a product of interpretation, and this is a function of the researcher himself.

All data analysis depends upon the manner in which the information has been measured. Standards of measurement vary immensely, from the vagueness

---

[1] One of the first commercial ventures in political science that used simulation was the "Inter-Nation Simulation Kit" developed by Harold Guetzkow and Cleo H. Cherryholmes and marketed by Science Research Associates, Chicago, Illinois, 1966.

of "big" and "little" to the precision of numerical scales. In a classic discussion about measurement, S. S. Stevens sets out four alternate schemes for assigning values to objects. Stevens points out that once a scheme has been selected, certain analytic operations are appropriate. The type of measurement selected to represent the information during the early stages of evaluation is consequently of great importance during the latter data analysis stages.

The growing popularity of behavioral forms of inquiry in political science is evidenced by the increasing use of quantification, statistical analysis, and mathematical reasoning, which stress objectivity, reliability, and precision. Although these techniques have much in common, they should not be confused with one another. Quantification means simply that data are stated in terms of quantities (numbers of things) rather than in qualities (types of things). Statistical analysis involves procedures whereby the data are submitted to numerical manipulation according to prescribed procedures. The goal in this manipulation is to determine significance from a probabilistic point of view. These procedures are governed by mathematical reasoning, which is a system of logic.

At one time many persons were attracted to the social sciences because relatively little was required in the way of mathematical skills. The growing use of behavioralism has changed the outlook of these disciplines with regard to quantification and mathematical reasoning because these methods increase objectivity and precision in data analysis. Consequently, political scientists have begun to "retool" in these areas and have found that mathematics is a superior means of communication and has a better system of logic than any other medium.

In a brief introduction to the use of statistics, Claire Selltiz, Marie Jahoda, Morton Deutsch, and Stuart W. Cook indicate some of the general types of statistical operations. They stress the difference between the manipulative phase and the interpretative phase of statistical analysis by pointing out that significance as determined by numerical manipulation is meaningful only within its own context and not necessarily of substantive importance. Further, this reading clearly points out the advantages and risks of quantitative-statistical analysis, a theme further discussed in the essay by Leslie Kish that follows.

One outgrowth of using quantitative-mathematical methods has been an increase in the laborious processes of data manipulation and computation. This growth of tedious work has been more than matched, however, by recent developments in the technology of electronic equipment, especially the computer. Max Gunther casts a sympathetic but critical eye toward computers and their uses in his article on computers and their limitations. He places the frequently heard cliché, "computers never make mistakes," in perspective by illustrating how a computer is dependent upon people. His conclusion that computers are devices to permit the individual to do more complex operations in a shorter period of time and with fewer errors extends the horizons of inquiry in all disciplines.

In the final analysis, the critical factor in determining the success of any research venture continues to be the researcher. The best techniques and devices alone can only provide facts and figures. The resourceful and informed investigator is the key to interpreting these facts and providing them with meaning. Without valid, reliable, and precise facts, however, the investigator must substitute intuition and guesswork for clear and reasoned data analysis. Thus reliable data *and* an intelligent, resourceful investigator are both essential elements in meaningful investigation.

Raymond L. Gold

Roles in Social Field Observations[1]

Buford Junker has suggested four theoretically possible roles for sociologists conducting field work.[2] These range from the complete participant at one extreme to the complete observer at the other. Between these, but nearer the former, is the participant-as-observer; nearer the latter is the observer-as-participant.

. . . .

[1] Read before the nineteenth annual meeting of the Southern Sociological Society, Atlanta, Georgia, April 13, 1956.
[2] Buford Junker, "Some Suggestions for the Design of Field Work Learning Experiences," in Everett C. Hughes, *et al.*, *Cases on Field Work* (hectographed by The University of Chicago, 1952), Part III-A.

Reprinted from *Social Forces*, 36 (March, 1958), 217–23, where it appeared under the title, "Roles in Sociological Field Observations." By permission of the author and the publisher.

My aim in this paper is to present extensions of Junker's thinking growing out of systematic interviews with field workers whose experience had been cast in one or more of these patterns of researcher-subject relationship. All of these field workers had gathered data in natural or nonexperimental settings. I would like in this paper to analyze generic characteristics of Junker's four field observer roles and to call attention to the demands each one places on an observer, as a person and as a sociologist plying his trade.

Every field work role is at once a social interaction device for securing information for scientific purposes and a set of behaviors in which an observer's self is involved.[3] While playing a field work role and attempting to take the role of an informant, the field observer often attempts to master hitherto strange or only generally understood universes of discourse relating to many attitudes and behaviors. He continually introspects, raising endless questions about the informant and the developing field relationship,

[3] To simplify this presentation, I am assuming that the field worker is an experienced observer who has incorporated the role into his self-conceptions. Through this incorporation, he is self-involved in the role and feels that self is at stake in it. However, being experienced in the role, he can balance role-demands and self-demands in virtually all field situations, that is, all except those to be discussed shortly.

with a view to playing the field work role as successfully as possible. A sociological assumption here is that the more successful the field worker is in playing his role, the more successful he must be in taking the informant's role. Success in both role-taking and role-playing requires success in blending the demands of self-expression and self-integrity with the demands of the role.

It is axiomatic that a person who finds a role natural and congenial, and who acts convincingly in it, has in fact found how to balance role-demands with those of self. If need be he can subordinate self-demands in the interest of the role and role-demands in the interest of self whenever he perceives that either self or role is in any way threatened. If, while playing the role, someone with whom he is interacting attacks anything in which he has self-involvement, he can point out to himself that the best way to protect self at the moment is to subordinate (or defer) self-expression to allow successful performance in the role. In other words, he uses role to protect self. Also, when he perceives that he is performing inadequately in the role he can indicate to himself that he can do better by changing tactics. Here he uses self as a source of new behaviors to protect role. The case of using role to protect self from perceived threat is one of acute self-consciousness, a matter of diminishing over-sensitivity to self-demands by introspectively noting corresponding demands of role. The case of using self to protect role from perceived threat is one of acute role-consciousness, a matter of diminishing over-sensitivity to role-demands by introspectively indicating that they are disproportionately larger than those of self. Both cases represent situations in which role-demands and self-demands are out of balance with each other as a result of perceived threat, and are then restored to balance by appropriate introspection.

Yet, no matter how congenial the two sets of demands seem to be, a person who plays a role in greatly varied situations (and this is especially true of a sociologist field observer) sometimes experiences threats which markedly impair his effectiveness as an interactor in the situation.

Where attempting to assess informational products of field work, it is instructive to examine the field worker's role-taking and role-playing in situations of perceived, but unresolved, threat. Because he defines success in the role partly in terms of doing everything he can to remain in even threatening situations to secure desired information, he may find that persevering is sometimes more heroic than fruitful.

The situation may be one in which he finds the informant an almost intolerable bigot. The field worker decides to stick it out by attempting to subordinate self-demands to those of role. He succeeds to the extent of refraining from "telling off" the informant, but fails in that he is too self-conscious to play his role effectively. He may think of countless things he would like to say and do to the informant, all of which are dysfunctional to role-demands, since his role requires taking the role of the other as an informant, not as a bigot. At the extreme of nearly overwhelming self-consciousness, the field worker may still protect his role by getting out of the situation while the getting is good. Once out and in the company of understanding colleagues, he will finally be able to achieve self-expression (i.e., finally air his views of the informant) without damaging the field role.[4]

Should the situation be such that the field worker finds the informant practically inscrutable (i.e., a "bad" informant), he may decide to persevere despite inability to meet role-taking and role-playing demands. In this situation he becomes acutely role-conscious, since he is hypersensitive to role-demands, hyposensitive to self. This partial breakdown of his self-process thwarts his drawing on past experiences and current observations to raise meaningful questions and perceive meaningful answers. At the extreme, a role-conscious field worker may play his role so mechanically and unconvincingly that the informant, too, develops role-and-self problems.

[4] An inexperienced field worker might "explode" on the spot, feeling that role and self are not congenial in this *or any other* situation. But an experienced field worker would leave such a situation as gracefully as possible to protect the role, feeling that role and self are not congenial in *this* situation only.

The following discussion utilizes these conceptions of role and self to aid in analyzing field work roles as "master roles" for developing lesser role-relationships with informants.[5] While a field worker cannot be all things to all men, he routinely tries to fit himself into as many roles as he can, so long as playing them helps him to develop relationships with informants in his master role (i.e., participant-as-observer, etc.).

COMPLETE PARTICIPANT

The true identity and purpose of the complete participant in field research are not known to those whom he observes. He interacts with them as naturally as possible in whatever areas of their living interest him and are accessible to him as situations in which he can play, or learn to play, requisite day-to-day roles successfully. He may, for example, work in a factory to learn about inner-workings of informal groups. After gaining acceptance at least as a novice, he may be permitted to share not only in work activities and attitudes but also in the intimate life of the workers outside the factory.

Role-pretense is a basic theme in these activities. It matters little whether the complete participant in a factory situation has an upper-lower class background and perhaps some factory experience, or whether he has an upper-middle class background quite divorced from factory work and the norms of such workers. What really matters is that he knows that he is pretending to be a colleague. I mean to suggest by this that the crucial value as far as research yield is concerned lies more in the self-orientation of the complete participant than in his surface role-behaviors as he initiates his study. The complete participant realizes that he, and he alone, knows that he is in reality other than the person he pretends to be. He must pretend that his real self is represented by the role, or roles, he plays in and out of the factory situation in relationships with people who, to him, are but informants, and this implies an interactive construction that has deep ramifications. He must bind the mask of pretense to himself or stand the risk of exposure and research failure.

In effect, the complete participant operates continually under an additional set of situational demands. Situational role-and-self demands ordinarily tend to correspond closely. For this reason, even when a person is in the act of learning to play a role, he is likely to believe that pretending to have achieved this correspondence (i.e., fourflushing) will be unnecessary when he can actually "be himself" in the role. But the complete observer simply cannot "be himself"; to do so would almost invariably preclude successful pretense. At the very least, attempting to "be himself"—that is, to achieve self-realization in pretended roles—would arouse suspicion of the kind that would lead others to remain aloof in interacting with him. He must be sensitive to demands of self, of the observer role, and of the momentarily pretended role. Being sensitive to the set of demands accompanying role-pretense is a matter of being sensitive to a large variety of overt and covert mannerisms and other social cues representing the observer's pretended self. Instead of being himself in the pretended role, all he can be is a "not self," in the sense of perceiving that his actions are meaningful in a contrived role.

The following illustration of the pretense of a complete participant comes from an interview with a field worker who drove a cab for many months to study big-city cab drivers. Here a field worker reveals how a pretended role fosters a heightened sense of self-awareness, an introspective attitude, because of the sheer necessity of indicating continually to himself that certain

---

[5] Lesser role-relationships include all achieved and ascribed roles which the field worker plays in the act of developing a field relationship with an informant. For example, he may become the "nice man that old ladies can't resist" as part of his over-all role-repertoire in a community study. Whether he deliberately sets out to achieve such relationships with old ladies or discovers that old ladies ascribe him "irresistible" characteristics, he is still a participant-as-observer who interacts with local old ladies as a "nice man." Were he not there to study the community, he might choose *not* to engage in this role-relationship, especially if being irresistible to old ladies is not helpful in whatever master role(s) brought him to town. (Cf. any experienced community researcher.)

experiences are merely part of playing a pretended role. These indications serve as self-assurance that customers are not really treating *him* as they seem to do, since he is actually someone else, namely, a field worker.

> Well, I've noticed that the cab driver who *is* a cab driver acts differently than the part-time cab drivers, who don't think of themselves as real cab drivers. When somebody throws a slam at men who drive only part of the year, such as, "Well, you're just a goddamn cab driver!," they do one of two things. They may make it known to the guy that they are not a cab driver; they are something else. But as a rule, that doesn't work out, because the customer comes back with, "Well, if you're not a cab driver what the hell are you driving this cab for?" So, as a rule, they mostly just rationalize it to themselves by thinking, "Well, this is not my role or the real me. He just doesn't understand. Just consider the source and drop it." But a cab driver who *is* a cab driver, if you make a crack at him such as, "You're just a goddamn cab driver!" he's going to take you out of the back seat and whip you.

Other complete participant roles may pose more or less of a challenge to the field worker than those mentioned above. Playing the role of potential convert to study a religious sect almost inevitably leads the field worker to feel not only that he has "taken" the people who belong to the sect, but that he has done it in ways which are difficult to justify. In short, he may suffer severe qualms about his mandate to get information in a role where he pretends to be a colleague in moral, as well as in other social, respects.

All complete participant roles have in common two potential problems; continuation in a pretended role ultimately leads the observer to reckon with one or the other. One, he may become so self-conscious about revealing his true self that he is handicapped when attempting to perform convincingly in the pretended role. Or two, he may "go native," incorporate the role into his self-conceptions and achieve self-expression in the role, but find he has so violated his observer role that it is almost impossible to report his findings. Consequently, the field worker needs cooling-off periods during and after complete participation, at which times he can "be himself" and look back on his field behavior dispassionately and sociologically.

While the complete participant role offers possibilities of learning about aspects of behavior that might otherwise escape a field observer, it places him in pretended roles which call for delicate balances between demands of role and self. A complete participant must continually remind himself that, above all, he is there as an observer: this is his primary role. If he succumbs to demands of the pretended role (or roles), or to demands of self-expression and self-integrity, he can no longer function as an observer. When he can defer self-expression no longer, he steps out of the pretended role to find opportunities for congenial interaction with those who are, in fact, colleagues.

PARTICIPANT-AS-OBSERVER

Although basically similar to the complete observer role, the participant-as-observer role differs significantly in that both field worker and informant are aware that theirs is a field relationship. This mutual awareness tends to minimize problems of role-pretending; yet, the role carries with it numerous opportunities for compartmentalizing mistakes and dilemmas which typically bedevil the complete participant.

Probably the most frequent use of this role is in community studies, where an observer develops relationships with informants through time, and where he is apt to spend more time and energy participating than observing. At times he observes formally, as in scheduled interview situations; and at other times he observes informally—when attending parties, for example. During early stages of his stay in the community, informants may be somewhat uneasy about him in both formal and informal situations, but their uneasiness is likely to disappear when they learn to trust him and he them.

But just when the research atmosphere seems

ripe for gathering information, problems of role and self are apt to arise. Should field worker and informant begin to interact in much the same way as ordinary friends, they tend to jeopardize their field roles in at least two important ways. First, the informant may become too identified with the field worker to continue functioning as merely an informant. In this event the informant becomes too much of an observer. Second, the field worker may over-identify with the informant and start to lose his research perspective by "going native." Should this occur the field worker may still continue going through the motions of observing, but he is only pretending.

Although the field worker in the participant-as-observer role strives to bring his relationship with the informant to the point of friendship, to the point of intimate form, it behooves him to retain sufficient elements of "the stranger" to avoid actually reaching intimate form. Simmel's distinction between intimate content and intimate form contains an implicit warning that the latter is inimical to field observation.[6] When content of interaction is intimate, secrets may be shared without either of the interactors feeling compelled to maintain the relationship for more than a short time. This is the interaction of sociological strangers. On the other hand, when form of interaction is intimate, continuation of the relationship (which is no longer merely a field relationship) may become more important to one or both of the interactors than continuation of the roles through which they initiated the relationship.

[6] "In other words, intimacy is not based on the *content* of the relationship. . . . Inversely, certain external situations or moods may move us to make very personal statements and confessions, usually reserved for our closest friends only, to relatively strange people. But in such cases we nevertheless feel that this 'intimate' *content* does not yet make the relation an intimate one. For in its basic significance, the whole relation to these people is based only on its general, unindividual ingredients. That 'intimate' content, although we have perhaps never revealed it before and thus limit it entirely to this particular relationship, does nevertheless not become the basis of its form, and thus leaes it outside the sphere of intimacy." K. H. Wolff (ed.), *The Sociology of Georg Simmel* (Glencoe, Illinois: The Free Press, 1950), p. 127.

In general, the demands of pretense in this role, as in that of the complete participant, are continuing and great; for here the field worker is often defined by informants as more of a colleague than he feels capable of being. He tries to pretend that he is as much of a colleague as they seem to think he is, while searching to discover how to make the pretense appear natural and convincing. Whenever pretense becomes too challenging, the participant-as-observer leaves the field to re-clarify his self-conceptions and his role-relationships.

Observer-as-Participant

The observer-as-participant role is used in studies involving one-visit interviews. It calls for relatively more formal observation than either informal observation or participation of any kind. It also entails less risk of "going native" than either the complete participant role or the participant-as-observer role. However, because the observer-as-participant's contact with an informant is so brief, and perhaps superficial, he is more likely than the other two to misunderstand the informant, and to be misunderstood by him.

These misunderstandings contribute to a problem of self-expression that is almost unique to this role. To a field worker (as to other human beings), self-expression becomes a problem at any time he perceives he is threatened. Since he meets more varieties of people for shorter periods of time than either the complete participant or the participant-as-observer, the observer-as-participant inclines more to feel threatened. Brief relationships with numerous informants expose an observer-as-participant to many inadequately understood universes of discourse that he cannot take time to master. These frustratingly brief encounters with informants also contribute to mistaken perceptions which set up communication barriers the field worker may not even be aware of until too late. Continuing relationships with apparently threatening informants offer an opportunity to redefine them as more congenial partners in interaction, but such is not the fortune of a field worker in this role. Consequently, using his prerogative to break off relationships with threatening informants, an observer-as-participant, more easily than the other two, can leave the field almost at will to regain the kind of role-and-self balance that he, being who he is, must regain.

COMPLETE OBSERVER

The complete observer role entirely removes a field worker from social interaction with informants. Here a field worker attempts to observe people in ways which make it unnecessary for them to take him into account, for they do not know he is observing them or that, in some sense, they are serving as his informants. Of the four field work roles, this alone is almost never the dominant one. It is sometimes used as one of the subordinate roles employed to implement the dominant ones.

It is generally true that with increasingly more observation than participation, the chances of "going native" become smaller, although the possibility of ethnocentrism becomes greater. With respect to achieving rapport in a field relationship, ethnocentrism may be considered a logical opposite of "going native." Ethnocentrism occurs whenever a field worker cannot or will not interact meaningfully with an informant. He then seemingly or actually rejects the informant's views without ever getting to the point of understanding them. At the other extreme, a field worker who "goes native" passes the point of field rapport by literally accepting his informant's views as his own. Both are cases of pretending to be an observer, but for obviously opposite reasons. Because a complete observer remains entirely outside the observed interaction, he faces the greatest danger of misunderstanding the observed. For the same reason, his role carries the least chance of "going native."

The complete observer role is illustrated by systematic eavesdropping, or by reconnaissance of any kind of social setting as preparation for more intensive study in another field role. While watching the rest of the world roll by, a complete observer may feel comfortably detached, for he takes no self-risks, participates not one whit. Yet, there are many times when he wishes he could ask representatives of the observed world to qualify what they have said, or to answer other questions his observations of them have brought to mind. For some purposes, however, these very questions are important starting points for subsequent observations and interactions in appropriate roles. It is not surprising that reconnaissance is almost always a prelude to using the participant-as-observer role in community study. The field worker, feeling comfortably detached, can first "case" the town before committing himself to casing *by* the town.

CONCLUSIONS

Those of us who teach field work courses or supervise graduate students and others doing field observations have long been concerned with the kinds of interactional problems and processes discussed above. We find such common "mistakes" as that of the beginner who over-identifies with an informant simply because the person treats him compassionately after others have refused to grant him an interview. This limited, although very real, case of "going native" becomes much more understandable to the beginner when we have analyzed it for him sociologically. When he can begin utilizing theory of role and self to reflect on his own assets and shortcomings in the field, he will be well on the way to dealing meaningfully with problems of controlling *his* interactions with informants.

Beyond this level of control, sophistication in field observation requires manipulating informants to help them play their role effectively. Once a field worker learns that a field relationship in process of being structured creates role-and-self problems for informants that are remarkably similar to those he has experienced, he is in a position to offer informants whatever kinds of "reassurances" they need to fit into their role. Certainly a field worker has mastered his role only to the extent that he can help informants to master theirs. Learning this fact (and doing something about it!) will eliminate nearly all excuses about "bad" or "inept" informants, since, willy-nilly, an informant is likely to play his role only as fruitfully or as fruitlessly as a field worker plays his.[7]

Experienced field workers recognize limitations in their ability to develop relationships in various roles and situations. They have also discovered that they can maximize their take of information by selecting a field role which permits them to adjust their own role-repertories to research objectives. Objectively, a selected role is simply an expedient device for securing a given level of information. For instance, a complete participant obviously develops relationships and frames of reference which yield a somewhat different perspective of the subject matter than that which any of the other field work roles would yield. These subjective and objective factors come together in the fact that degree of success in securing the level of information which a field role makes available to a field worker is largely a matter of his skill in playing and taking roles.

Each of the four field work roles has been shown to offer advantages and disadvantages with respect to both demands of role and self and level of information. No attempt has been made in this report to show how a sociological conception of field work roles can do more than provide lines of thought and action for dealing with problems and processes of field interaction. Obviously, however, a theory of role and self growing out of study of field interaction is in no sense limited to that area of human activity. Learning to take and play roles, although dramatized in the field, is essentially the same kind of social learning people engage in throughout life.

In any case, the foregoing discussion has suggested that a field worker selects and plays a role so that he, being who he is, can best study those aspects of society in which he is interested.

[7] In a recent article on interviewing, Theodore Caplow also recognizes the key role played by the field worker in structuring the field relationship. He concludes, "The quality and quantity of the information secured probably depend far more upon the competence of the interviewer than upon the respondent." "The Dynamics of Information Interviewing," *American Journal of Sociology*, 62 (September, 1956), p. 169. Cf. also the studies by Junker, "Some Suggestions for the Design . . ." and Raymond L. Gold, "Toward a Social Interaction Methodology for Sociological Field Observation," unpublished Ph. D. Dissertation, University of Chicago, 1954.

William J. Goode and *Paul K. Hatt*

The Interview:
A Data Collection Technique

Just as sampling procedures have developed in complexity and precision far beyond common-sense mental operations, yet are still based on activities common to all men, so is

Reprinted from *Methods in Social Research* by W. J. Goode and Paul K. Hatt, copyright 1952 by McGraw-Hill Book Company, Inc., where it appeared under the title "The Interview." By permission of the author and the publisher.

interviewing the development of precision, focus, reliability, and validity in another common social act—conversation. When parents attempt to find out what "really happened in school" by questioning children, they are carrying out an interview. Perhaps most readers of this book have been through a "job interview," in which they were asked an embarrassing series of questions designed to find out "What can you do?" Almost everyone has seen a "whodunit" film, in which the master detective carries out a number of interviews with the murder suspects. He "probes" more deeply if he believes the answer does not tell the whole story. He asks a series of questions, designed to cross-check a set of earlier answers. He may ask innocent questions, in order to make the murder suspect relax his guard. The prospective purchaser of real estate becomes an interviewer, also, when he questions the salesman about the property, or returns to the neighborhood later in order to question other residents of the area.

Everyone, then, has been interviewer and interviewee at some time or another, and all have listened to interviews. Some of these have been efficiently performed, while others have failed to elicit the information desired. A few have antagonized the interviewee, while others have become the beginning of a fast friendship. Some have been trivial in nature, and others have been of great significance. It is common to feel, after such interviews, that something different should have been said. Or self-congratulation follows some particularly shrewd question which cleared up an important ambiguity. Once in a while, also, it is recognized that the person who spoke with us "felt a lot better" for having talked about his troubles.

Like other social activities, interviewing has many facets. There are many types of interviews, and their purposes are many. Nevertheless, interviewing can be studied in order to develop skill. Although interviewing is easier for some than for others, everyone can improve his technique by learning to avoid certain types of errors, by developing an alertness to ambiguities and deceptions, and by becoming aware of the purpose of the interview, as well as the interaction between interviewer and respondent.

It is of particular importance that the modern social investigator develop his skill. Increasingly, the social scientist has turned from books to social phenomena in an effort to build the foundations of science. It is true, of course, that speech adds a further complex dimension to research, which the physical scientist does not have to probe. The rock cannot speak. But, as Max Weber once noted, this dimension is also a source of information. If it is not to be ignored, tools for its exploitation must be developed. One can maintain, of course, that every phase of any research is crucial. Errors at any stage may weaken or destroy the validity of the investigation. Yet the interview is, in a sense, the foundation upon which all other elements rest, for it is the data-gathering phase.

Its importance is further seen in the gradual recognition, location, and control of interviewer bias, since the interviewer is really a tool or an instrument. One interviewer may not penetrate the mask of refusal which a potential respondent offers. Another will be given a cordial reception. One interviewer will meet with cliché answers, moderate in tenor and logical in structure, from a certain respondent. Another may find that the same respondent is quite violent in his answers and in his emotion pays little attention to logic. These differences may be extreme cases, but all may be encountered. Important differences between interviewers are generally found, raising the fundamental question of *interviewer reliability*: "To what extent can the answers so obtained be repeated?"

Interviewing has become of greater importance in contemporary research because of the reassessment of the *qualitative interview*. Social scientists of the turn of the century used this type of interview almost exclusively. The interview was likely to be rather unstructured in character and more in the nature of a probing conversation. Guided by a shrewd, careful observer, this could be a powerful instrument for obtaining information. However, it was also an unstandardized instrument. The investigator could not offer definitive proof that his data were as described. The interview was of the character of the anthropological interview, in which no other interviewer was expected to check on the information and the problem of reliability was not often raised.

The development of highly structured schedules was seen as one possible solution to the problem of standardization. Its most complete development, of course, is the polling interview, in which the same questions are asked of every respondent. They were to be asked in the same form, in the same order, with no deviation from respondent to respondent. In this fashion, it was possible to obtain certain items of information for each respondent. This facilitated comparative analysis between individuals or subclasses. However, *depth* was usually sacrificed in order to gain this standardization. As a consequence, there is a movement back to the qualitative interview through the use of the *interview guide*, which requires certain items of information about each respondent but allows the interviewer to rephrase the question in keeping with his understanding of the situation. This permits the interviewer to express the question in such a fashion that the respondent can understand it most easily. Further, the interviewer may probe more deeply when the occasion demands. This permits a more adequate interpretation of the answers to each question. In addition, the development of content analysis and qualitative coding permits some standardization of answers not of the "yes-no" type. Thus, one of the basic objections to the qualitative interview has been partially removed.

On the other hand, this method requires an even *higher level of interviewing quality*. The greater the amount of discretion allowed the interviewer, the more necessary is a high level of competence. The application of more rigid sampling controls takes from the interviewer the choice of respondent. If, however, his interviewing report contains information quite different from that obtained by other interviewers, the problem of "sampling" is reraised—this time, a *sampling of the responses* of the interviewee to these particular questions. If the responses are entirely different from one interview to another, the adequacy of the data is always in question. Consequently, the development of interviewing skills as well as interviewer controls to a high level is of great importance.

Interviewing as a Social Process

Neither reliability nor depth can be achieved, however, unless it is kept clearly in mind that interviewing is fundamentally a process of social interaction. Its primary purpose may be research, but this is its purpose *for the investigator. For the respondent*, its foundation and meaning may be different. Even if both have research as an interest, the process of obtaining information is so structured by its character as social interaction that considerable attention to this aspect is required.

Let us first look at the element in social interaction which is most difficult to define, that of *insight* or intuition. This is an unfortunate term, since for many it possesses overtones of vagueness, subjectivity, and even mysticism. Yet no such connotations are intended here. Reference is rather made to the fact that some of the individuals in a social group seem to understand the dislikes and likes of the rest better than others do. They can predict more accurately what the others will say, and respond more precisely to their intended meaning. They know when one feels offended, and what lies behind the casual comments of another.

It is commonplace to feel, when on close terms with a friend, that a casual word, gesture, or look conveys a complete message or story. Yet this is not usual between mere acquaintances, and this describes the importance of what may be called *subliminal cues*. That is, everyone betrays his emotions in various ways. As we become accustomed to friends, we learn, consciously or unconsciously, the tiny behavioral accompaniments of these emotions. Those cues which are not recognized consciously, which are below the threshold of perception, are called *subliminal*. A good poker player wins as much by his guesses as to the plans and emotions of his fellow players as by his knowledge of the cards themselves. Indeed, he will play the cards on the basis of these guesses about his opponents. Sometimes guesses are based on a conscious recognition of these cues; other guesses, equally good, which seem to be based on no such recognition, spring from such unconscious observations. If insight refers to such procedures, then it is clear that it can be acquired. To improve his "insight," the student of social relations should attempt consciously to:

1. Develop an alertness to the fact that there *are* many subliminal cues, and that one can learn to "read" them.

2. Attempt to bring these cues to a conscious level, so that comparisons can be made with the hunches of other observers and interviewers.

3. Systematically check the predictions made from these hunches, to see which are correct.

The process of social interaction in the interview is complicated by the fact that the *interviewee* also has insight. This means that the interviewer must not only attempt to be conscious of the real meaning of the answers made by the interviewee; he must also be aware of the fact that his respondent is, in turn, guessing at the motives of the interviewer, responding to the embarrassment of the latter, even to the lack of insight on his part. At times the respondent will give more information because he feels the interviewer "already knows." He responds, then, to the image of himself which he believes the interviewer possesses. This is of real importance when the interviewer must "probe" in order to test or check another answer.

The interviewer must, therefore, become alert to what he is bringing to the interview situation: his appearance, his facial and manual gestures, his intonation, his fears and anxieties, his obtuseness and his cleverness. How do these affect the interviewee? Over some of these characteristics, he may have some control. Others, however, are so much a part of his personality that he can discipline them only slightly. The result will be that every interviewer will meet with some interviewees with whom no rapport will develop, and no adequate interview situation can exist. However, being alert to these characteristics allows him at least to change those elements which are under his control, even if only in the restricted context of the interview situation.

A concrete example lies in one of the most common questions asked by the beginning interviewer: How should he dress? If he is to interview lower class people, for example, should he attempt to dress shabbily? Will he get better answers if he dresses in overalls, or even in poorly cut suits? Further, should he indicate, by using lower class "grammar," that he is a member of that class?

The answers to this series of questions are not entirely certain, but some general rules are apparent. The basic rule derives from the *social role* of the interviewer. Whatever else he may be, he is a researcher in this particular situation, and most of his decisions follow from that role definition. There are research situations in which other considerations enter, but in general this status is a middle-class one. The interviewer will find that "overdressing" is as incongruous with this status as is wearing overalls. Indeed, the latter costume may arouse some disbelief that the interviewer is really a representative of an established research organization. The external characteristics of his functional role include such items as adequate grammar, alertness, confidence and seriousness, and clothing whose aim is neither to attract the opposite sex nor to arouse pity. This is not advice to "steer a middle course." Rather, the interviewer's actions, gestures, speech, and dress should divert attention from himself; in this situation, it is the respondent who is important. Just as extreme dress will arouse attention, so will exaggerated mannerisms or overprecise speech.

These externals are not, be it noted, matters of individual personality and taste, but are the indexes by which the respondent himself will make preliminary judgments. Consequently, they may determine whether the interview will be obtained at all. The social researcher is rapidly becoming a definite status in the society, which means concretely that the student who would pass muster must "act the part." His range of choice in these matters is limited by the public's image of his activities. And, of course, since the interviewer is almost always a representative of some organization, he is limited further by its position in the area. His contribution to the total project will be a negative one if he obtains the interview but manages to arouse antagonism or suspicion toward the organization itself.

This is not to say that for all research situations an apparently middle-class role or behavior will adequate. Several studies have indicated that a greater range and intensity of attitude are more likely to be expressed when the interviewer is closer to the class and ethnic position of the respondent. This is most especially true, of course, when the opinions to be expressed are somewhat opposed to general public opinion. Thus, to take an extreme case, white interviewers would have a more difficult time in obtaining a true set of attitudes from Negro respondents in Mobile than would Negro interviewers. Similarly, in a town torn by union strife, a very obviously white-collar interviewer might meet with considerable suspicion and might find that many respondents express a suspiciously high proportion of promanagement attitudes.

These facts follow, of course, from the general notion that the respondent, too, has insight and will judge the interviewer by his external characteristics, both gross and subtle. The situation, however, is one to be taken account of in the research design itself, so that the most adequate interviewers are chosen for the particular job. A highly trained interviewer can break through most of these barriers. A poorly trained one will not be adequate even if his class position is superficially in conformity with the group being interviewed. Furthermore, even the average interviewer can learn to become alert to suspicion or reservations on the part of the respondent and to deal with them in an adequate manner.

A common response to a request for an interview is a housewife's, "Oh, I'm much too busy right now. Come back some other time." The interviewer must be able to decide whether the respondent is really too busy or is merely using this claim as a way of avoiding the interview. In some cases, the puzzled or suspicious look on the face of the housewife will tell the interviewer that he should take a few minutes to explain what he is doing and why he is doing it. Even a few casual remarks about the neighborhood, the weather, or his understanding of the house work itself may break through these barriers. In some cases, he may ask a few questions about how to go to his next respondent's house, so as to make known to his interviewee that he is engaged in a quite ordinary activity. It is *not* sufficient merely to ask the respondent if she will set a time for an early appointment. She may be willing to do that, also, in order to avoid the present situation, but there is no guarantee that she will appear. The interviewer, then, must learn to "read" this situation carefully before accepting her claim. She may be obviously leaving the house or may be dressed for housework and annoyed at the interruption. On the other hand, she may really be asking for further reassurance.

Often, when the interviewer has a list of specific respondents whom he must interview personally, his first contact is not with the interviewee but with a member of the family or a friend. In these cases, he must remember that these people must be understood as well, if he is to persuade the respondent to give the interview. A husband or fiancé may be suspicious and refuse permission. A friend of the family may decide that the interviewer is a salesman or a bill collector, and he may give false information or prejudice the respondent so that the latter will not permit the interview. This situation may be very delicate, and particularly so if the subject of the interview is to be explained in detail only to the respondent. The problem must be met before the interviewer goes into the field, so that an adequate answer can be given. However, each situation must be understood as it occurs, so that active cooperation can be obtained from those who may bar the way to the respondent. For example, in one research situation, it was learned that in many lower class areas the statement "I am looking for Mr. Jones" frequently aroused suspicion. Alert to expressions of suspicion such as "What for?" or "Who are you?" or "I don't know the man"—the latter statement made after some hesitation or a long "sizing up" of the interviewer —the decision was made to avoid the question altogether. Instead, the interviewers began their first contact with an expression of smiling, near assurance, "Mr. Jones?" This led more often to a truthful denial of the identity and an offer of information concerning the whereabouts of Mr. Jones. Without this alertness on the part of the interviewer, however, a number of respondents would have been lost because of noncooperation on the part of the those who knew the respondents and who had it in their power to misinform the interviewer or refuse access to the ultimate respondent.

Herbert McClosky

Survey Research in Political Science

NATURE AND TYPES OF SURVEYS

A survey may be defined as any procedure in which data are systematically collected from a population, or a sample thereof, through some form of direct solicitation, such as face-to-face interviews, telephone interviews, or mail questionnaires. Surveys, of course, are not the only devices political scientists can use to collect systematic data on populations. They can also consult ecological reports for information on consumption, industrialization, and other indices of "modernization," state and county publications for aggregative data on party registration and voting preferences, or census and other government reports for figures on population changes, income levels, and number and size of voting

Reprinted from *Survey Research in the Social Sciences*, Charles Y. Glock (ed.), (New York: Russell Sage Foundation, 1967), pp. 131–37, where it appeared under the title "Survey Research in Political Science." By permission of the author and the publisher, the Russell Sage Foundation.

districts. The utility of these sources, however, is severely limited. Aggregative data, for example, can tell us little about the political attitudes and motivations of the population or its subgroups; official election returns can furnish only a crude estimate of which groups have voted for whom;[1] and comparisons of gross voting figures in various elections cannot specify which voters have shifted their preference for a candidate or party from one election to the next. These and scores of similar questions can usually be answered only through surveys. Surveys can, of course, be justified "only when the desired information cannot be obtained more easily and less expensively from other sources."[2]

The survey investigator may be interested in the individuals or groups who compose his "sample" not for themselves but for what they "represent." Usually, in fact, he chooses for study persons who collectively possess in miniature the same characteristics as the "universe" of persons he wishes to investigate. That universe may be the adult population of the entire nation

[1] Warren S. Robinson, "Ecological Correlation and the Behavior of Individuals," *American Sociological Review*, 15 (June, 1950), 351–57; and Austin Ranney, "The Utility and Limitations of Aggregate Data in the Study of Electoral Data," in Austin Ranney (ed.), *Essays on the Behavioral Study of Politics* (Urbana, Ill.: University of Illinois Press, 1962), pp. 91–102, at p. 91.

[2] Angus Campbell and George Katona, "The Sample Survey: A Technique for Social Science Research," in Leon Festinger and Daniel Katz (eds.), *Research Methods in the Behavioral Sciences* (New York: Dryden Press, 1953), p. 16.

or of a smaller unit. It may consist of active party members, political officer holders, or some other special group, such as state legislators, judges, or young people attending high school. There is no inherent restriction on either the size or composition of the universe. These are determined by the investigator's interest, and by the questions he wishes to answer. The universe may be small enough to permit the investigator to interview all of its members (this would be the case, for example, if one were doing a survey of national committeemen of the American parties); or it may be so large and geographically far-flung that the sample selected for study will number only a tiny fraction of the whole. Not all samples employed in political studies need to be "representative" in the sense of being cross-section miniatures of the general population; ordinarily they need only to mirror the characteristics of the particular universe being studied. A cross-section sample of the general population is essential only if one is actually investigating the general population. If one is studying political leaders, it would be wasteful to draw a sample of the general population in order to cull the small number of leaders to be found among them.

The accuracy with which a sample reflects the characteristics of the universe, and the number and type of characteristics represented in the sample, can vary somewhat with the nature of the inquiry. In general, a sample must more perfectly reflect the characteristics of the universe being studied if the investigator wishes to describe that universe than if his main concern is to discover or test relationships among variables. A scholar seeking to ascertain the number of Democratic voters in the general population will need a sample whose characteristics approximate quite closely those of the general population itself. This requirement becomes still more urgent if he also intends to predict the magnitude of the Democratic vote in an election, for even a small misrepresentation in the sample may lead to an erroneous forecast of the outcome. If, however, he is seeking to discover what the approximate correlation is between, say, belief in democracy and personality characteristics, he may be able to get by with a less perfect sample, for the correlation between these variables is not likely to be severely altered by the over-representation of certain groups—providing, of course, the errors are not extremely large.

Fulfilling the requirements of representative sampling may in some political studies be hindered by the fact that the characteristics of the universe are not known, and no practicable procedure may be available for ascertaining them. For example, no adequate description of the universe of persons active in politics is presently available, and an investigator who wishes to sample this universe cannot be certain that he has achieved an appropriate likeness. In such cases he may attempt to reduce systematic bias and undesired over-representation by casting a wide net and drawing his sample from many party units at different levels of activity in widely separated localities. Given sufficient funds, he might also employ a procedure known as "double sampling," in which a preliminary survey would first be carried out merely to locate and learn the characteristics of the party actives throughout the nation—merely, in other words, to determine the nature of the universe to be sampled.

The discovery that one can, by scientific sampling, reproduce in miniature and with remarkable accuracy the characteristics of a given universe is the *sine qua non* of the survey procedure, and holds out immense opportunities for the study of politics. Since political science is concerned with institutional and group phenomena involving thousands or even millions of people, procedures are needed for making the observation of such multitudes manageable. Without sampling, the study even of a local election, or of special groups such as Negroes, farmers, and political leaders, would require countless observations of many thousands of people. Through sampling, however, these and dozens of other groups can be reproduced in miniature, reduced to manageable proportions, and studied directly, closely, and in detail. The potential value of this procedure is incalculable, for it has now become possible simultaneously to study one or many political sub-cultures, to collect detailed information on millions of people by sampling only one or two thousand of them, to gather evidence systematically on a large number of questions and hypotheses and, by virtue of these possibilities, to advance the scientific study of politics in a way that could not have been imagined even fifty years ago.

Before the development of survey methods and scientific sampling, the political analysis of mass responses rested upon the observational skills and intuition of the individual investigator. He was often compelled to gather evidence on a hit or miss basis, relying on published articles or books, on such letters or diaries as he could find, and on discussions with persons whose opinions seemed to him worthwhile. Whether the political responses yielded by these sources were truly representative of a larger universe or were opinions unique to the individuals who expressed them was rarely known. Generalizations about slave owners or debtors or Northerners or those who voted for Wilson were, as a result, inadequately grounded and lacking in warranty. As retrospective empirical studies of past elections and other political events indicate, many historical inferences formerly regarded as "settled" turn out, in the light of contemporary survey knowledge, to be suspect.[3] They reflect not only the biases of the investigator, but the fortuitous character of his data-gathering procedures.

With the aid of surveys, some of these difficulties can now be mitigated. In principle, we can accurately compare and contrast individuals and groups from every part of the society, occupants of every kind of political role, persons or groups holding every type of political belief, political leaders and followers, Northerners and Southerners, Democrats and Republicans, rich and poor, debtors and creditors, supporters and opponents of various public policies, and every combination of these that interest us. We are thus able to test generalizations that until now have been loosely grounded in impressions gathered from anecdotal and other unsystematic evidence. This potentiality is essential for the development of a scientific understanding of politics.

Survey methods can be used in the comparative study of institutions, nations, and political practices as well as in the study of individuals and groups. One might, for example, discover much about the practices of bureaucratic agencies by surveying samples of their officials. One might learn how their decisions are actually made, whether their procedures differ according to the size of the agency, the degree of "modernization," and the types of training received by their functionaries, how they relate in practice to the legislative branches, and whether executive policies are in fact dominated by permanent civil servants. Comparable studies could be done on the institutional practices of judicial bodies, legislatures, or political parties—all with a degree of scientific refinement rarely possible heretofore.

Just as political surveys differ in the types of universes they investigate, so do they also vary in their design and purpose. The term "survey" usually conjures up the image of an interviewer canvassing house to house, asking direct opinion questions and receiving direct answers. While this is the classic pattern of the public opinion poll, many other kinds of political surveys are possible. The primary concern of some surveys has been to describe the beliefs of the general population or certain of its segments, while others have sought to test hypotheses and to investigate the relations among variables. Some have used randomly drawn samples, while others have deliberately over-sampled certain groups in order to study them more closely. Some have merely ascertained surface opinions, while others have probed deeply into the underlying attitude structure and personality characteristics of the respondents. Some have focused upon beliefs and others upon actions, some upon opinion-makers and others upon opinion-consumers. Most surveys have been one-shot affairs, collecting all of

---

[3]David B. Truman, "Research in Political Behavior," *American Political Science Review*, 46 (December, 1952), 1003–1006.

their data at a single point in time, but others have employed a "panel" of respondents who are interviewed and reinterviewed at different points in time in order to assess their shifts in opinion and party preference, or their response to changing political events.

The adaptability of the survey method is also illustrated by the range of subjects to which surveys may be addressed. They can, for example, be employed either for omnibus political studies or for studies that are narrowly restricted in subject matter. An investigator may be interested in a single political outcome, such as voting turnout, but if he is to explore adequately even this one variable, he will need information on many different kinds of people. He will require samples of regular voters, occasional voters, and nonvoters, and on each of these he will doubtless want information on their group, psychological, political, and intellectual characteristics. He can collect such data in a single survey, and can use the results to explore his respondents' behavior either intensively or extensively. He can use the information yielded by the survey to construct indices and scales that will permit him to examine, from various points of view and with considerable thoroughness and refinement, the relationships in which he is interested. Although a survey utilizing large samples may seem an uneconomical way to collect data on a single dependent variable, it often turns out to be the most efficient and economical method possible. Among the most striking examples of this are the handful of large-scale voting surveys that have in three decades taught us more about the act of voting than was learned in all previous history.[4] These survey studies have shown that even the apparently simple act of voting is in reality the behavioral manifestation of extraordinarily complex forces.

Surveys offer the further advantage of furnishing information simultaneously on more than one set of dependent and independent variables. A survey that is primarily focused on voting can also collect information on a number of related behaviors, any one of which may be treated as a dependent variable and explored in its own right. Party affiliation, level of political participation, and political ideology may, in relation to voting, be considered as independent variables; but from other perspectives they can obviously be treated as dependent variables. A large-scale survey that measures and collects data on many dimensions at once permits the investigator an extraordinary measure of flexibility not only in the manipulation of variables but in the selection of hypotheses he may wish to test. He can observe the same dimension from different perspectives and he can control or vary particular factors so that their relation to certain behavioral outcomes can be assessed more precisely.

Omnibus surveys also permit the investigator to undertake new lines of inquiry or even new studies that were not expressly provided for in the original design. If certain lines of analysis prove fruitless, the investigator often has on hand the information he needs to shift the direction of the inquiry and to test alternative explanations. Every survey of any magnitude contains far more data than the original investigator is likely to use. Indeed, recent scholarship has shown that survey data can often be exploited by investigators who had nothing to do with the original study but who find that it contains information they can readily adapt to their own inquiries. Secondary analysis of the data accumulated by opinion polls has often provided more significant results than were yielded by the initial analysis.

The impression is common, though erroneous, that political surveys have been used almost exclusively to assess public opinion. Although political surveys began as opinion polls, public opinion is only one of many subjects to which they are now addressed. Surveys are being adapted to

---

[4]Paul F. Lazarsfeld, Bernard R. Berelson and Hazel Gaudet, *The People's Choice: How the Voter Makes Up His Mind in a Presidential Campaign* (New York: Duell, Sloan and Pearce, 1944); Bernard R. Berelson, Paul F. Lazarsfeld and William W. McPhee, *Voting: A Study of Opinion Formation in a Presidential Campaign* (Chicago: Chicago University Press, 1954); Angus Campbell, *et al.*, *The American Voter* (New York: John Wiley & Sons, 1960).

almost every area of political study and to many different types of problems. They have been used to skim off the opinions of respondents and to interview them in depth, focusing intensively on certain aspects of behavior or on complex systems of belief. While surveys, of course, cannot usually investigate these matters as thoroughly as they can be explored in repeated psychiatric interviews, they can employ clinical personality keys, attitude scales, and other sophisticated measuring devices to achieve a fairly detailed profile of the personality and attitude characteristics of respondents.

Flexibility is also possible in the design of the survey questionnaire and procedure. The questionnaire may be open-ended or focused, loosely or highly structured, intensive or extensive, probing or concerned with surface responses, addressed to a single question or to many questions. It may, in its level of language and conceptualization, be fashioned either for particular classes of people or for people of many different kinds. The procedure may consist entirely of face-to-face interviews, telephone interviews, mail questionnaires, questionnaires delivered by interviewers that are self-administered by respondents and then returned by mail (or alternatively picked up, when completed, by interviewers), or some combination of these. Each method has its advantages and disadvantages.

. . . .

Establishing the Facts

The primary function of a political survey is to collect and certify the facts. What do voters believe about the candidates? What are their opinions on issues? Do individuals at different levels of the society differ in their political attitudes? Before surveys came into use, even such simple questions often could not be answered accurately, or the accuracy of the answer could not be certified by reference to an objective procedure.

The ability to establish the facts is, of course, a *sine qua non* of any scholarly field of inquiry. Yet traditional political and historical studies have tended to pay less attention than they should to the task of certifying the evidence needed to confirm their conclusions. Many of their claims, therefore, are intuitive or conjectural rather than factual—the products of partial, anecdotal, and haphazard observation. They are given, for example, to infer a nation's climate of opinion from a relatively small number of books, articles, speeches, editorials, or legislative acts. They are likely to take isolated events—a demonstration, a violent incident, a sharp conflict, a protest by a vociferous group—as signifying major trends in political belief and preference. When one observes political phenomena anecdotally, one's eye is drawn to whatever is dramatic and exciting, which means, often, to whatever is idiosyncratic and atypical. The tendency to overstress the unusual and to ignore the everyday occurrence or the modal opinion is characteristic not only of journalists but also of scholars who fail to adopt rigorous research techniques.

The habit of generalizing from mere impressions or isolated instances may still be observed in current social and political analysis. Contemporary writing, for example, is ridden with assertions that modern man is alienated, conformist, escaping from freedom, frustrated about his status, other-directed, psychologically disoriented by reason of rapid social mobility, class conscious, hungry for meaning, overwhelmed by a debilitating sense of aimlessness and drift, bothered by conflicting norms and frantic for guidance, separated from meaningful work, oppressed by mechanization, and dehumanized by the anonymities of mass society. I cite these not to contest their validity but only to observe that the facts they allege have not been adequately established and that the magnitude and pervasiveness of the conditions they profess to find have rarely been investigated by their proponents in the thorough, empirical, systematic manner they deserve. Yet many of the facts needed to assess the warranty of such allegations are accessible through survey methods. One can use surveys to map a range and variety of responses bearing on these matters. In place of sweeping generalizations about modern man's anxiety or aimlessness or conformity, one might substitute fairly precise estimates of the frequency and extent of these states. Exactly what proportion of men and women feel this way? Do the old exhibit these responses more than the young? The uneducated more than the educated? Workers more than managers? Rural residents more than urban residents? Non-believers more than those who attend church? Persons in conflicting status roles more than those in congruent roles? One can, if one chooses, also ascertain through cross-cultural comparisons of survey results the essential facts concerning the relation of these attitude states to the level of industrialization and state of political culture.

While surveys are not the only procedures for searching out vital political and social facts, they can often furnish information that is both more precise and more varied than the information supplied by other sources. Surveys, for example, can substantially augment aggregative types of data such as election statistics. Official voting returns can tell us who won an election, which districts voted for which candidates, and whether the turnout was large or small. But they cannot tell us what motivated the voters, why they divided as they did, how much they knew about issues, and whether they were significantly affected by the personalities of the candidates. A careful assessment of the facts on these questions requires a survey of some type. Even if a shrewd observer could, by intuitive observation, correctly divine the answers to such questions, a survey would still be needed to confirm his conclusions since, inevitably, an equally shrewd observer will have been led by *his* intuitive observation to a very different set of conclusions. Even if surveys were employed solely to confirm "what everybody knows" (a widely repeated charge that is patently false), they would still be essential. As it turns out, "what everybody knows" is almost invariably contradicted by an opposing and equally "well-known" conclusion. Unless we can verify which of the two "well-known" claims is correct, we can never resolve the conflict and separate truth from error.

It has become fashionable in some intellectual circles to deplore "fact-grubbing" and to exalt "theory." But the polarity implied by this distinction is spurious. While theory, of course, is vital to scientific advancement, facts are the building blocks of every science, essential to both the construction and testing of theory. Their verification, therefore, is essential to the progress of a science. In the study of politics, survey methods are often indispensable for this purpose.

Increasing the Range and Amount of Information

Surveys also make it possible to collect in a single effort vast amounts of information on large numbers of people. Their potentiality in this regard is probably greater than that of any alternative method. Studies using aggregative data, for example, may supply information on many people, but the range of the information is usually narrow and little or nothing can be learned about any individual. Laboratory or field experiments are in some respects superior to surveys in their power to confirm or disconfirm hypotheses, but they are generally compelled for reasons of cost, manpower, and the nature of the research design, to limit themselves to a small number of subjects and a severely restricted set of variables. Documentary studies of political phenomena can encompass many people and many topics, but are deficient in the ability to portray those phenomena accurately in all their range and variety. They can, furthermore, supply detailed personal information only on the few individuals on whom biographical data are available. Anthropological studies of communities come closest to surveys in their capacity for collecting large amounts and varieties of detailed data on large numbers of people. Many of these studies, however, are in reality crude surveys, less systematic and quantitative than the standard survey, but superior in their ability to observe how men act, how they relate to each other, and how they are affected by institutional practices.

In a single political survey that utilizes a one- or two-hour interview of a sample of, say, 1500 persons, skilled interviewers can collect vast amounts of data on personal and social background factors (e.g., age, sex, education, religion, income, occupation, residence, marital status, organizational memberships) and on such political matters as party preference, political interest and participation, nature and frequency of political discussions, family political background, opinions on issues, attitudes on conservatism, internationalism, equalitarianism, and democratic beliefs and practices. Even psychological information can be collected by using personality scales, questions about primary group involvement, measures of behavioral conformity and non-conformity, and the like. Much can also be learned about the reaction of individuals and groups to current political developments, their response to conflicting political choices, and their predictions about political trends.

Even these questions and answers do not exhaust the data-gathering potential of the survey procedure. Various questions (and their answers) can be combined and recombined into indices and scales that will furnish information on entirely new variables. Consider a simple illustration: if we ask a respondent for his own party preference and that of his father, we can, by combining the answers, simultaneously learn whether the respondent has shifted from or remained loyal to his father's party. This datum may be useful in answering a number of questions that interest political scientists. Do people shift their party preference because of changes in economic status, life style, or geographic location? Are they more likely to shift if their families lack cohesiveness or possess conflicting status characteristics? Are there personality differences between shifters and non-shifters? Is the rejection of the parental party preference an aspect of a generalized rebellion against parents? Thus, even this simple example opens the possibility of exploring many questions that were not specifically asked in the initial questionnaire. Our illustration contains only two variables that were combined into a third, but indices can be constructed that consist of three, four, or even more elements. A questionnaire containing 25 questions, hence, may yield measures several times that number.

The number of findings yielded by a survey can be further augmented by manipulating variables in the course of the analysis so as to vary, in effect, the conditions under which behavior occurs. The sample, for example, can be divided by education and the less educated compared with the more educated. Or college graduates with high incomes can be compared with college graduates of low income. Numerous other combinations involving occupation, religion, age, rural-urban residence, party membership, and so forth, are equally possible. Nor is the survey investigator restricted to an analysis of the so-called "marginals," i.e. the figures resulting from the *prima facie* answers of the total sample to the questions used. The marginals reported in public opinion polls, for example, represent only the first approximation of the knowledge uncovered by the survey. Much more can be learned, some of it of great significance and subtlety, by comparing various subgroups of the population.

New knowledge may also result from unexpected findings that are sometimes turned up by surveys. Such findings usually set off a train of questions and conjectures that lead the investigator to redirect his inquiry. How are the unanticipated findings to be explained? Do they appear to be related to some special circumstance, to the unsuspected influence of some variable, or to the interaction effect produced by the joining of two variables? The survey method frequently provides the information needed to explore and to resolve such questions.

Using survey methods to study politics furnishes more and better answers to old questions as well as new answers to new questions. It is characteristic of a science that each improvement in research techniques inspires new questions that had either never been thought of before or were dismissed for want of a method for finding the answers. Surveys have encouraged political scientists to ask and to investigate numerous questions they would scarcely have raised a few decades ago, including, for example, questions about the effects of personality on leader preference, the relation of cognitive habits to political attitudes, the group process by which political beliefs are transmitted and sustained, the degree of correspondence between leader and follower opinions, and scores of other problems that involve social and psychological mechanisms. As answers to these questions are found, they inspire, in turn, new questions, and sometimes new research techniques as well. There is thus a continual interplay between advances in research technique and the acquisition of scientific knowledge.[5]

[5] Evron M. Kirkpatrick, "The Impact of the Behavioral Approach on Traditional Political Science," in *Essays on the Behavioral Study of Politics*. . . , p. 15.

Barry Anderson

The Social Science Experiment

The experiment may be regarded as a model of the scientific method. It is a miniaturization of the scientific method in that, in its simplest form, it is a procedure for arriving at a single descriptive statement. It is an idealization of the scientific method in that the descriptive statements it produces are statements of causal relationship and in that the procedure for arriving at these statements embodies the principle of controlled observation more fully than any other procedure.

According to the principle of controlled observation, before one can make a descriptive statement that a change in variable A produces a change in variable B, all variables other than

From *The Psychology Experiment: An Introduction to the Scientific Method* by Barry F. Anderson. © 1966 by Wadsworth Publishing Company, Inc., Belmont, California. Reprinted by permission of the publisher, Brooks/Cole Publishing Company.

A must be discounted as possible causes of the change in B. In an experiment, one can immediately discount a great number of variables because, by the very nature of an experiment, these variables are controlled. *An experiment is a situation in which one observes the relationship between two variables by deliberately producing a change in one and looking to see whether this alteration produces a change in the other.* One kicks the television set, for example, to see whether this action will restore the picture. Because one changes this variable oneself, he is in a position to know that a great many other variables have not changed and, therefore, could not possibly have caused any observed change in the television picture. If it comes back on, one can be quite certain that opening the door had nothing to do with it because the door was not opened; that turning the lights off had nothing to do with it because the lights were not turned off; and that raising the room temperature had nothing to do with it because the room temperature was not raised. The list of controlled variables could be extended indefinitely.

The variable which the experimenter changes directly is referred to as the *independent variable* (IV), so named because its value is independently manipulated by the experimenter. The variable which the experimenter examines to see whether it is affected by changes in the IV is called the *dependent variable* (DV), so named because its value is expected to be dependent on the value of the IV. Thus, one might apply heat (IV) to sulfur to see whether it will burn (DV), put fertilizer (IV) on mushrooms to see whether they will grow faster (DV), or make noise (IV) to see whether this will affect a learner's rate of learning (DV).

In psychology, IVs are almost always stimulus variables, and DVs are response variables. The psychologist is trying to understand behavior, and his usual way of manipulating behavior, so that he can see how it works, is by varying the things the organism sees, hears, or in some other way senses.

The values on the IV define the *experimental conditions*. If one is interested in the effect of noise level on studying, for example, there are any number of different values on the noise-level variable one might use: silence, a quiet radio, a loud radio, and a very loud radio, to mention only a few. In the *simple experiment*, . . . just two values on the IV—and hence two experimental conditions—are used. In the example, these might be a "silence condition" and a "loud-radio condition". . . .

Notice that the *fewest* experimental conditions an experiment can have is two. Observations under a single condition can result only in state description. For process description, one must observe changes; and for a change to exist on the IV, there must be at least two conditions.

One of the experimental conditions is often called the *control condition*, and the other simply the *experimental condition*. The experimental condition is the one in which the treatment of interest—heat, fertilizer, noise—is administered. The control condition is one which is exactly like the experimental condition on most variables other than the variable of interest—the IV. Those variables on which the experimental and control conditions are identical can be discounted as possible causes of the change on the DV because they are controlled; hence, the name "control condition." The fact that experimental and control conditions are referred to collectively as experimental conditions may seem confusing, but the meaning of the term "experimental condition" that is intended is usually quite clear from its context.

According to the principle of repeated observations, if one wishes to generalize a descriptive statement to apply to an entire population, the observation should be repeated on several members of the population. This introduces another variable into the experiment, the *sampling variable* (SV). The sampling variable refers to the particular objects studied. These might be samples of a chemical in a chemistry experiment, plots of land in an agricultural experiment, or Ss[1] in a psychology experiment. The sampling variable is so called because, usually, not all of the objects to which the experiment's conclusions will be applied are included in the experiment itself; only a limited sample is studied. Thus, every experiment has an IV, a DV, and an SV.

The main features of an experiment can be conveniently represented in the following kind of diagram:

IV: NOISE LEVEL

	Silence	Loud Radio
S1	80	70
S2	70	70
S3	90	80
S4	80	85
S5	65	60
S6	75	70

DV: Comprehension Test Score

The IV and the two values on it are identified at the top of the diagram. The Ss, which are numbered arbitrarily, are indicated down the left side. The numbers in the cells refer to values on the DV, those in each row having been mea-

[1](*Editor's note*) In this context the letter "S" is used to refer to the people who act as subjects in experiments.

sured on the S indicated at the left of that row, and those in each column having been measured under the experimental condition indicated at the top of that column. The DV is identified to the right of the rest of the diagram.

THE NATURAL EXPERIMENT

In order to observe a relationship, one variable must be changed and another observed. In an experiment, the experimenter produces the changes; in a *natural experiment*, nature produces the changes. This may seem like a small difference, but it has far-reaching implications.

To make the difference clearer, let us consider an example. Suppose that one is interested in finding out whether skipping breakfast makes a person irritable during the day. One could seek relevant evidence by way of an experiment or a natural experiment. In an experiment, one would manipulate the breakfast variable *oneself* by having some Ss eat breakfast and other Ss eat no breakfast; later in the day, one would make observations of both groups on the irritability variable. In a natural experiment, one would allow *nature* to manipulate the breakfast variable by simply looking for Ss who happened to have eaten breakfast and Ss who happened to have skipped breakfast. One manipulates nothing in the natural experiment; one simply looks. Hence, properly speaking, there is no IV or DV in a natural experiment, only SVs; one samples situations instead of producing them.

If one found a difference in irritability between the "breakfast" Ss and the "no breakfast" Ss in the experiment, one could safely conclude that eating breakfast affects the irritability variable. If one found such a difference in the natural experiment, however, one would not know whether the breakfast variable was related to irritability or whether some *third variable*, which varied along with these, actually produced the observed relationship. Maybe those Ss who skipped breakfast did so because they were irritable in the first place; or because their wives had decided to sleep late that morning, and that made them irritable; or because they were out of breakfast food, and that made them irritable.

A serious problem with natural experiments, then, is that one can never be sure whether an obtained relationship is causal or noncausal. It is always possible that a third variable, C, is causally related to both A and B and has produced the observed relationship between them—in which case the A-B relationship is itself noncausal. . . .

In the purest case of the natural experiment, all variables are free to vary, and so there are always many variables which could act as third variables to produce a noncausal relationship. In the purest case of the experiment, the problem of the third variable never comes up, for, in line with the principle of controlled observation, all variables except the IV are controlled or in some other way discounted as possible causes of the change on the DV.

Controls can, in fact, be introduced into natural experiments. The procedures for establishing such controls will not be discussed in this book, but one important point should be made about the differing nature of control in experiments and natural experiments. *The natural experiment begins with a situation in which all variables are free to vary, and controls, if they are introduced, are introduced one variable at a time.* In marked contrast, *the experiment begins with a situation in which all variables are controlled, and variation is introduced one variable at a time.*

Some variables are difficult or impossible for the experimenter to manipulate, however. Thus, natural experiments are possible in many cases where true experiments cannot possibly be carried out, for nature makes changes on many variables over which man has no control. Indeed, a number of highly respectable sciences are, of necessity, bound to the use of natural experiments. Astronomy, oceanography, and geology are among these, for it is beyond man's present powers to manipulate the mass of a planet, the temperature of the ocean, or isostatic forces in the earth's crust. In many areas of psychology as well, the manipulation of variables is either not possible or not ethical, and the natural experiment is the only acceptable alternative. For example, one does not raise children in isolation in order to determine the effects of such isolation on their social behavior.

Even where a choice exists, the natural experiment is sometimes preferable to the true experiment. Especially in the early stages of research in an area, it is wise to observe the operation of variables in their natural settings to determine the relative importance of these variables before studying them under the more controlled conditions of the laboratory. Variables which can be made to produce large effects in the laboratory are sometimes relatively unimportant in nature because their effects are swamped by variables which were controlled in the laboratory setting. Conversely, variables which cannot be made to produce large effects in the laboratory are sometimes of considerable importance in nature. The fear induced by nature on the battlefield or in death row, for example, is far greater than that induced by an experimenter using the strongest electric shock he dares.

Thus, the natural experiment will always have an important place in science. The person who uses the natural experiment should be aware of its limitations, however, so that he may avoid the many pitfalls attendant upon its use. The limitations of the natural experiment are most easily seen when it is contrasted with the true experiment.

. . . .

STEPS IN CONDUCTING AN EXPERIMENT. . .

1 Getting the Idea

One does not run an experiment until one has an idea to test, a question to ask. . . .

2 Searching the Literature

Reading the relevant literature can be an excellent source of ideas. It can also be an important help in designing the experiment. Finding out how other researchers working on similar problems have operationally defined their IVs and DVs can save a considerable amount of time. Finding out the length of the lists, number of trials, exposure time, or rate of presentation used by others can reduce the number of preliminary runs required to adjust your experiment to the right level of difficulty for your Ss. Finally, a knowledge of related results will often suggest important variables that should be controlled. . . .

3 Designing the Experiment

Once you have an idea, you must make it operational. You must decide exactly how you are going to manipulate your IV and exactly how you are going to measure you DV; what variables you will control; whether you will use an identical-Ss, matched-Ss, or independent-Ss design; and how you are going to present your stimuli and record your responses. You must prepare instructions for your Ss. In addition, it it highly advisable to decide, at the time that you are designing the experiment, how you are going to analyze your data, for, if you do not, you may design an experiment whose results you will be unable to analyze. Finally, in designing an experiment, it is important to anticipate the various ways the results might come out and the conclusions you would want to be able to draw in each case, so that you can make provisions for gathering data or instituting controls relevant to as many of these conclusions as possible. All too often an experimenter realizes, once he has begun to write up his research paper, that he would be able to make such-and-such a statement *if* he had only gotten such-and-such data or instituted such-and-such controls. Once the experiment is over, it is too late to attend to these matters; such matters must be attended to at the time that the experiment is being designed. Conclusions are not something which are tacked on at the end of an experiment, but something which, from the very beginning, are woven into the fabric of the experimental design. . . .

4 Instructing the Ss

Instructions are to motivate and direct. Simply explaining the purpose of the experiment to S will often stimulate him to do his best, though, of course, such information should not be given to him when it would distort the results of the experiment. The task should be clearly explained, so that S will know what he is expected to do. The instructions should be presented in simple language, and important parts should be repeated. When the task is at all complex, it is advisable to instruct S on part of his task and have him practice that part before instructing him on the rest. When the experiment consists of several different tasks, it is likewise better to instruct S for each task just before he is to perform it, rather than to present all the instructions at the beginning of the experiment. Instructions should ordinarily be written out and read verbatim to S by the experimenter, to ensure that all Ss in each condition receive identical instructions. Faulty instructions can introduce serious errors into an experiment. For this reason, it is of the utmost importance that the instructions be thoughtfully prepared and adequately pretested.

5 Presenting the Stimuli

Many kinds of stimuli are possible, from shock to mazes to propaganda, and only the most general considerations relating to the presentation of stimuli can be dealt with here. Stimuli are most commonly visual or auditory. Simple visual stimuli are often presented in the form of lights, and simple auditory stimuli in the form of bells and buzzers. Complex visual stimuli, such as patterns or written words, are often presented by means of a slide projector or a memory drum (a mechanical device for serial presentation); and complex auditory stimuli, such as music or spoken words, by means of a tape recorder. For experimenters with minimal equipment resources, complex visual stimuli can usually be quite adequately presented on 3-by-5 cards, which S turns over and views one at a time on signals from the experimenter; and verbal auditory stimuli can be read from a script by the experimenter. Stimuli should be made as simple as is consistent with the hypothesis to be tested, for unnecessary complexity only makes the results more difficult to interpret by making it difficult to determine which aspects of the stimuli S responded to.

6 Recording the Responses

The response required of S during the experiment should be as simple as possible, consistent with the hypothesis, for requiring an unnecessarily complex response only makes it more difficult for S to give proper attention to the task required by the hypothesis. A written record of S's responses should be made, by the experimenter or by S himself, at the time they are made. A kind of response which often proves to be quite valuable is S's own introspective report, after the experiment, of what he thinks the experiment was about and what he was trying to do in the experiment. Introspective reports help in providing a check on the adequacy of the instructions and in serving as a source of new ideas.

7 Running a Pilot Study

Before the experiment proper is run, it is wise to try the procedure out on a few Ss. A small trial experiment which is run before the experiment proper is referred to as a *pilot study*. A pilot study is useful for trying out the instructions, for making sure that the equipment works, for adjusting the conditions of the experiment so that the task is neither too easy nor too difficult, and for enabling the experimenter himself to master the routine of the experiment.

8 Running the Experiment

If the experiment has been well thought out beforehand, running it should be a straightforward carrying out of the plan. It is usually helpful, in running experiments of any complexity, for the experimenter to prepare himself a checklist to follow in administering the conditions of the experiment; it is all too easy to forget to administer a condition, to administer a condition out of order, or to administer a condition to the wrong S.

9 Analyzing the Data

There are two problems involved in analyzing the data. One is to simplify the great mass of numbers resulting from the experiment; graphs and descriptive statistics, such as averages, are used for this task. . . . The other problem is to make inferences from the few Ss tested to all the people to whom the conclusion is intended to apply; for this, statistical inference is used. . . .

10 Interpreting the Results

Many considerations are involved in interpreting the results of an experiment: the design of the experiment, the statistics used, [and] findings of other investigators. . . .

11 Writing the Research Paper

Science is a social enterprise, and the results of a particular experiment are of no scientific value until they are communicated to other members of the scientific community. . . .

Richard E. Dawson

Data Generation Through Simulation

Simulation is of increasing importance to social and behavioral scientists. The term appears frequently in a wide variety of social science literature and in reports of social science experiments. Earlier, the term was used almost exclusively in the discourses of various branches of engineering. In recent years, however, the use of the term and a more explicit use of the method it designates have spread into military science, industrial engineering, business and management training and research, economics, psychology, political science and sociology.

Richard E. Dawson, "Simulation in the Social Sciences" in Harold Guetzkow (ed.), *Simulation in Social Science: Readings.* © 1962. Reprinted by permission of Prentice-Hall, Inc., Englewood Cliffs, New Jersey.

The current employment of simulation in social science is related to the development of system analysis, the social psychological study of groups, the use of more formal models, the employment of various mathematical techniques and the availability of high speed electronic computers. Simulation, as the study of systems through the construction and operation of models, has followed the use of these different but closely related techniques into the social sciences. This essay will discuss the meaning, types, uses, values and problems of simulation as a tool for research, training and teaching in the social sciences.

I. What Is Simulation?

The term *simulation* has also been used in a variety of different, yet closely related, senses. Like other terms currently used in social science literature, it has a technical meaning and two or more popular meanings. In popular usage the term sometimes refers to the assumption of the appearance of something without having its reality. For instance, electric light may be termed simulated sunlight. It possesses many of the illuminating properties of sunlight but is not actually sunlight. In other instances, simulation refers more specifically to the assumption of false appearances for the sake of deception. This second usage of the term has been adopted in the biological sciences. *A Dictionary of Scientific Terms* defines simulation as: "Assumption of features or structures intended to deceive enemies, as forms of leaf and stick insects, and all varieties of protective coloration."[1] As currently used to depict a method of research in the social sciences, the term is employed without the connotation of deception.

Simulation is certainly not a new device. Its use probably precedes recorded history. In a very broad sense of the term, it can be argued that man has been simulating objects ever since he first began to draw and carve representations of objects on tree trunks and on the stone walls of cliffs and caves. In this very broad meaning of the word any construction of a "model," whether symbolic (pictorial, verbal, mathematical) or physical might be termed simulation. In this sense the classical dialogues of Plato, the 15th century art of Leonardo Da Vinci and the abstract art of the twentieth century might all be termed simulation, inasmuch as they are attempts to portray or reproduce by means of words, stone or canvas their authors' conception of various aspects of human life or physical objects.

The earliest practical use of simulation was the construction of physical models of real objects, particularly for work in designing tools and other objects. As Conway *et al.* point out: "Simulation has for a good many years been a useful device to the designer. He has tank-tested hull shapes, tested airfoils and scale models of airframes in wind-tunnels, and built pilot versions of chemical processing plants."[2] The construction of a model of or simulation of the real object, permits the designer to test the whole or specified aspects of the object he wants to build on the replica and thus avoid mistakes and waste in the construction of the real object.

In the general, more popular use of the term, all model building might be termed simulation. But we are interested in a more specific meaning when referring to the use of simulation as a research, training and teaching device in the study of human systems. The social or behavioral scientist is primarily concerned with the study of psychological and social processes. Simulation, as a social science research technique, refers to the construction and manipulation of an *operating* model, that model being a physical or symbolic representation of all or some aspects of a social or psychological process. Simulation, for the social scientist, is the building of an operating model of an individual or group process and experimenting on this replication by manipulating its variables and their interrelationships.

II. Simulations Are Models

Since simulation is so closely related to the construction of and the use of models, it seems appropriate to discuss briefly the use of models in social science. A model of something—a physical object, a living organism or a social system—is a physical or symbolic representation of that object, designed to incorporate or reproduce those features of the real object that the researcher deems significant for his research problem. The term model, as used here, refers to a scientific tool. It does not connote that the representation is an ideal or a "good model," worthy of emulation. Brody points out that "Developing a model involves abstracting from reality those components and relationships which are hypothesized as crucial to what is being modeled."[3] The choice of essential aspects of the reality being modeled depends upon the purposes for which the model

---

[1] I. F. Henderson and W. D. Henderson, *A Dictionary of Scientific Terms* (Edinburgh: Oliver and Boyd Ltd., 1960), p. 497.

[2] R. W. Conway, B. M. Johnson and W. L. Maxwell, "Some Problems of Digital Systems Simulation," *Management Science*, 6, No. 1 (1959), 92.

[3] Richard A. Brody, "Political Games for Model Construction in International Relations," The Program of Graduate Training and Research in International Relations, Department of Political Science, Northwestern University, June, 1961 (unpublished mimeo); to be published in Harold Guetzkow, *et al.*, *The Use of Simulation for Teaching and Research in International Relations* (Englewood Cliffs, N.J.: Prentice-Hall, Inc., 1963), p. 2.

is being constructed. In some cases—like a model ship to be displayed in a bottle—the object of the model construction is to reproduce as many of the details of the real object as possible, only on a smaller scale. In other cases—like the hydraulic model of an economic system,[4]—whether the model has the physical appearance of the real object is of little importance. The important factor is that the components and variables being investigated through the model respond in a manner comparable to that of the behavior of the real system.

Models can be constructed in several different media. Until recently nearly all description and analysis in the social sciences has been verbal, pictorial or diagrammatic. Plato's *Republic* and Toynbee's *Study of History* use verbal models for the description and analysis of social systems. Economists and sociologists have made much use of pictorial or diagrammatic models in the form of supply and demand charts and organizational authority flow charts.

As the social sciences have developed and the social scientists have increased their understanding of the phenomena they study, it has been possible to translate physical, verbal and pictorial models into mathematical models. In the construction of mathematical models, properties of the real objects or systems are abstracted by measurement and expressed in a set of mathematical equations, as Simon has done in his *Models of Man*.[5] Quantitative measures supplement the qualitative distinctions of verbal and pictorial models, and general descriptions of the relationships between variables are replaced by more precise equations. Before mathematical models can be developed, the relationships among units and variables must be structured so that the latter correspond to the rules of relationship which constitute the mathematical methods being used. When appropriate mathematical models can be constructed, the scientist has a powerful tool to aid him in understanding the behavior of the object or system he is investigating. The abstractness of the symbol system in a mathematical model makes possible the recognition of similarities and congruences between various models and thus between the realities they represent. Many of the simulation exercises used in social sciences are done wholly or in part with mathematical models.[6]

As was intimated above, simulation as employed by social science makes use of models constructed in such a way that they may become operative or functioning. Operating models are representations of behaving systems that attempt to reproduce processes in action. As such, operating models provide information about variable, component and relationship changes within a system over time. These models or parts of them can be expressed in physical, chemical and biological media and through verbal, pictorial or mathematical languages. In most instances operating models will involve combinations of these several modes of representation.

III. Purposes for Which Simulation Is Employed

As the above discussion has suggested, simulation can be used by social scientists for a wide variety of purposes. Basically the social scientist simulates

---

[4]Irving Fisher, *Mathematical Investigations in the Theory of Value and Prices* (New Haven: Yale University Press, 1925), p. 44.

[5]H. A. Simon, *Models of Man* (New York: John Wiley and Sons, Inc., 1957).

[6]For more extensive discussion of the relationship between models and simulation see Brody, "Political Games for Model . . ."; Martin Shubik, "Simulation, Its Uses and Potential, Part 1," (unpublished mimeo), Anticipation Project Expository and Development Paper, No. 3, General Electric Corp., Operations Research and Synthesis Consulting Service, May 4, 1959; and Guy H. Orcutt, "Simulation of Economic Systems," *The American Economic Review*, 50, No. 5 (1960), 893–907. For a general discussion of the use of models in social science see Anatol Rapoport, "Uses and Limitations of Mathematical Models in Social Sciences," and May Brodbeck, "Models, Meaning, and Theories," both in Llewellyn Gross (ed.), *Symposium on Sociological Theory* (Evanston, Ill.: Row, Peterson and Co., 1959).

to investigate and to learn about the behavior of individual and group processes. Such learning may be used for at least four very useful purposes: (1) design; (2) the development of a body of knowledge; (3) training; and (4) teaching.

To aid in the designing of new and improving the design of old materials or systems is probably the oldest use to which simulation has been put. It has been widely used for design purposes in the physical sciences, especially in engineering, for some time. The experiments with aircraft models in wind tunnels illustrate how simulation is useful in choosing between alternative designs. By manipulating the pressure and direction of the wind, as well as size, structure and materials of the model plane, in the simulated situation, the experimenter is able to learn something of how various designs would work under different conditions in the real atmosphere. This same principle is quite useful in designing and refining various social systems. For instance, in the simulation of an air defense system . . . the researchers seek to experiment with the system in simulated situations to learn how the system can be best designed to cope with real problems.

Simulation is also useful in helping the social scientist build a body of knowledge concerning various social or individual systems. The object of science is the formulation of theories that explain and predict behavior. Simulation is a very useful device for the exploration of verbal theories and the testing of hypotheses. In many instances, especially in the study of social and psychological phenomena, it is undesirable or even impossible to conduct experiments upon real systems. By successfully simulating the significant variables, it is possible to explore such phenomena by experimenting with the simulated system.[7]

Simulation has value as a teaching device. The hydraulic model of an economic system portrays quite vividly the general nature of a complex economic system, an analytical system often too "abstract" for many to understand through verbal description. Guetzkow's inter-nation simulation and the Carnegie Tech Management Game are used to help teach college students about the behavior of complex social systems by having the students make decisions, handle data and experience consequences in the simulated systems, comparable to those which occur in the real system.

The training purposes of simulation are closely related to the teaching functions. In teaching, the student is presumed to learn about the system. In training, it is hoped that the trainees will learn how to fill various operating roles in the real system by participating in comparable roles in simulated situations. As Orcutt observes of the training function of simulation, ". . . the trainee is able to get some feel of what he would experience in the real situation and some indication of the likely outcome of various actions and responses on his part."[8] . . .

IV. THE VOCABULARY OF SIMULATION

A variety of simulation techniques has been developed and employed by social scientists, and a number of terms are currently in use designating different approaches, purposes and techniques. Among the more frequent terms used in the social science literature in connection with simulation are man-machine simulation, gaming, Monte Carlo techniques, digital computer simulation, analog computer simulation, machine simulation and real-time simulation. At present, there is disagreement and sometimes even confusion as to the meanings and distinctions between some of these terms; especially concerning the relationship between simulation, gaming and

---

[7]For a discussion of the use of simulation for theory exploration and hypothesis testing see Harold Guetzkow "A Use of Simulation in the Study of International Relations," in Harold Guetzkow (ed.), *Simulation in Social Science: Readings* (Englewood Cliffs, N.J.: Prentice-Hall, Inc., 1962), pp. 82–93.

[8]Orcutt, "Simulation of Economic Systems . . .," 895.

Monte Carlo techniques. Commenting on the use of these terms Thomas and Deemer say:

> It is difficult to agree on a common terminology for current operational gaming, however, because its antecedents suggest diverse usages. Traditional war gaming, Monte Carlo computation, parlor games, and the von Neumann-Morgenstern theory of games all have contributed ideas and words. Beyond this historically inspired confusion, diverse local usages spring up at individual establishments to meet the needs of particular problems.[9]

Although Thomas and Deemer are speaking more specifically to the field of operational gaming in operations research, their comments are applicable to the current situation in social science simulation.

Some writers attempt to make clear distinctions between simulation, gaming and Monte Carlo techniques.[10] Others fuse the techniques, either regarding them as synonymous or seeing one or two of the processes as subcategories of the other one or two.[11] Distinctions between simulation (or man-machine simulation), gaming and Monte Carlo techniques have been suggested based on the general purposes to be served by the process,[12] the role of human beings vs. machines,[13] the element of competition,[14] the tightness or looseness of interactions within the simulated system, and the type of language in which the simulation process is expressed.[15] Although there are values to be gained from classifying different types of simulation procedures and related techniques according to any of these criteria, these distinctions are generally offered in connection with the solution of specified problems in particular areas of concern. When one attempts to apply these categorical distinctions between simulation, gaming and Monte Carlo techniques to a wider range of problems, they tend to become unnecessarily restrictive. Given the general disagreement in the literature as to how classifications should be made, the ambiguity of some of these classification schemes, and the young but rapidly developing state of simulation in the social sciences, it seems most useful to regard simulation as a general term referring to constructing and operating on a model that replicates behavioral processes; gaming as a type of simulation; and Monte Carlo techniques as a process used in some simulation operations.

It is rewarding to discuss the terms which have been applied in describing simulations. Some social scientists distinguish between simulations which are conducted entirely on machines (pure-machine), and simulations which make use of human decision makers with or without the additional use of machines (man-machine or all-man simulation).

Pure-Machine Simulation

It seems useful to identify two subgroups within the pure-machine simulation category; (a) physical analog simulation—those operations using some sort of physical model; and, (b) those

---

[9]Clayton J. Thomas and Walter L. Deemer, Jr., "The Role of Operational Gaming in Operations Research," *Operations Research*, 5, No. 1 (1957), 3.

[10]See Thomas and Deemer, "The Role of Operational". and Martin Shubik, "Simulation of Industry," *The American Economic Review*, 50, No. 5 (1960), 910–12.

[11]See Charles D. Flagle, "Simulation Techniques," in Charles D. Flagle, et al. (eds.), *Operations Research and Systems Engineering* (Baltimore: The Johns Hopkins Press, 1960), p. 446; R. P. Rich, "Simulation as an Aid in Model Building," *Operations Research*, 3 (1955), 15–19; N. M. Smith, Jr., "A Rationale for Operational Gaming," paper presented to 8th National Meeting of the Operations Research Society, Ottawa, Canada, January 10, 1956; and R. A. Brody, "Political Games for Model."

[12]Shubik, "Simulation, Its Uses and Potential," Part 1.

[13]Brody, "Political Games for Model," p. 6.

[14]E. W. Martin, "Teaching Executives Via Simulation," *Business Horizons*, 2, No. 2 (1959), 101.

[15]Thomas and Deemer, "The Role of Operational Gaming," 4–7.

simulations employing only a mathematical model. In physical analog simulation, problems are solved by manipulating parts of one physical system constructed as a model of another physical or analytical system. . . .

In the nonphysical analog simulation the model being operated on is constructed in symbolic form. The model is operated by manipulating the various symbols and programs which replicate the variables and components of the system. In most cases this operation is carried out on a digital electronic computer. . . .

Simulations Including Human Actors

Simulations in which human decision-makers act and interact within the simulated system are currently used by social scientists, especially for training and teaching purposes. . . . Both what is labeled "man-machine simulation" and those simulations commonly called "gaming" fall in this category. In man-machine simulation and games both the actions of human actors (participating as system decision-makers, not as experimenters) and computing machines simulate a social or psychological system. This type of simulation may be used to train the participants to serve in various capacities such as top management positions, to teach students about the operations of some social system and to help the scholar investigate behavior in a real system by manipulating activities in a simulated system. The AMA Top Management Game[16] is an example of a game or man-machine simulation of business situations developed to train executives in the making of top level decisions. The Northwestern inter-nation simulation and the Carnegie Tech Management Game . . . are examples of man-machine simulation or gaming, used both to teach students about inter-nation and business systems and to investigate hypotheses concerning international relations and business organizations. In the RAND simulation of an air defense system, . . . the man-machine simulation was employed to help the researchers learn about the performance of an air defense organization.

Games

Since the term "game" or "gaming" is often used in describing some simulations in which human participants serve as decision-makers acting within the system, it seems desirable to describe briefly the development of those operations labeled gaming. Although some writers maintain that there is a distinction between simulation and gaming, this writer does not consider simulation and gaming to be two separate techniques. He considers gaming to be a term sometimes applied to some simulations in which human actors participate within the simulated system, generally in a competitive situation.

E. W. Martin, Jr., defines a game as "a technical term denoting a simulation in which the results for one group depend on the actions of their competitors."[17] Martin Shubik, who distinguishes between simulation and gaming, suggests that in gaming the human decision-makers learn from "playing." "When a model (a man-machine model) is used for gaming the individuals in the system are presumed to learn from the play."[18] Teaching, however, is not the only function served by the man-machine simulations used in gaming, as was pointed out above. J. M. Kibbee says of one type of gaming, "Any business game could be used for some form of research; the obverse of the coin is that any simulation could be used for training."[19]

---

[16]F. M. Ricciardi, *et al.*, *Top Management Decision Simulation: The AMA Approach* (New York: American Management Association, Inc., 1958).

[17]Martin, "Teaching Executives Via Simulation."
[18]Shubik, "Simulation, Its Uses and Potential," 19.
[19]J. M. Kibbee, "Management Control Simulation," in Donald G. Malcolm, *et al.* (eds.), *Management Control Systems* (New York: John Wiley and Sons, Inc., 1960), p. 304.

There are three main streams of gaming of interest to social scientists: war gaming, business or management gaming and political gaming. The first of these gaming devices to be developed was war gaming, which can be traced back many centuries—possibly to the ancient game of chess. War games for purposes of military training and analysis were developed during the nineteenth century, and they were employed by both sides in the two World Wars. The Naval War College has defined war games as "a generic term describing the means for simulating the play of systematic strategic or tactical operations of opposing forces. It may include twosided board maneuvers, chart maneuvers, electronic maneuver board games, tactical games, or strategic war games."[20] Both business and political games trace their genesis to the military use of the technique. The recent application of gaming to the problems of political science and management science has been stimulated by the use of mathematical formalizations and social laboratory techniques in the study of social and psychological phenomena, and by the development of high speed electronic computers.

Business and management games are now being used by some large corporations in the training and selection of management personnel. The first important such game to be developed was the AMA Top Management Decision Game, developed by the American Management Association in 1956.[21] In more recent years the use of management and business games has spread to academic institutions, where they are used both for teaching and research purposes. In addition to teaching the techniques of management and business operation, management games have potential for research in economic systems, organizational theory, psychology, industrial relations and problems of production and marketing.[22]

Political games, generally simulating international situations, have been developed in several academic and research institutions in the past few years. The earliest attempts at formulating political games were made by the RAND Corporation starting about 1954. Their work grew out of RAND's earlier work with war games.[23] Later, political games were developed at Northwestern University and then at M. I. T.[24] These games have been used for teaching and research purposes. It would seem, however, that they could also be employed for the training of decision-makers and for policy formulation analysis.

In addition to terms that point out differences in the degree to which men and/or machines are involved in a simulation, two other terms are widely used in describing the differences in the time dimension and in the extent to which detailed mechanisms are supplemented by stochastic representation. These may now be presented in concluding this section on the types and techniques of simulation.

Real-Time Simulation

One of the significant advantages of many of the simulation processes is their ability to replicate years of activity in a very short period of time or to slow down time for detailed study of specific situations. In some simulation situations, however, especially those used for training purposes, it is advantageous for the simulation to be carried out in ordinary clock time. That is, the

---

[20]"Brief History of War Gaming," The United States War College, Part IV.

[21]For an extensive discussion of management games see Kalman J. Cohen and Eric Rhenman, "The Role of Management Games in Education and Research," *Management Science*, 7, No. 2 (1961), pp. 131–166.

[22]See Kalman J. Cohen, *et al.*, "The Carnegie Tech Management Game," in H. Guetzkow (ed.), *Simulation in the Social Sciences: Readings* . . ., pp. 104–123.

[23]Social Science Division, "Experimental Research on Political Gaming" (Santa Monica: RAND Corp., Rand Report P-1540-RC, Nov. 10, 1958).

[24]See Lincoln P. Bloomfield and Norman J. Padelford, "Three Experiments in Political Gaming," *The American Political Science Review*, 53 (1959), pp. 1105–1115.

simulated activity must take as long as the real activity it is replicating. Real-time simulation is especially useful in simulators used to train persons for positions where prompt, accurate decisions are at a premium.

The Monte Carlo Method

Another technique incorporated in the operation of some simulations is the Monte Carlo method. The name *Monte Carlo* is given to the process of simulating with models that include probability distributions. The Monte Carlo method is defined by W. E. Alberts as "a computational method or techniques of introducing data of a random or probabilistic nature into a model. Its purpose is to reproduce data in the same manner as would occur in a real life situation."[25] In some simulation situations in which a large number of variables are involved and in which the value of the variables and components is not constant, it is necessary to model the behavior of the system in probability terms. The Monte Carlo method is employed in such instances. Daniel D. McCracken says of this method:

> The Monte Carlo method, in general, is used to solve problems which depend in some important way upon probability—problems where physical experimentation is impracticable and the creation of an exact formula is impossible. Often the process we wish to study consists of a long sequence of steps, each of which involves probability, as for instance the travels of the neutron through matter. We can write formulas for the probabilities at each collision, but we are often not able to write anything useful for the probabilities for the entire sequence. . . .[26]

[25] W. E. Alberts, "Report to the Eighth A.I.I.E. National Convention on the Systems Simulation Symposium," in D. G. Malcolm (ed.), *Report of Systems Simulation Symposium* (Baltimore: Waverly Press, Inc., 1958), p. 4.
[26] Daniel D. McCracken, "The Monte Carlo Method," *Scientific American*, 192, No. 5 (1955), p. 94.

V. Advantages and Problems in Simulation

Up to this point we have discussed what simulation is, types and techniques of simulation, and the general purposes for which it can be used. Simulation is, like other research techniques, merely a tool, and only one of many available to the social scientist. It is not a magical cure-all for the study of all problems faced by social scientists. It must be evaluated against other research techniques. These criteria may be used in making an evaluation: (1) Applicability: Will the technique adequately solve the problems involved in the research or training exercise? (2) Cost: Are the costs in terms of time, money, equipment required and effort expended less than those for any other technique yielding comparable results? (3) Simplicity and communicability: Is it the least complex and most comprehensible technique, considering the persons who might be using it and those to whom the results might be communicated?

One of the most significant advantages of simulation is that it permits the experimenter to study process in ways that nature prohibits. The simulation can be run many times with the values of the parameters being modified between runs and the changes in outputs observed. This makes possible the effective study of operating models containing many different components, variables and interrelationships. The experimenter, in short, exercises a great amount of control through which he can study and evaluate outcomes resulting from a variety of alternative conditions and relationships. Conway, Johnson and Maxwell sum up this advantage: "Simulation is often described as a means of incorporating a fourth dimension—time—in what have previously of necessity been static methods of analysis."[27]

Simulation also permits the researcher, teacher

[27] Conway, *et al.*, "Some Problems of Digital Systems Simulation . . .," p. 95.

or trainer to compress or expand real time. He can simulate the operations of a system over a period of years in a matter of minutes, or he can slow down the process so that he can more carefully analyze or demonstrate what is going on in specific areas. In other situations the simulation can be used to reproduce real time.

Experimenting with a simulated system, instead of the real system, permits the social scientist to study problems that would be impractical or altogether impossible to study in real life. It is impossible for a political scientist studying international relations to experiment with the real inter-nation system of decision-makers, capabilities and organizations, manipulating variables like nuclear weapons to see what results such changes might have on international tensions. However, when significant aspects of the internation system are simulated by physical analogs, mathematical formulas and/or human decision-makers the variables in the model can be manipulated and properties concerning the real system may be inferred. The ability to experiment in this fashion is particularly useful in the social sciences because moral and physical factors often prohibit experimenting with real people and real social systems. Helmer and Rescher call experimentation on simulated models "pseudo-experimentation." Speaking of the value of this process they comment, "Generally it may be said that in many cases judicious pseudo-experimentation may effectively annul the oft-regretted infeasibility of carrying out experiments proper in the social sciences by providing an acceptable substitute which, moreover, has been tried and proved in the applied physical sciences."[28]

Many simulation processes are relatively free from complex mathematics, making them more widely comprehensible than other more complex systems of formal mathematical analysis. The lack of dependency upon complex mathematical analysis not only has the advantage of making simulation comprehensible to the mathematically unsophisticated, but it can also be used in studying situations where mathematical methods capable of considering all of the desired factors are not available.

The central problem inherent in all simulation processes, and in all model building as well, is that of adequate reproduction of the real system. In simulation the researcher, teacher or trainer is trying to learn or teach about a real system by working with a model of it. If the simulator does not validly model the necessary attributes of the real system, the results found in solving problems in the simulated environment cannot successfully indicate the behavior of the real system. This means that the researcher must know a great deal about the real system before he can presume to simulate it, and that he must have reliable means (mathematical, physical or human) of reproducing it. If the replication of the system and the means of operating it are not valid, the experimenter will find the use of simulation dysfunctional rather than useful.

One of the liabilities in using simulation is its high cost. Because simulation generally requires the use of large machines and/or a number of trained participants, the cost of simulation is often quite high. Cost in most instances, however, is a relative consideration. The cost of a simulation operation for the solution of a problem must be evaluated against both the cost and the results of using other techniques. . . . In some instances it might be decided that the cost of simulation prohibits its use and other less costly and/or less satisfactory techniques must be employed. In other situations, relative costliness cannot be considered because simulation is the only acceptable way to handle the problem.

Simulation is a useful tool when the researcher knows enough about the real system or process adequately to reproduce its behavior in an operating model, and when the problem cannot be solved successfully by simpler techniques. The user of simulation must also be careful in selecting the best type of simulation process (pure-

---

[28]Olaf Helmer and Nicholas Rescher, "On the Epistemology of the Inexact Sciences," *Management Science*, 6, No. 1 (1959), p. 49.

machine, man-machine, physical analog, Monte Carlo, and so on) for the specific problem he seeks to investigate.

VI. CONCLUSION

Simulation, although a relatively new tool for the social sciences, has already proven useful for training and research concerning human organizations and psychological and social processes. . . . [S]ocial or behavioral scientists have, in the past several years, begun to use simulation for a wide range of purposes. The increasingly frequent mention of the term and its use in the social science literature indicate that the technique has been found to be useful, especially for teaching, training and the generation of research hypotheses. All of this suggests an increase in the employment of simulation in the future. As the proficiency and availability of electronic computers increases, as more empirical data become available to the social scientist, as mathematical and socio-psychological techniques are improved and as the social scientist continues in his emphasis upon the study of human systems and processes, it seems reasonable to assume that the popularity and usefulness of simulation will increase.

S. S. Stevens

Levels of Measurement

A CLASSIFICATION OF SCALES OF MEASUREMENT

Paraphrasing N. R. Campbell[1] . . . we may say that measurement, in the broadest sense, is defined as the assignment of numerals to objects or events according to rules. The fact that numerals can be assigned under different rules leads to different kinds of scales and different kinds of measurement. The problem then becomes that of making explicit (a) the various rules for the assignment of numerals, (b) the mathematical properties (or group structure) of the resulting scales, and (c) the statistical operations applicable to measurements made with each type of scale.

Scales are possible in the first place only because there is a certain isomorphism between what we can do with the aspects of objects and the properties of the numeral series. In dealing with the aspects of objects we invoke empirical operations for determining equality (classifying), for rank-ordering, and for determining when differences and when ratios between the aspects of objects are equal. The conventional series of numerals yields to analogous operations: We can identify the members of a numeral series and classify them. We know their order as given by convention. We can determine equal differences, as $8 - 6 = 4 - 2$, and equal ratios, as $8/4 = 6/3$. The isomorphism between these properties of the numeral series and certain empirical operations which we perform with objects permits the use of the series as a *model* to represent aspects of the empirical world.

The type of scale achieved depends upon the

[1] See "The Final Report," British Association for the Advancement of Science on the problem of measurement, 1940, p. 340.

Reprinted from *Science*, Vol. 103, No. 2684 (June 7, 1946), 677–80, where it appeared under the title, "On the Theory of Scales of Measurement." By permission of S. S. Stevens and the publisher, *Science*.

character of the basic empirical operations performed. These operations are limited ordinarily by the nature of the thing being scaled and by our choice of procedures, but, once selected, the operations determine that there will eventuate one or another of the scales listed in Table 1.[2] . . .

Table I[3]

Scale	Basic Empirical Operations	Permissible Statistics (invariantive)
Nominal	Determination of equality	Number of cases Mode Contingency correlation
Ordinal	Determination of greater or less	Median Percentiles
Interval	Determination of equality of intervals or differences	Mean Standard deviation Rank-order correlation Product-moment correlation
Ratio	Determination of equality of ratios	Coefficient of variation

It will be noted that the column listing the basic operations needed to create each type of scale is cumulative: to an operation listed opposite a particular scale must be added all those operations preceding it. Thus, an interval scale can be erected only provided we have an operation for determining equality of intervals, for determining greater or less, and for determining equality (not greater and not less). To these

[2] A classification essentially equivalent to that contained in this table was presented before the International Congress for the Unity of Science, September, 1941. The writer is indebted to the late Professor G. D. Birkhoff for a stimulating discussion which led to the completion of the table in essentially its present form.

[3] (*Editor's note*) A portion of Table 1 as it appeared in its original form has been omitted.

operations must be added a method for ascertaining equality of ratios if a ratio scale is to be achieved. . . .

The last column presents examples of the type of statistical operations appropriate to each scale. This column is cumulative in that *all* statistics listed are admissible for data scaled against a ratio scale. . . .

NOMINAL SCALE

The *nominal scale* represents the most unrestricted assignment of numerals. The numerals are used only as labels or type numbers, and words or letters would serve as well. Two types of nominal assignments are sometimes distinguished, as illustrated (a) by the "numbering" of football players for the identification of the individuals, and (b) by the "numbering" of types or classes, where each member of a class is assigned the same numeral. Actually, the first is a special case of the second, for when we label our football players we are dealing with unit classes of one member each. Since the purpose is just as well served when any two designating numerals are interchanged, this scale form remains invariant under the general substitution or permutation group (sometimes called the symmetric group of transformations). The only statistic relevant to nominal scales of Type A is the number of cases, e.g. the number of players assigned numerals. But once classes containing several individuals have been formed (Type B), we can determine the most numerous class (the mode), and under certain conditions we can test, by the contingency methods, hypotheses regarding the distribution of cases among the classes.

The nominal scale is a primitive form, and quite naturally there are many who will urge that it is absurd to attribute to this process of assigning numerals the dignity implied by the term measurement. Certainly there can be no quarrel with this objection, for the naming of things is an arbitrary business. However we christen it, the use of numerals as names for classes is an example of the "assignment of numerals according to rule." The rule is: Do not assign the same numeral to different classes or different numerals to the same class. Beyond that, anything goes with the nominal scale.

Ordinal Scale

The *ordinal scale* arises from the operation of rank-ordering. Since any "order-preserving" transformation will leave the scale form invariant, this scale has the structure of what may be called the isotonic or order-preserving group. A classic example of an ordinal scale is the scale of hardness of minerals. Other instances are found among scales of intelligence, personality traits, grade or quality of leather, etc.

As a matter of fact, most of the scales used widely and effectively by psychologists are ordinal scales. In the strictest propriety the ordinary statistics involving means and standard deviations ought not to be used with these scales, for these statistics imply a knowledge of something more than the relative rank-order of data. On the other hand, for this "illegal" statisticizing there can be invoked a kind of pragmatic sanction: In numerous instances it leads to fruitful results. While the outlawing of this procedure would probably serve no good purpose, it is proper to point out that means and standard deviations computed on an ordinal scale are in error to the extent that the successive intervals on the scale are unequal in size. When only the rank-order of data is known, we should proceed cautiously with our statistics, and especially with the conclusions we draw from them.

Even in applying those statistics that are normally appropriate for ordinal scales, we sometimes find rigor compromised. Thus, although it is indicated in Table 1 that percentile measures may be applied to rank-ordered data, it should be pointed out that the customary procedure of assigning a value to a percentile by interpolating linearly within a class interval is, in all strictness, wholly out of bounds. Likewise, it is not strictly proper to determine the mid-point of a class interval by linear interpolation, because the linearity of an ordinal scale is precisely the property which is open to question.

Interval Scale

With the *interval scale* we come to a form that is "quantitative" in the ordinary sense of the word. Almost all the usual statistical measures are applicable here, unless they are the kinds that imply a knowledge of a 'true' zero point. The zero point on an interval scale is a matter of convention or convenience, as is shown by the fact that the scale form remains invariant when a constant is added.

This point is illustrated by our two scales of temperature, Centigrade and Fahrenheit. Equal intervals of temperature are scaled off by noting equal volumes of expansion; an arbitrary zero is agreed upon for each scale; and a numerical value on one of the scales is transformed into a value on the other by means of an equation of the form $x' = ax + b$. Our scales of time offer a similar example. Dates on one calendar are transformed to those on another by way of this same equation. On these scales, of course, it is meaningless to say that one value is twice or some other proportion greater than another.

Periods of time, however, can be measured on ratio scales and one period may be correctly defined as double another. The same is probably true of temperature measured on the so-called Absolute Scale.

Most psychological measurement aspires to create interval scales, and it sometimes succeeds. The problem usually is to devise operations for equalizing the units of the scales—a problem not always easy of solution but one for which there are several possible modes of attack. Only occasionally is there concern for the location of a 'true' zero point, because the human attributes measured by psychologists usually exist in a positive degree that is large compared with the range of its variation. In this respect these attributes are analogous to temperature as it is encountered in everyday life. Intelligence, for example, is usefully assessed on ordinal scales which try to approximate interval scales, and it is not necessary to define what zero intelligence would mean.

Ratio Scale

Ratio scales are those most commonly encountered in physics and are possible only when there exist operations for determining all four relations: equality, rank-order, equality of intervals, and equality of ratios. Once such a scale is erected, its numerical values can be transformed (as from inches to feet) only by multiplying each value by a constant. An absolute zero is always implied, even though the zero value on some scales (e.g., Absolute Temperature) may never be produced. All types of statistical measures are applicable to ratio scales, and only with these scales may we properly indulge in logarithmic transformations such as are involved in the use of decibels.

Foremost among the ratio scales is the scale of number itself—cardinal number—the scale we use when we count such things as eggs, pennies, and apples. This scale of the numerosity of aggregates is so basic and so common that it is ordinarily not even mentioned in discussions of measurement.

It is conventional in physics to distinguish between two types of ratio scales: *fundamental* and *derived*. Fundamental scales are represented by length, weight, and electrical resistance, whereas derived scales are represented by density, force, and elasticity.

These latter are *derived* magnitudes in the sense that they are mathematical functions of certain fundamental magnitudes. They are actually more numerous in physics than are the fundamental magnitudes, which are commonly held to be basic because they satisfy the criterion of *additivity*. Weights, lengths, and resistances can be added in the physical sense, but this important empirical fact is generally accorded more prominence in the theory of measurement than it deserves. The so-called fundamental scales are important instances of ratio scales, but they are only instances. As a matter of fact, it can be demonstrated that the fundamental scales could be set up even if the physical operation of addition were ruled out as impossible of performance. Given three balances, for example, each having the proper construction, a set of standard weights could be manufactured without it ever being necessary to place two weights in the same scale pan at the same time. The procedure is too long to describe in these pages, but its feasibility is mentioned here simply to suggest that physical addition, even though it is sometimes possible, is not necessarily the basis of all measurement. Too much measuring goes on where resort can never be had to the process of laying things end-to-end or of piling them up in a heap.

Ratio scales of psychological magnitudes are rare but not entirely unknown. The Sone scale . . . is an example founded on a deliberate attempt to have human observers judge the loudness ratios of pairs of tones. The judgment of equal intervals had long been established as a legitimate method, and with the work on sensory ratios, started independently in several laboratories, the final step was taken to assign numerals to sensations of loudness in such a way that relations among the sensations are reflected by the ordinary arithmetical relations in the numeral series. As in all measurement, there are limits imposed by error and variability, but within these limits the Sone scale ought properly to be classed as a ratio scale.

. . . We may venture to suggest by way of conclusion that the most liberal and useful definition of measurement is, as one of its members advised, "the assignment of numerals to things so as to represent facts and conventions about them." The problem as to what is and is not measurement then reduces to the simple question: What are the rules, if any, under which numerals are assigned? If we can point to a consistent set of rules, we are obviously concerned with measurement of some sort, and we can then proceed to the more interesting question as to the kind of measurement it is. In most cases a formulation of the rules of assignment discloses directly the kind of measurement and hence the kind of scale involved. If there remains any ambiguity, we may seek the final and definitive answer in the mathematical group-structure of the scale form: In what ways can we transform its values and still have it serve all the functions previously fulfilled? We know that the values of all scales can be multiplied by a constant, which changes the size of the unit. If, in addition, a constant can be added (or a new zero point chosen), it is proof positive that we are not concerned with a ratio scale. Then, if the purpose of the scale is still served when its values are squared or cubed, it is not even an interval scale. And finally, if any two values may be interchanged at will, the ordinal scale is ruled out and the nominal scale is the sole remaining possibility.

This proposed solution to the semantic problem is not meant to imply that all scales belonging to the same mathematical group are equally precise or accurate or useful or "fundamental." Measurement is never better than the empirical operations by which it is carried out, and operations range from bad to good. Any particular scale, sensory or physical, may be objected to on the grounds of bias, low precision, restricted generality, and other factors, but the objector should remember that these are relative and practical matters and that no scale used by mortals is perfectly free of their taint.

Claire Selltiz, Marie Jahoda, Morton Deutsch, and *Stuart W. Cook*

Quantitative Descriptions of Data

Let us suppose that we have asked a thousand people, who have been selected as a sample of the adult population of England, a series of questions to gather information about their movie-going habits. Assume that we have developed a number of sets of categories and have coded the responses of each individual. The coding may be considered a method of summarizing the responses of each individual; if we were concerned only about each individual, this might be as far as we would go in the analysis. However, the purposes of our research are broader than this. We wish to know more than that a

Reprinted from Chapter 11 of *Research Methods in Social Relations,* revised edition, by Claire Selltiz, Marie Jahoda, Morton Deutsch, and Stuart W. Cook. Copyright 1951, © 1959 by Holt, Rinehart and Winston, Inc. Reprinted by permission of Holt, Rinehart and Winston, Inc.

given government clerk in London goes to the movies four times a month and that an innkeeper in Yarmouth goes only once a month. Our inquiry is directed toward providing information about the adult population of England.[1]

[1]Even if we were interested primarily in individual cases, we would still find a characterization of the total population useful for study of the individual. A fact about an individual frequently takes on its significance in relation to facts about the population of which the individual is a part. Thus, if one knows that it is customary behavior for members of a group to go to the movies twice a month, we have a frame of reference against which to interpret the individual who sees five movies a month. Our picture of this same individual might be different if the average member of the group went to the movies ten times a month. Such *normative* interpretation—the evaluation of the individual's behavior in the light of the group standard—is the most common type of interpretation. This is not to deny the value of *ipsative* interpretation, in which the frame of reference is the individual himself. Thus, the fact that an individual reports going to the movies five times a month may take on varying significance depending upon whether, in the past, it was customary for him to go ten times or not to go at all. Additional meaning may be given by other intra-individual comparisons—for example, by comparing the amount of time the individual spends at the movies with the amount of time he spends reading books. For a statistical approach to the ipsative study of the individual, see A. L. Baldwin, "Statistical Problems in the Treatment of Case Histories," *Journal of Clinical Psychology,* 6 (1950), 6–12; and R. B. Cattell and L. B. Luborsky, "T-technique Demonstrated as a New Clinical Method for Determining Personality and Symptom Structure," *Journal of General Psychology,* 42 (1950), 3–24.

As a necessary step in characterizing this population, we must describe or summarize the data we have obtained on the sample we have studied. Tabulation is a part of this step. In addition, we must estimate the reliability of generalizations from the obtained data to the total population. Statistical methods are used to fulfill both these functions. The term *descriptive statistics* is often applied to characterize the methods employed in summarizing the obtained data, and *sampling statistics* to characterize the methods utilized in making and evaluating generalizations from the data. . . .

Description of the Data

In giving an adequate description of a mass of data, we usually wish to do one or another, or several, of the following things:

1. *To Characterize What Is "Typical" in the Group.*

We wish to know, for example, how many movies, on the average, the people in our sample attend, or, perhaps, we want to know what kind of movies most of them prefer. In the terminology of statistics, we wish to get some indication of the *central tendency*. There are various measures of central tendency, each of which makes assumptions about the nature of the data. If those assumptions are not met, then the measure of central tendency may be misleading. Thus, an *arithmetic mean* or average (the sum of individual scores divided by the number of individuals) implies equality of intervals, or an interval scale. It is appropriate to such data as the number of times people go to the movies each month, since here the scale is that of number itself, with equal intervals between every two whole numbers.[2] On the other hand, suppose that we had asked the respondents to rate each of several types of movie (comedies, documentaries, etc.) on a scale that attempted to gauge preference in terms of frequency of seeing such films; for example, "I go to see almost all the films of this type that are shown, . . . I occasionally go to see films of this type, . . . I seldom go to see films of this type, . . . I avoid films of this type." If we were to assign the numbers 1 to 4 to these four scale positions and average the rating for each type of film, the average would have no clear meaning, since there is no reason to believe that the scale positions are equally distant from one another. In such a case, it would be preferable to employ the *median* (the point on the scale above which—and correspondingly, below which—50 per cent of the cases lie). Or suppose that we had simply asked each respondent to check which of various types of film he preferred. Here the data would be in the form of a nominal scale, since various types of film have no relation of order to one another. In this case, the only appropriate method for measuring the central tendency would be the *mode* (the score that occurs with the greatest frequency —in this case, the type of film mentioned by the greatest number of respondents).

The various measures of central tendency not only make different assumptions about the nature of the data but provide somewhat different kinds of information. The arithmetic mean may be thought of as the point on the scale around which the cases balance: one case at one extreme of the distribution, for example, may be counterbalanced by one or more cases at the other extreme. If the cases are not symmetrically distributed around this point, the arithmetic mean may be very misleading in certain respects. Thus, it takes a very large number of cases of low income to counterbalance one case of a person whose annual income comes to over a million

---

[2]*Number of times* a person goes to the movies is, of course, on a ratio scale, and *a fortiori* on an interval scale. But, strictly speaking, even this scale violates one of the assumptions involved in the use of the arithmetic average. This scale is *discrete*; i.e., only integers are possible. What, for instance, does it mean to say that a person has gone to the movies 1.763 times? The arithmetic average presupposes a *continuous* scale; i.e., one in which all fractional values are possible. Sometimes, however, the attribution of continuity to a discrete scale is a convenient fiction.

dollars; as a consequence, the arithmetic mean of income is far above the income of the vast majority of the population. It does not follow that knowledge of the mean income is pointless information; whether it has point depends on what you want to do with it. The median also implies a concept of balance, but, since it takes account only of the ordinal positions of scores rather than of their absolute values, one millionaire *is* counterbalanced by one pauper.

The use of the mode dispenses with the idea of a point of balance. In fact, it is possible to have several modes in one distribution; for example, frequency of movie-going may be *bimodal*, with a large number of people going once a week, another large number going less than once a month. As a rule, *multimodal* distributions such as this come about as a result of the intermingling of distinct populations. In our illustration, for example, going to the movies once a week may be the mode for people under 25; going less than once a month, the mode for an older group. However, it is not always easy to identify the basis of distinction.

It is also possible for a distribution of scores to have no mode at all, every score occurring about as frequently as every other. In this case, as in the case of bimodal and multimodal distributions, it is perhaps misleading to speak of *central tendency*, a term suggesting that the cases tend to cluster around some point more or less in the middle of the distribution. But even in these instances, the arithmetic mean and the median may be useful in bringing out aspects of the distributions which they describe.

2. To Indicate How Widely Individuals in the Group Vary.

We might wish to know, for example, whether the people in our sample are similar in their film preferences, so that most people prefer films of a given type, or whether there is great diversity. Or we might wish to know whether there is much variation in the frequency of movie-going among the sample being studied.

There are many measures of interindividual variation. The purpose of each of them is to indicate how similar or how different the individuals in the group are with respect to a given characteristic. Some of the common measures are the range, the average deviation, the standard deviation, and the quartile deviation.

As with measures of central tendency, each of the measures of variability makes assumptions about the nature of the measurements and gives somewhat different kinds of information. The *range* shows the extremes of variation in the group; it might show, for example, that at least one person never goes to the movies, while at least one other goes every day. Obviously, the range is affected by extreme cases, and therefore may be misleading as a picture of the group as a whole. To avoid this difficulty, other measures of variation focus on the limits within which half of the group fall, or on the average distance of individuals from the group mean. The *quartile deviation* shows the points within which the central half of the cases fall; like the median, it assumes that the data correspond to an ordinal scale. The *average deviation* and the more frequently used *standard deviation* are measures of the average distance of individuals from the group mean; like the mean, they assume that the data correspond to an interval scale.

3. To Show Other Aspects of How the Individuals Are Distributed with Respect to the Variable Being Measured.

For example, is the number of people who do not go to the movies at all about the same as the number who go three times a month? Or do relatively many people go three times a month, while relatively few do not go at all and relatively few go six or more times a month? If you plotted the figures on a graph, using the frequency of movie-going on the horizontal axis and the number of people reporting each frequency on the vertical axis, what would be the shape of the resulting graph? Is it rectangular (that is, are there equal numbers of people at each point on the scale, resulting in a graph that takes the form of a straight line across the page); or is it a bell-shaped or "normal" distribution; or is it an asymmetrical curve, with a piling up of cases at one side or the other; or does it have more than one mode (that is, is there a piling up of cases at two or more points along the scale, with relatively few cases in between)?[3]

Knowing the shape of the distribution curve is fundamental to the use of efficient statistical methods, since the more efficient methods make specific assumptions about the nature of the distribution curve. It is common to assume that the distribution curve for any variable is normal, but this may not be so. If it is not, one may see whether other known distribution curves fit the data or whether one can transform the raw data by mathematical manipulation to a known distribution.

4. To Show the Relation of the Different Variables in the Data to One Another.

We wish to know, for example, whether, within our sample, the frequency of movie-going or the preference for different types of movies seems related to income, to sex, to age, etc. That is, we wish to know whether a variation in one characteristic is associated with or paralleled by variations in another characteristic. There are several methods of determining the relationship between variables. None of these methods in itself, however, permits the conclusion that an association or correlation between variables in one's data is indicative of a causal relationship. . . .

5. To Describe the Differences Between Two or More Groups of Individuals.

For example, we may wish to compare the movie-going habits of those members of our sample who live in communities of less than 10,000 (whom

[3] For a fuller discussion of these aspects of distribution, look up *skewness* and *kurtosis* in almost any statistics textbook. An unusually thorough and not overly technical discussion of various types of distribution may be found in J. G. Smith and A. J. Duncan, *Sampling Statistics and Applications* (New York: McGraw-Hill Book Company, 1945).

we shall, for convenience, call rural residents) and of those members who live in communities of 10,000 or more (whom we shall call urban residents). This is, of course, a special case of showing the relationship between two variables. We distinguish it solely because in much social research interest is focused on the comparison of groups. Although such comparisons most commonly involve measures of central tendency, they need not be limited to this; they may include comparison of measures of variation within the groups or of the relations among variables in the two groups. It might be, for example, that the average frequency of movie-going is similar in the two groups but that there is greater variation among urban than among rural residents, or vice versa.

GENERALIZATION TO THE POPULATIONS FROM WHICH THE SAMPLES WERE DRAWN

Suppose that we have studied samples of rural and urban Englishmen and that our results show differences between the two samples. One may ask whether the differences that have been obtained reflect true differences between rural and urban Englishmen, or whether the two samples might have differed to this extent by chance even though the total rural and urban populations are alike in their movie-going habits. Through statistical procedures, one is able to answer such a question in terms of a statement of probability.

When we are contrasting samples or studying the differences between experimental and control groups, we usually wish to test some hypothesis about the nature of the true difference between the larger populations represented by the samples. Most commonly, in the social sciences, we are still concerned with relatively crude hypotheses (for example, that urban residents go to the movies more often than rural residents); we are usually not in a position to consider more specific hypotheses (for example, that they go to the movies *twice* as often). Suppose our data show that our sample of urban Englishmen attend an average of three motion pictures a month and that our sample of rural Englishmen attend an average of only two motion pictures. Clearly, the findings within our samples are in line with the hypothesis: urban residents attend the movies more often than rural residents. But we know that the findings based on our samples are not very likely to be exactly the same as the findings we would obtain if we had interviewed all the adults in England. . . . Now we want to estimate whether, if we had interviewed the total population, we would still have found more frequent movie attendance on the part of urban residents. This we do, ordinarily, by testing the *null hypothesis*—in this case, the hypothesis that in the English population as a whole, rural and urban residents do *not* differ in frequency of movie-going. Various statistical techniques (called statistical *tests of significance*) have been devised, which tell us the likelihood that our two samples might have differed as much as they do by chance even if there were *no* difference between urban and rural Englishmen as a whole.[4]

It may seem odd that, when interested in one hypothesis (that there *is* a difference between the two populations represented by our samples), we should test its opposite (that there is *no* difference). But the reason is not too difficult to follow. Since we do not know the true state of affairs in the population, all we can do is make inferences about it on the basis of our sample findings. If we are comparing two groups, there are obviously two possibilities: either the two populations are alike, or they are different. Suppose that our *samples* from the two populations are different on a particular measure or attribute. Clearly, this would be likely to happen if the two populations from which the samples are drawn do in fact differ on that attribute. However, it does not in itself constitute evidence that they *do* differ, since there is always the possibility that the samples do not correspond exactly to the populations they are intended to represent. We must consider the possibility that the element of chance which is involved in the selection of a sample may have given us samples which differ from each other even if the two populations do not differ. Thus the crucial question is: Is it likely that we would have come up with samples that differ to this extent if the two populations were actually alike? This is the question the test of the null hypothesis answers; it tells us what the chances are that two samples differing to this extent would have been drawn from two populations that are in fact alike.[5] Only if the statistical test indicates that it is improbable that two samples differing to this extent could have been drawn from similar populations can we conclude that the two populations probably differ from each other.

Suppose, however, that our findings show no difference between the two samples; let us say that in our samples, both rural and urban Englishmen attend the movies, on an average, two and one-half times a month. Can we then conclude that the total populations of rural and urban Englishmen are alike in frequency of movie-going? Not with any certainty. Just as there is the possibility that samples may *differ* when the populations are alike, so there is the possibility that samples may be *alike* when in fact the populations differ. Thus, if our two samples are alike, all we can conclude is that we have no

---

[4]Which method of testing significance is appropriate depends on the nature of the measurements used and the distribution of the characteristic. Most of the tests commonly described in statistics textbooks assume that the measurements are in the form of at least interval scales and that the distribution of the characteristic is normal. These conditions, however, are seldom met in social research. In recent years, a number of statistical tests have been developed that do not rest on these assumptions; they are called *nonparametric* or *distribution-free* statistics. For a presentation of tests of this type, see S. Siegel, *Nonparametric Statistics for the Behavioral Sciences* (New York: McGraw-Hill Book Company, 1956).

[5]It should be kept in mind that all statistical tests of significance, and thus all generalizations from samples to populations, rest on the assumption that the samples are not biased—that is, that the cases to be included in the samples have been selected by some procedure that gives every case in the population an equal, or at least a specifiable, chance of being included in the sample. If this assumption is not justified, significance tests become meaningless.

evidence that the populations differ—in other words, that the idea that the two populations are alike is tenable.

But to go back to the case where the two samples differ. We can affirm that the two populations they represent probably differ if we can reject the null hypothesis—that is, if we can show that the obtained difference between the two samples would be unlikely to appear if the two populations were in fact the same. It is, however, in the nature of probability that even highly improbable events can sometimes happen. Thus, we can never be absolutely certain of our generalizations to the total population. Whenever we reject the null hypothesis, there is some chance that we are wrong in doing so.

However, since we are always dealing with inferences and probabilities, there is always also some chance that if we *accept* the null hypothesis, we are wrong in doing so. That is, even if our statistical test indicates that the sample differences might easily have arisen by chance even if the two populations are alike, it may nevertheless be true that the populations differ.

In other words, we are always confronted with the risk of making one of two types of error. We can reject the null hypothesis when, in fact, it is true; that is, we may conclude that there is a difference between the two populations when, in fact, they are alike. This is commonly referred to as the *Type I error*. Or, on the other hand, we can accept the null hypothesis as tenable when, in fact, it is false; that is, we may conclude that the two populations are alike when, in fact, they are different. This is referred to as the *Type II error*.

The risk of making the Type I error is determined by the *level of significance* we accept in our statistical testing. Thus, if we decide that we will conclude that the populations truly differ whenever a test of significance shows that the obtained difference between two samples would be expected to occur by chance not more than 5 times in 100 if the two populations were in fact alike, we are accepting 5 chances in 100 that we will be wrong in rejecting the null hypothesis.

We can reduce the risk of a Type I error by making our criterion for rejecting the null hypothesis more extreme; for example, by rejecting the null hypothesis only if the statistical test indicates that the sample difference might have appeared by chance only once in a hundred times, or once in a thousand times, or once in ten thousand times. Unfortunately, however, the chances of making Type I and Type II errors are inversely related. The more we protect ourselves against the risk of making a Type I error (that is, the less likely we are to conclude that two populations differ when in fact they do not), the more likely we are to make a Type II error (that is, to fail to recognize population differences which actually exist). Once we have determined the degree of Type I risk we are willing to run, the only way of reducing the possibility of Type II error is to take larger samples and/or to use statistical tests that make the maximum use of available relevant information.[6]

The inverse relationship of the risks of the two types of error makes it necessary to strike a reasonable balance. In the social sciences, it is more or less conventional to reject the null hypothesis when the statistical analysis indicates that the observed difference would not occur more than 5 times out of 100 by chance alone. If the statistical analysis indicates that the difference between the two samples might have appeared by chance more than 5 times out of 100, the null hypothesis is not rejected. But these conventions are useful only when there is no other reasonable guide. The decision as to just how the balance between the two kinds of error should be struck must be made by the investigator. In some instances, it is obviously more important to be sure of rejecting a hypothesis when it is false than to fail to accept it when it is true. In other cases,

---

[6]For a discussion of the extent to which different types of statistical test offer protection against Type II errors, consult a statistics textbook that discusses this matter; for example, W. J. Dixon and F. J. Massey, Jr., *Introduction to Statistical Analysis* 2nd edition (New York: McGraw-Hill Book Company, 1957), chap. 15; *or* H. M. Walker and J. Lev, *Statistical Inference* (New York: Holt Publishing Co., 1953), pp. 60–67.

the reverse may be true. This may be seen clearly in an example from everyday life, outside the sphere of statistical analysis. In many countries it is considered more important to reject a hypothesis of guilt when it is false than to fail to accept this hypothesis when it is true; a person is considered not guilty so long as there is reasonable doubt as to his guilt. In other countries, the acceptance of a false hypothesis of guilt is deemed less costly than the rejection of this hypothesis if it were true; a person charged with a crime is considered guilty until he has demonstrated his lack of guilt.

In much research, of course, there is no clear basis for deciding whether a Type I or a Type II error would be more costly, and so the investigator makes use of the conventional level for determining statistical significance. However, there are some studies in which one type of error would clearly be more costly than the other. Suppose that in a certain school it has been suggested that a new method of teaching arithmetic would be more effective; suppose also that this method would require expensive equipment for each arithmetic class. An experiment is set up to test whether the new method would in fact lead to better learning of arithmetic. Two groups of children are randomly selected; they may also be matched in arithmetic ability and achievement and in other relevant respects. One is taught by the new method, one by the old. Since the new method requires expensive equipment, the school system would not wish to adopt it unless there were considerable assurance of its superiority; in other words, it would be costly to make a Type I error and conclude that the new method is better when in fact it is not. On the other hand, if there were no difference in the expense of the two methods, a Type I error would not be especially costly, whereas a Type II error might lead to failure to adopt the new method when in fact it is superior.

Let us go back to the fact that any generalization from samples to populations is simply a statement of statistical probability, of the chances that a given difference between samples reflects a true difference between the populations. Let us say that we have decided to work with a 5 per cent level of confidence. This means that we will reject the hypothesis of *no* difference between the populations only if a sample difference as large as the one we have found can be expected to occur by chance 5 times or less in 100; if such a difference can be expected more than 5 times in 100, we will accept the null hypothesis. Is there any way of estimating whether or not our finding represents one of the 5 times that such a difference might have appeared by chance? On the basis of an isolated finding, there is not; but we can draw further inferences by examining patterns within our findings.

Suppose that we are interested in testing the effects of a series of lectures about the United Nations on attitudes toward that body. We have set up a careful study design, with randomly selected experimental and control groups, perhaps also matched in terms of initial attitudes, with precautions to assure the frankness of responses, etc. Now suppose that we use as our measure of attitudes toward the United Nations only one item—say, attitudes toward the establishment of a U.N. police force. We find that those who have attended the lectures are more favorable on this question than those who have not, and a statistical test indicates that the difference would not have appeared by chance due to random sampling fluctuations more than 5 times in 100. However, this has the corollary that it *might* have appeared by chance 5 times in 100, and we have no way of knowing whether this is one of those 5 times. Let us say, however, that we have asked 20 different questions that are reasonable indicators of attitude toward the United Nations. If we are using a confidence level such that we accept as significant a difference that might have occurred by chance 5 times in 100, then, if we had 100 questions, we might expect to find, by chance, statistically significant differences on 5 of them; out of 20 questions, we might expect to find, by chance, a difference on 1 of them. But suppose we find that on 12 of our 20 questions those who have attended the lectures are more favorable than those who have not. We may feel much safer in concluding that there is a true difference in attitudes, even though on each question the statistical test indicates that the difference might have arisen by chance 5 times in 100.

What if, out of the 20 questions, only the one about a U.N. police force shows a statistically significant difference between the two groups? This difference might well have occurred by chance; on the other hand, it may be that the lectures actually did influence opinions on this point though on no other. Unless our hypotheses specifically predicted that the lectures would be more likely to affect beliefs about an international police force than any of the other 19 items, we are not justified in making this latter interpretation, no matter how convincingly we may argue (after we have the finding) that the content of the lectures was such that they were especially likely to affect beliefs about an international police force. For if it had happened that the one item that showed a change dealt with the veto power in the Security Council, we might (in the absence of specific predictions about which items would be most affected) discover in the lectures material especially likely to change beliefs about the veto power. In other words, given any two variables that show a statistically significant relationship, an investigator usually finds it possible to propose an explanation for the relationship. However, the critical test of an obtained relation is not the *ex post facto* rationales and explanations for it, but rather the ability to predict it or to predict other relationships on the basis of it.[7] Thus, our unpredicted finding of a difference in attitudes toward a U.N. police force, even though "statistically significant," cannot be considered as established by the study we have carried out. It may, however, provide a fruitful hypothesis for a future study.

This point is clear enough as an abstract principle, but its implications are frequently ignored in analyzing and reporting data. Sometimes an investigator "runs wild" with an IBM machine and relates every variable to every other variable; it is then to be expected that a specified proportion of the relationships will appear statistically significant. Or he may first inspect the data for relationships and analyze only those that appear to be statistically significant. In either case, if he reports only those relationships that turned out to be statistically significant, the report may be entirely misleading, since the reader has no basis for judging whether these relationships form a consistent pattern within the findings or whether they are such a small proportion of the total number of relationships that it is more reasonable to assume their chance occurrence.

Since statistical statements are always statements of probability, we can never rely on statistical evidence alone for a judgment of whether or not we will accept a hypothesis as true. Confidence in the interpretation of a research result requires not only statistical confidence in the reliability of the finding (i.e., that the differences are not likely to have occurred by chance) but, in addition, evidence about the validity of the presuppositions of the research.[8] This evidence is necessarily indirect. It comes from the congruence of the given research findings with other knowledge about which there is considerable assurance. Even in the most rigorously controlled investigation, the establishment of confidence in the interpretations of one's results or in the imputation of causal relationships requires repetition of research and the relating of the findings to those of other studies.

It is important to recognize that even when statistical tests and the findings of a number of studies suggest that there is indeed a consistent difference between two groups, or a consistent relationship between two variables, this still does

---

[7] As pointed out . . . "prediction" in this context is not limited in meaning to the forecasting of future occurrences; it includes also the announcing of past or current events prior to knowledge of their occurrence.

[8] Some of the common presuppositions are that the measures are relevant to the variables included in the hypothesis and that the effect of extraneous variables has been controlled or taken into account. . . .

not constitute evidence of *the reason* for the relationship. If we want to make causal inferences —that is, to say that one variable or event has led to another—we must meet assumptions over and above those required for establishing the existence of a relationship. . . .

One further point should be mentioned. The fact that a result is *statistically* significant does not necessarily mean that it is *socially* or *psychologically* significant. Many statistically significant differences are trivial. For example, given enough cases,[9] an average difference in intelligence between men and women of less than one IQ point may be statistically significant, but it is difficult to see any real import in the finding. On the other hand, there are cases where a small but reliable difference has great practical importance. For example, in a large-scale survey designed to give information about a population, a difference of one half of 1 per cent may represent hundreds of thousands of people, and knowledge of the difference may be important for policy decisions. One must constantly be concerned with the social and psychological meaning of one's findings as well as their statistical significance.

---

[9] The larger a sample, the less likely are findings based on it to deviate from the true state of affairs in the population. Tests of significance therefore take into account the number of cases in the sample. The larger the samples, the greater the likelihood that a given difference between them is statistically significant. . . .

Leslie Kish

Problems in Statistical Analysis[1]

. . . The researcher designates the explanatory variables on the basis of substantive scientific theories. He recognizes the evidence of other *sources of variation* and he needs to separate these from the explanatory variables. Sorting all sources of variation into four classes seems to me a useful simplification. Furthermore, no confusion need result from talking about sorting and treating "variables," instead of "sources of variation."

I. The *explanatory* variables, sometimes called the "experimental" variables, are the objects of the research. They are the variables among which the researcher wishes to find and to measure some specified relationships. They include both the "dependent" and the "independent" variables, that is, the "predictand" and "predictor" variables.[2] With respect to the aims of the research all

---

[1] This research has been supported by a grant from the Ford Foundation for Development of the Behavioral Sciences. It has benefited from the suggestions and encouragement of John W. Tukey and others. But the author alone is responsible for any controversial opinions.

Reprinted from *American Sociological Reivew* Vol. 24, No. 3 (June, 1959), 328–38, where it appeared under the title, "Some Statistical Problems in Research Design." By permission of the author and the publisher, the American Sociological Association.

[2] Kendall points out that these latter terms are preferable. See his two-part paper M. G. Kendall, "Regression, Structure and Functional Relationship," *Biometrika*, 38 (June, 1951), 12–25; and 39 (June, 1952), 96–108; and M. G. Kendall and W. R. Buckland, *A Dictionary of Statistical Terms* (London: Oliver and Boyd, 1957), prepared for the International Statistical Institute with assistance of UNESCO. I have also tried to follow in IV below his distinction of "variate" for random variables from "variables" for the usual (non-random) variable.

other variables, of which there are three classes, are extraneous.

II. There are extraneous variables which are *controlled*. The control may be exercised in either or both the selection and the estimation procedures.

III. There may exist extraneous uncontrolled variables which are *confounded* with the Class I variables.

IV. There are extraneous uncontrolled variables which are treated as *randomized* errors.
. . . Randomization may be regarded as a substitute for experimental control or as a form of control.

The aim of efficient design both in experiments and in surveys is to place as many of the extraneous variables as is feasible into the second class. The aim of randomization in experiments is to place all of the third class into the fourth class; in the "ideal" experiment there are no variables in the third class. And it is the aim of controls of various kinds in surveys to separate variables of the third class from those of the first class; these controls may involve the use of repeated cross-tabulations, regression, standardization, matching of units, and so on.

The function of statistical "tests of significance" is to test the effects found among the Class I variables against the effects of the variables of Class IV. . . . The scientist must make many decisions as to which variables are extraneous to his objectives, which should and can be controlled, and what methods of control he should use. He must decide where and how to introduce statistical tests of hypotheses into the analysis.

As a simple example, suppose that from a probability sample survey of adults of the United States we find that the level of political interest is higher in urban than in rural areas. A test of significance will show whether or not the difference in the "levels" is large enough, compared with the sampling error of the difference, to be considered "significant." Better still, the confidence interval of the difference will disclose the limits within which we can expect the "true" population value of the difference to lie.[3]

. . . If the test of significance rejects the null hypothesis of no difference, *several* hypotheses remain in addition to that of a simple relationship between urban *versus* rural residence and political interest. Could differences in income, in occupation, or in family life cycle account for the difference in the levels? The analyst may try to remove (for example, through cross-tabulation, regression, standardization) the effects due to such variables, which are extraneous to his expressed interest; then he computes the difference, between the urban and rural residents, of the levels of interest now free of several confounding variables. This can be followed by a proper test of significance—or, preferably, by some other form of statistical inference, such as a statement of confidence intervals. . . .

Agreement on these ideas can eliminate certain confusion, exemplified by Selvin in a recent article:

> Statistical tests are unsatisfactory in nonexperimental research for two fundamental reasons: it is almost impossible to design studies that meet the conditions for using the tests, and the situations in which the tests are employed make it difficult to draw correct inferences. The basic difficulty in design is that sociologists are unable to randomize their uncontrolled variables, so that the difference between "experimental" and "control" groups (or their analogs in nonexperimental situations) are a mixture of the effects of the variable being studied and the uncontrolled variables or correlated biases. Since there is no way of knowing, in general, the sizes of these correlated biases and their directions, there is no point

---

[3] The sampling error measures the chance fluctuation in the difference of levels due to the sampling operations. The computation of the sampling error must take proper account of the actual sample design, and not blindly follow the standard sample random formulas. See Leslie Kish, "Confidence Intervals for Complex Samples," *American Sociological Review*, 22 (April, 1957), 154–65.

in asking for the probability that the observed differences could have been produced by random errors. The place for significance tests is after all relevant correlated biases have been controlled. . . . In design and in interpretation, in principle and in practice, tests of statistical significance are inapplicable in nonexperimental research.[4]

. . . Actually, much research—in the social, biological, and physical sciences—must be based on nonexperimental methods. In such cases the rejection of the null hypothesis leads to several alternate hypotheses that may explain the discovered relationships. It is the duty of scientists to search, with painstaking effort and with ingenuity, for bases on which to decide among these hypotheses.

As for Selvin's advice to refrain from making tests of significance until "after all relevant" uncontrolled variables have been controlled—this seems rather far-fetched to scientists engaged in empirical work who consider themselves lucky if they can explain 25 or 50 per cent of the total variance. The control of all relevant variables is a goal seldom even approached in practice. To postpone to that distant goal all statistical tests illustrates that often "the perfect is the enemy of the good."[5]

. . . .

SOME MISUSES OF STATISTICAL TESTS

Of the many kinds of current misuses this discussion is confined to a few of the most common. There is irony in the circumstance that these are committed usually by the more statistically inclined investigators; they are avoided in research presented in terms of qualitative statements or simple descriptions.

First, there is "hunting with a shot-gun" for significant differences. Statistical tests are designed for distinguishing results at a predetermined level of improbability (say at $P = .05$) under a specified null hypothesis of random events. A rigorous theory for dealing with individual experiments has been developed by Fisher, the Pearsons, Neyman, Wold, and others. However, the researcher often faces more complicated situations, especially in the analysis of survey results; he is often searching for interesting relationships among a vast number of data.

. . .

Perhaps the problem has become more acute

---

[4] Hanan C. Selvin, "A Critique of Tests of Significance in Survey Research," *American Sociological Review*, 22 (October, 1957), 527. In a criticism of this article, McGinnis shows that the separation of explanatory from extraneous variables depends on the type of hypothesis at which the research is aimed. Robert McGinnis, "Randomization and Inference in Sociological Research," *American Sociological Review*, 23 (August, 1958), 408–14.

[5] Selvin performs a service in pointing to several common mistakes: (a) The mechanical use of "significance tests" can lead to false conclusions. (b) Statistical "significance" should not be confused with substantive importance. (c) The probability levels of the common statistical tests are not appropriate to the practice of "hunting" for a few differences among a mass of results. However, Selvin gives poor advice on what to do about these mistakes; particularly when, in his central thesis, he reiterates that "tests of significance are inapplicable in nonexperimental research," and that "the tests are applicable only when all relevant variables have been controlled." I hope that the benefits of his warnings outweigh the damages of his confusion.

I noticed three misleading references in the article. (a) In the paper which Selvin appears to use as supporting him, Herman Wold, "Causal Inference From Observational Data," *Journal of the Royal Statistical Society*, (A) 119, Part I (January, 1956), 39, specifically disagrees with Selvin's central thesis, stating that "the need for testing the statistical inference is no less than when dealing with experimental data, but with observational data other approaches come to the foreground." (b) In discussing problems caused by complex sample designs, Selvin writes that "such errors are easy enough to discover and remedy" (p. 520), referring to Kish, "Confidence Intervals for Complex Samples,". . . . On the contrary, my article pointed out the seriousness of the problem and the difficulties in dealing with it. (c) "Correlated biases" is a poor term for the confounded uncontrolled variables and it is not true that the term is so used in literature. Specifically, the reference to Cochran is misleading, since he is dealing there only with errors of measurement which may be correlated with the "true" value. See William G. Cochran, *Sampling Techniques* (New York: John Wiley and Sons, Inc., 1953), p. 305.

now that high-speed computers allow hundreds of significance tests to be made. There is no easy answer to this problem. We must be constantly aware of the nature of tests of null hypotheses in searching survey data for interesting results. After finding a result improbable under the null hypothesis the researcher must not accept blindly the hypothesis of "significance" due to a presumed cause. Among the several alternative hypotheses is that of having discovered an improbable random event through sheer diligence. Remedy can be found sometimes by a reformulation of the statistical aims of the research so as to fit the available tests. Unfortunately, the classic statistical tests give clear answers only to some simple decision problems; often these bear but faint resemblance to the complex problems faced by the scientist. . . .

Second, statistical "significance" is often confused with and substituted for substantive significance. There are instances of research results presented in terms of probability values of "statistical significance" alone, without noting the magnitude and importance of the relationships found. These attempts to use the probability levels of significance tests as measures of the strengths of relationships are very common and very mistaken. The function of statistical tests is merely to answer: Is the variation great enough for us to place some confidence in the result; or, contrarily, may the latter be merely a happenstance of the specific sample on which the test was made? This question is interesting, but it is surely *secondary*, auxiliary, to the main question: Does the result show a relationship which is of substantive interest because of its nature and its magnitude? Better still: Is the result consistent with an assumed relationship of substantive interest?

The results of statistical "tests of significance" are functions not only of the magnitude of the relationships studied but also of the numbers of sampling units used (and the efficiency of design). In small samples significant, that is, meaningful, results may fail to appear "statistically significant." But if the sample is large enough the most insignificant relationships will appear "statistically significant."

Significance should stand for meaning and refer to substantive matter. The statistical tests merely answer the question: Is there a big enough relationship here which *needs* explanation (and is not merely chance fluctuation)? The word *significance* should be attached to another question, a substantive question: Is there a relationship here *worth* explaining (because it is important and meaningful)? As a remedial step I would recommend that statisticians discard the phrase "test of significance," perhaps in favor of the somewhat longer but proper phrase "test against the null hypothesis" or the abbreviation "TANH."

Yates, after praising Fisher's classic *Statistical Methods*, makes the following observations on the use of "tests of significance":

> Second, and more important, it has caused scientific research workers to pay undue attention to the results of the tests of significance they perform on their data, particularly data derived from experiments, and too little to the estimates of the magnitude of the effects they are investigating.
>
> Nevertheless the occasions, even in research work, in which quantitative data are collected solely with the object of proving or disproving a given hypothesis are relatively rare. Usually quantitative estimates and fiducial limits are required. Tests of significance are preliminary or ancillary.
>
> The emphasis on tests of significance, and the consideration of the results of each experiment in isolation, have had the unfortunate consequence that scientific workers have often regarded the execution of a test of significance on an experiment as the ultimate objective. Results are significant or not significant and this is the end of it.[6]

[6] Frank Yates, "The Influence of *Statistical Methods for Research Workers* on the Development of the Science of Statistics," *Journal of the American Statistical Association*, 46 (March, 1951), 32–33.

For presenting research results statistical estimation is more frequently appropriate than tests of significance. The estimates should be provided with some measure of sampling variability. For this purpose confidence intervals are used most widely. In large samples, statements of the standard errors provide useful guides to action. These problems need further development by theoretical statisticians.[7] . . .

Third, the tests of null hypotheses of *zero* differences, of no relationships, are frequently weak, perhaps trivial statements of the researcher's aims. In place of the test of zero difference (the nullest of null hypotheses), the researcher should often substitute, say, a test for a difference of a specific size based on some specified model. Better still, in many cases, instead of the tests of significance it would be more to the point to measure the magnitudes of the relationships, attaching proper statements of their sampling variation. The magnitudes of relationships cannot be measured in terms of levels of significance; they can be measured in terms of the difference of two means, or of the proportion of the total variance "explained," of coefficients of correlations and of regressions, of measures of association, and so on. These views are shared by many, perhaps most, consulting statisticians—although they have not published full statements of their philosophy. Savage expresses himself forcefully: "Null hypotheses of no difference are usually known to be false before the data are collected; when they are, their rejection or acceptance simply reflects the size of the sample and the power of the test, and is not a contribution to science."[8]

Too much of social research is planned and presented in terms of the mere existence of some relationship, such as: individuals high on variate *x* are also high on variate *y*. The *exploratory* stage of research may be well served by statements of this order. But these statements are relatively weak and can serve *only* in the primitive stages of research. Contrary to a common misconception, the more advanced stages of research should be phrased in terms of the quantitative aspects of the relationships. Again, to quote Tukey:

> *There are normal sequences of growth in immediate ends.* One natural sequence of immediate ends follows the sequence: (1) Description, (2) Significance statements, (3) Estimation, (4) Confidence statement, (5) Evaluation. . . . There are, of course, other normal sequences of immediate ends, leading mainly through various decision procedures, which are appropriate to development research and to operations research, just as the sequence we have just discussed is appropriate to basic research.[9]

At one extreme, then, we may find that the contrast between two "treatments" of a labor force results in a difference in productivity of 5 per cent. This difference may appear "statistically significant" in a sample of, say, 1000 cases. It may also mean a difference of millions of dollars to the company. However, it "explains" only about one per cent of the total variance in productivity. At the other extreme is the far-away land of completely determinate behavior, where every action and attitude is explainable, with nothing left to chance for explanation.

The aims of most basic research in the social sciences, it seems to me, should be somewhere between the two extremes; but too much of it is presented at the first extreme, at the primitive level. This is a matter of over-all strategy for an entire area of any science. It is difficult to make this judgment off-hand regarding any specific piece of research of this kind: the status of research throughout the entire area should be

---

[7] D. R. Cox, "Some Problems Connected with Statistical Inference," *Annals of Mathematical Statistics*, 29 (June, 1958), 357–72.

[8] Richard J. Sabage, "Nonparametric Statistics," *Journal of the American Statistical Association*, 52 (September, 1957), 332–33.

[9] John W. Tukey, "Unsolved Problems of Experimental Statistics," *Journal of the American Statistical Association*, 39 (December, 1954), 712–13.

considered. But the superabundance of research aimed at this primitive level seems to imply that the over-all strategy of research errs in this respect. The construction of scientific theories to cover broader fields—the persistent aim of science—is based on the synthesis of the separate research results in those fields. A coherent synthesis cannot be forged from a collection of relationships of unknown strengths and magnitudes. The necessary conditions for a synthesis include an *evaluation* of the results available in the field, a coherent interrelating of the *magnitudes* found in those results, and the construction of models based on those magnitudes.

Max Gunther

Computers: Their Built-In Limitations

"Oh, My God!" croaked a network-TV director in New York. He seemed to be strangling in his turtleneck shirt. It was the evening of Election Day, 1966, and the director's world was caving in. Here he was, on the air with the desperately important Election Night coverage, competing with the two enemy networks to see whose magnificently transistorized, fearfully fast electronic computer could predict the poll results soonest and best. Live coverage: tense-voiced, sweating announcers, papers flapping around, aura of unbearable suspense. The whole country watching. And what happens? The damned computer quits.

Oh, my God. The computer rooms disintegrated in panic. Engineers leaped with trembling screwdrivers at the machine's intestines. The director stared fisheyed at a mathematician. A key-punch girl yattered terrified questions at a programmer. Young Madison Avenue types rushed in and out, uttering shrill cries. And the computer just sat there.

The story of this ghastly evening has circulated quietly in the computer business ever since. You hear it in out-of-the-way bars and dim corners of cocktail parties, told in hoarse, quavering tones. It has never reached the public at large, for two reasons. One reason is obvious: Those concerned have sat on it. The second reason is less obvious and much more interesting.

When the initial panic subsided, the director freed some of his jammed synapses and lurched into action. He rounded up mathematicians, programmers, political experts, research girls and others. And he rounded up some hand-operated adding machines. "All right," he said, "we'll simplify the calculations and do the whole thing by hand. This may be my last night on TV; but, by God, I'll go on the air with *something*!"

And so they perspired through the long, jangling night. The network's election predictions appeared on the screen just like its competitors'. The director and his aides gulped coffee, clutched burning stomachs and smoked appalling numbers of cigarettes. They kept waiting for an ax to fall. Somebody was bound to notice something wrong sooner or later, they thought. The hand-cranked predictions couldn't conceivably be as good as the computerized punditry of the competition. Maybe the hand-cranked answers would be totally wrong! Maybe the network would become the laughingstock of the nation! Maybe. . . . Oh, my God!

Reprinted from *Playboy*, Vol. 14, No. 10 (October, 1967), 94ff., where it appeared under the title, "Computers: Their Built-In Limitations." By permission of the author and the publisher. Copyright © 1967 by HMH Publishing Co., Inc.

Well. As history now tells us, the entire poll-predicting razzmatazz was the laughingstock of the nation that November. None of the three networks was wronger than the other two. When the half-gutted director and his fellow conspirators skulked out of bed the next morning and focused smoldering eyes on their newspapers, they at last recognized the obscure little facts that had saved their professional lives:

An electronic computer, no matter how big or how expensive or how gorgeously bejeweled with flashing lights or how thoroughly crammed with unpronounceable components, is no smarter than the men who use it. Its answers can never be better than the data and formulas that are programed into it. It has no magical insights of its own. Given inadequate data and inexact formulas, it will produce the same wrong answers as a man with an aching head and an adding machine. It will produce them in a more scientific-looking manner, that's all.

Over the past ten years, it has been fashionable to call these great buzzing, clattering machines "brains." Science-fiction writers and Japanese moviemakers have had a lovely time with the idea. Superintelligent machines take over the world! Squish people with deadly squish rays! Hypnotize nubile girls with horrible mind rays, baby! It's all nonsense, of course. A computer is a machine like any other machine. It produces numbers on order. That's all it can do.

Yet computers have been crowned with a halo of exaggerated glamor, and the TV election-predicting circus is a classic example. The Columbia Broadcasting System got into this peculiar business back in 1952, using a Remington Rand Univac. The Univac did well. In 1956, for instance, with 1/27 of the popular vote in at 9:15 P.M., it predicted that Dwight Eisenhower would win with 56 percent of the votes. His actual share turned out to be 57.4 percent, and everybody said, "My, my, what a clever machine!" The Univac certainly was a nicely wrought piece of engineering, one of the two or three fastest and most reliable then existing. But the credit for insight belonged to the political experts and mathematicians who told the Univac what to do. It was they, not the machine, who estimated that if Swampwater County went Democratic by X percent, the odds were Y over Z that the rest of the state would go Democratic by X-plus-N percent. The Univac only did the routine arithmetic.

Which escaped attention. By the 1960s, the U.S. public had the idea that some kind of arcane, unknowable, hyperhuman magic was soldered into computers—that a computerized answer was categorically better than a hand-cranked answer. As the TV networks and hundreds of other businesses realized, computers could be used to impress people. A poll prediction looked much more accurate on computer print-out paper than in human handwriting. But, as became clear at least to a few in 1966, it's the input that counts. Honeywell programing expert Malcolm Smith says: "You feed guesswork into a computer, you get beautiful neat guesswork back out. The machine contains no Automatic Guess Rectifier or Factualizing Whatchamacallit."

The fact is, computers are monumentally dense—"so literal," says Smith, "so inflexible, so flat-footed dumb that it sometimes makes you want to burst into tears." Smith knows, for he spends his life trying to make the great dimwits cogitate. To most people, however, computers are metallic magic, wonderful, tireless, emotionless, infallible brains that will finally solve mankind's every problem. Electronic data processing (EDP) is the great fad of the 1960s and perhaps the costliest fad in history. Companies big and small, universities, Government agencies are tumbling over one another in a gigantic scramble for the benefits of EDP. They believe EDP represents, at last, instant solutions to problems they've wrestled with for decades: problems of information flow, bookkeeping, inventory control. And they're hounded by dreams of status. To have a computer is "in." Even if you're a scruffy little company that nobody ever heard of, you must have a computer. Businessmen meeting at conventions like to drop phrases such as "My EDP manager told me" and "Our programing boys think," and watch the crestfallen looks of uncomputerized listeners.

It's a great business to be in. Computer makers shipped some 8000 machines in 1965 and 13,700 (3.75 billion dollars' worth) in 1966. There are over 30,000 computers at work in the country today and there will be (depending on whose guess you listen to) as many as 100,000 by 1975. It's a boom business in which young salesmen can buy Cadillacs and Porsches, while their college classmates in other professions are still eating canned beans in one-and-a-half-room flats. The salesmen don't need any unusual qualifications to strike it rich: just a two- or three-year apprenticeship, a sincere hard handshake, a radiating awareness of belonging to an elite group and a good memory for a polysyllabic vocabulary. (You don't sell machines; you sell "systems" or "systems concepts," or "integrated functional solid-state logic systems concepts." They seem to cost more that way.)

The salesmen are all business. They sell machines on a severely pragmatic level, maybe exaggerating their products' worth sometimes but, in general, avoiding any unbusinesslike talk about "superbrains." Computer manufacturers as a whole, in fact, avoid such talk. To their credit, they have struggled from the beginning to keep things in perspective, have publicly winced when imaginative journalists compared computers with that odd gray mushy stuff inside the human skull. "Don't call them brains! Please, *please* don't call them brains!" shouted IBM scientist Dr. Arthur Samuel at a reporter once. "But listen," said the reporter, "don't they——" "No, they don't!" howled Samuel. "Whatever you're going to say they do, they don't!" (Samuel, now at Stanford University, had won unwanted fame for programing an IBM machine to play checkers.) "Computers are just extremely fast idiots," says logician-mathematician Richard Bloch, a former Honeywell vice-president now working with Philadelphia's Auerbach Corporation. Bloch, a lean, dark, ferociously energetic man who smokes cigars incessantly, first tangled with the machines in the early 1940s, when he helped run Harvard University's historic Mark 1. "On second thought, 'idiots' is the wrong word. It suggests some innate thinking capacity gone wrong. Computers have no thinking capacity at all. They're just big shiny machines. When will people learn that machines don't think?"

Maybe never, though men like Bloch never tire of saying so. "A computer can multiply umpteen umpteen-digit numbers a second," says Bloch, "but this is only blind manipulation of numbers, not thinking. To think about a problem, you've got to understand it. A computer *never* understands a problem."

Arthur Samuel, for instance, tells about an early checkers-playing experiment. A British computer was given a simple set of rules in arithmetical form. Among other things, it was told that a king is worth three points, an uncrowned piece one point. It played an ordinary undistinguished game until its human opponent maneuvered a piece within one move of being crowned. Then the machine seemed to go mad.

Somewhere in its buzzing electrical innards, a chain of "reasoning" something like this took place: "Oh, my goodness! If my opponent gets his piece into the king's row, he'll gain a three-point king where he had only a one-point man before. In effect, this means I'll lose two points. What'll I do? (*buzz, buzz* . . .) Ah! I'll sacrifice one of my uncrowned pieces. The rules say he must take my piece if I offer it, and this will force him to use his move and prevent him from getting his man crowned. I'll have lost only one point instead of two!"

So the cunning computer sacrificed a man. The human player took it. The situation was now exactly the same as it had been before, so the computer slyly sacrificed another man. And so on. Piece by piece, the unthinking machine wiped itself out.

The computer had proved itself able to manipulate some of the arithmetical and logical formulas of checkers. But it had failed in one supremely important way. It simply didn't understand the game. It didn't grasp what no human novice ever needs to be told: that the basic object of a game is to win.

The trouble with computers is that they *seem* to be thinking. While cars, lawn mowers and other machines perform easily understood physical tasks, computers seem to be working with abstract thoughts. They aren't, of course; they are only switching electric currents along preordained paths. But they produce answers to questions, and this gives them a weird brainlike quality.

People expect too much of them, as a result, and this seriously worries some scientists. The late Norbert Wiener, coiner of the term "cybernetics," was particularly worried about the increasing use of computers in military decision making. Referring to machines that can manipulate the logical patterns of a game without understanding it, he once wrote that computers could win some future nuclear war "on points . . . at the cost of every interest we have at heart." He conjured up a nightmarish vision of a giant computer printing out "WAR WON: ASSIGNMENT COMPLETED . . ." and then shutting itself down, never to be used again, because there were no men left on earth to use it.

Secretary of Defense Robert S. McNamara has hinted at similar worries in the years-long argument about our famous (but so far nonexistent) Nike-X missile-defense system. Neither full-fledged hawk nor dove, McNamara favors a leisurely and limited building of Nike bases. He wants the U.S. to have some defense against a possible Russian or Chinese ballistic-missile attack, but he fears that an all-out missile-building program will involve us in a ghastly game of nuclear leapfrog with the Soviets—the two sides alternately jumping ahead of each other in countermeasures and counter-countermeasures until the radioactive end. One trouble with missile and anti-missile systems, as McNamara once expressed it to a group of reporters, is that "the bigger and more complex such systems get, the more remote grows man's control of them." In a nuclear-missile war, so many things would happen so fast, so much data would have to be interpreted in so limited a time that human brains could not possibly handle the job. The only answer for both the U.S. and Russia in a missile arms race would be increasing reliance on automatic control—in other words, on computers.

The last war might, in fact, be a war between computers. It would be a coldly efficient war, no doubt. A logical war: Score 70,000,000 deaths for my side, 60 megadeaths for your side; I'm ahead; your move, pal. How could we convey to the machines our totally illogical feelings about life and death? A country is made of people and money, and the people may properly be asked to give their lives for their country, yet a single human life is worth more than all the money in the world. Only the human brain is flexible enough to assimilate contradictions such as this without blowing a fuse.

A large modern computer can literally perform more arithmetic in an hour than can a football stadium full of human mathematicians in a lifetime, and it makes sense to enlist this lightning-fast electronic help in national defense. "But," said Norbert Wiener shortly before he died in 1964, "let us always keep human minds in the decision loop somewhere, if only at the last 'yes' or 'no.'"

The U.S. Ballistic Missile Early Warning System (BMEWS) is an example of the kind of setup that worried Wiener. Its radar eyes scan sky and space. Objects spotted up there are analyzed automatically to determine whether they are or aren't enemy missiles. The calculations performed by computers—distance of the objects, direction, checkoff against known craft—take place in fractions of a second, far faster than human thought. It all works beautifully most of the time, and this has led some enthusiasts to suggest going one step further in automation. "If BMEWS can spot enemy missiles by itself," they say, "why not hook up one more wire and have BMEWS launch *our* missiles?" But U.S.

military chiefs have so far agreed with Norbert Wiener. There is a subtlety in the human brain that no computer seems likely ever to duplicate.

A few years ago, an officer was monitoring a BMEWS computer station in the Arctic. It was night. The rest of the staff was in bed. Suddenly, the computer exploded into action. Lights flashed, a printer chattered, tape reels whirled. The officer gaped, horrified. The machine was signaling a massive missile attack.

There are self-checking devices and "redundant" networks in the BMEWS, as in any other large computer system, and the officer had no logical reason to suspect a mechanical breakdown. There could be little doubt that the computer was actually reporting what its far-flung radar eyes saw. The officer's orders were clear: In an event like this, he must send a message that would mobilize military installations all over the United States. Global war was only minutes away.

The officer hesitated. Questioned later, he couldn't explain why. He could only say, "It didn't *feel* right." And he gambled time to wake other staff members. One of them dashed outdoors to look at the cold, clear, starlit Arctic sky, ran back indoors, examined the computer's print-out, conferred with the others. Standing there in that antiseptic room full of shiny electronic equipment, the small knot of men made what may have been the most important decision in all the history of the world to date. They decided to wait.

They waited 30 awful seconds. The missile attack came no closer.

The officer's feeling had been correct. This was no missile attack. Unaccountably, through a freakish tangle of circumstances that should never have happened and could not have been predicted and was not fully unraveled until weeks later, the computer and its eyes had locked onto something quite without menace: earth's friendly companion and goddess of love, the moon, peacefully coming up over the horizon. If computers alone had handled the affair, the earth might now be a smoldering radioactive cinder.

Because of a man and his slow, strange human brain and its unfathomable intuition, we are all still here.

When a computer makes a mistake, it's likely to be a big one. In a situation where a man would stop and say, "Hey, something's wrong!" the machine blindly rushes ahead because it lacks the man's general awareness of what is and isn't reasonable in that particular situation; such as the time when a New York bank computer, supposed to issue a man a dividend check for $162.40, blandly mailed him one for $1,624,000; or the time when a computer working for a publishing company shipped a Massachusetts reader six huge cartons neatly packed with several hundred copies of the same book; or the time when an IBM machine was constructing a mathematical "model" of a new Air Force bomber that would fly automatically a few dozen feet off the ground. Halfway through the figuring, it became apparent that the computer was solemnly guiding its imaginary aircraft along a course some five feet below the ground. ("Goddamn it," roared General Curtis LeMay at one of the scientists, "I asked for an airplane, not a plow!") Or the time when——

Well, everybody makes mistakes. In general, society is most worried about mistakes made by war computers in the BMEWS style, for the potential result of a mistake in this field is the end of the world. Fearful imaginings such as *Fail-Safe* have expressed this fear, and most U.S. military planners share the fear and are cautious in their approach to computers. But no such colossal danger haunts computer users in science and business; and in these two fields, the great dumb machines have been pushed willy-nilly into all kinds of applications—some more sensible than others. A New York management-consultant firm, McKinsey and Company, exhaustively studied computer installations in 27 big manufacturing companies four years ago and found that only nine were getting enough benefits to make the machines pay.

"Sometimes computers are used for prestige purposes, sometimes as a means of avoiding human responsibility," says computer consultant John Diebold. Diebold, at 41, is a millionaire and an internationally sought-after expert on "automation" (a term he coined in the early 1950s). "Scientists and executives have discovered that it's impressive to walk into a meeting with a ream of computer print-out under your arm. The print-out may be utter nonsense, but it looks good, looks exact, gives you that secure, infallible feeling. Later, if the decision you were supposed to make or the theory you were propounding turns out to be wrong, you simply blame the computer or the man who programed it for you."

Professor David Johnson of the University of Washington is another well-known computer consultant who worries about what he calls "the mindless machines." He is amused by the fact that his engineering students seek status by using IBM cards as bookmarks—just as, 20 years from now, they will seek it by buying IBM machines for their companies. He praises computers for their ability to manipulate and organize huge masses of data at huge speeds. But, "What the computer does," he says, "is to allow us to believe in the myth of objectivity." The computer "acts without excessive hesitation, as if it is sure, as if it knows. . . ." A man who isn't sure can often make people think he is, simply by coming up with a bundle of factual-looking print-out. He hides his own bad brainwork, says Professor Johnson, by "sprinkling it with *eau de computer*."

Worse, Professor Johnson says, the growing availability of computers tends to make some researchers in scientific institutions avoid problems that don't lend themselves to machine handling: Problems involving human values, problems of morality and aesthetics, subtle problems that can't be translated into arithmetic and punched into those neat little snip-cornered cards —all these get left out of the calculations. The tendency is to wrench reality around and hammer it into a nice square shape so the inflexible machines can swallow it. Professor Johnson glumly cites the case of a computer-headed robot recently developed by a major agricultural-research center to pick tomatoes. It clanks along briskly, picking the juicy red fruits faster than a whole gang of human workers. The only trouble is, its blind, clumsy fingers break the tomatoes' skins. The agricultural scientists are now trying to solve the problem. By making the robot more gentle? No, by developing thicker-skinned tomatoes.

"It simply isn't accurate to call these machines 'clever,'" says Robert Cheek, a chief of the Westinghouse Tele-Computer Center near Pittsburgh. This is one of the biggest computer installations in the world, designed to handle Westinghouse's huge load of corporate clerical and accounting work, and it generates science-fictionish visions of an office of the (if the cliché may be pardoned) future. It's an entire modernistic building housing almost nothing but computing equipment. Clerks and secretaries who once populated it have been crowded out, and now it smells like the inside of a new car. Bob Cheek, a slight, mild man, looks small and lonely as he paces among the square whining monsters; and it is tempting to imagine that the machines have subjugated him as their slave. Actually, he is little more awed by this great aggregation of computing power than by an electric toaster. "Artificial intelligence?" he will say in response to the question he has heard too often. And he will look at his machines, think of the man-hours required to make them work, take off his glasses, rub his weary eyes and chuckle sourly.

Logician Richard Bloch is an example of high human intelligence. He learned chess at the age of three and is now, among other things, a Life Master bridge player and a blackjack shark. He once tried to teach a Honeywell computer to play bridge. "The experiment gave me new respect for the human brain," he recalls wryly. "The brain can act on insufficient, disorganized data. A bridge novice can start to play—badly but not stupidly—after an hour or so of mediocre instruction, in a half-drunken foursome. His brain makes generalizations on its own, reaches conclusions nobody ever told it to reach. It can absorb badly thought-out, unspecific instructions such as, 'If your hand looks pretty good, bid such and such.' What does 'pretty good' mean? The brain can feel it out. Now, you take a computer——"

Bloch pauses to chew moodily on his cigar. "A computer won't move unless you tell it every single step it must take, in excruciating detail. It took me more than a hundred pages of densely packed programing before I could even get the damned machine to make the first bid. Then I gave up."

The fact is, human thinking is so marvelous and mysterious a process that there is really not much serious hope of imitating it electronically—at least, not in this century. Nobody even knows how the brain works. Back in the late 1950s, during the first great soaring gush of enthusiasm over computers, journalists and some scientists were saying confidently that the brain works much like a very small, very complex digital computer—by means of X trillion tiny on-or-off switches. It remained only for IBM, Honeywell and Rem Rand to devise a monstrous mile-high machine with that many switches (and somehow figure out a way to supply its enormous power needs and somehow cool it so it wouldn't melt itself), and we'd have a full-fledged brain. But this was only another case of wrenching reality around to fit machinery. There is no reliable evidence that the brain works like an EDP machine. In fact, evidence is now growing that the basic components of human thought may be fantastically complicated molecules of RNA (ribonucleic acid), which seem to store and process information by means of a little-understood four-letter "code."

The human brain is uncanny. It programs itself. It asks itself questions and then tells itself how to answer them. It steps outside itself and looks back inside. It wonders what "thinking" is.

No computer ever wondered about anything. "It's the speed of computers that gives the false impression that they're thinking," says Reed Roberts, an automation expert who works for a New York management-consultant firm, Robert Sibson Associates. "Once a man has told a machine how to process a set of data, the machine will do the job faster than the man's brain could; so fast, in fact, that you're tempted to suspect the machine has worked out short cuts on its own. It hasn't. It has done the job in precisely the way it was told, showing no originality whatever."

For instance, you can program a machine to add the digits of each number from 1 to 10,000 and name every number whose digits add up to 9 or a multiple of 9. The machine will print out a list instantly—9, 18, 27—acting as though it has gone beyond its instructions and cleverly figured out a short cut. This is the way a man would tackle the problem. Instead of routinely adding the digits of every number from 1 to 10,000, he'd look for a formula. His brain would generalize: "Every time you multiply a number by 9, the result is a number whose digits add up to 9 or a multiple of 9. Therefore, I can do my assignment quickly just by listing the multiples of 9 and ignoring all other numbers." Is this what the computer did? No. With blinding speed but monumental stupidity, it laboriously tried every number, from 1 to 10,000, one by one.

In this example lies one of the main differences between thought and EDP. The human brain collects specific bits of data and makes generalizations out of them, organizes them into patterns. EDP works the other way around. A human programmer starts the machine out by giving it generalizations—problem-solving methods or "algorithms"—and the machine blindly applies these to specific data.

It is by no means easy to program a computer, and one of the great problems of the 1960s is a severe shortage of people who know how to do it. There are now some 150,000 professional programmers in the country, and computer owners are pitifully crying for at least 75,000 more. One estimate is that 500,000, all told, will be needed by the early 1970s.

The shortage is understandable. Computer programming is self-inflicted torture. The problem is to make a mindless machine behave rationally. Before you can tell the machine how to solve a problem, you must first figure out how your own brain solves it—every step, every detail. You watch your brain as it effortlessly snakes its way along some line of reasoning that loops back through itself, and then you try to draw a diagram showing how your brain did it, and you discover that your brain couldn't possibly have done it—yet you know it did. And there sits the computer. If you can't explain to yourself, how are you ever going to explain to *it*?

Aptitude tests for would-be programmers contain questions that begin, "If John is three years older than Mary would have been if she were three and a half times as old as John was when. . . ." This is the kind of human thought that must precede the switching on of a computer. The machine can't add two plus two unless there are clever, patient human brains to guide it. And even then it can't: All it can do is add one and one and one and one and come up with the answer—instantaneously, of course. *No* computer can multiply; all it can do is add, by ones, too fast for human conception. Nor can any computer divide; it can only subtract, again by ones. Feed it problems in square roots, cube roots, prime numbers, complex mathematical computations with mile-long formulas—it can solve them all with incredible rapidity. How? Essentially, by adding or subtracting one, as required, as often as required, to come up at once with an accurate answer it might take a team of mathematicians a thousand years to obtain—and another thousand to check for accuracy. It never invents its own mathematical short cuts. If it uses short cuts, they must be invented and programed into it by human thinkers.

A computer's only mental process is the ability to distinguish between is and isn't—the presence or absence of an electric current, the *this way* or *that way* of a magnetic field. In terms of human thought, this kind of distinction can be conceived as one and zero, yes and no. The machine can be made to perform binary arithmetic, which has a radix (base) of 2 instead of our familiar 10 and which is expressed with only two digits, 1 and 0. By stringing together yeses and noes in appropriate patterns, the machine can also be made to manipulate logical concepts such as "and," "or," "except when," "only when," and so on.

But it won't manipulate anything unless a man tells it how. Honeywell, whose aggressive EDP division has recently risen to become the nation's second-biggest computer maker, conducts a monthly programing seminar in a Boston suburb for top executives of its customer companies to help them understand what their EDP boys are gibbering about. The executives learn how to draw a "flow chart," agonizingly breaking down a problem-solving method into its smallest steps. They translate this flow chart into a set of instructions in a special, rigid, stilted English. (OPEN INPUT OMAST INVCRD. OPEN OUTPUT NMAST INVLST.) They watch a girl type out this semi-English version on a key-punch machine, which codes words and numbers in the form of holes punched into cards. These cards are fed into the computer, and another translation takes place. A canned "compiler" program (usually fed into the machine from a magnetic-tape unit) acts as an interpreter, translates the semi-English into logical statements in binary arithmetic. The computer finally does what the novice programmers have told it to do—if they've told it in the right way. The machine understands absolutely no deviations from its rigid language. Leave out so much as a comma, and it will either stop dead or go haywire. (At Cape Kennedy recently, a computer-guided rocket headed for Brazil instead of outer space because a programmer had left out a hyphen.) Finally, the executives head back to Boston's Logan International Airport, soothe their tired brains with ethyl alcohol with an olive or a twist, and morosely agree that nobody is so intractably, so maddeningly dense as a computer.

But they are glad to have learned. They've made a start toward finding out what goes on inside those strange square machines in the plant basement; and with that knowledge, they'll have a defense against a Machiavellian new kind of holdup that their Honeywell instructors have warned them about. It has happened more and more often and recently happened in one of the country's biggest publishing houses. Almost all the company's clerical work was computerized: inventory, billing, bookkeeping, payroll. With the corporate neurons thus inextricably tangled into the computer, the chief programmer went to the president and smilingly demanded that his salary be doubled. The president fired him on the spot—and shortly afterward realized the full enormity of what he had done. Nobody in the company, nobody in the whole world except the chief programmer knew what went on in the computer or how to make it do its work. The programs were too complex—and the computer, having no intelligence, could offer no explanations. As the horrified president now discovered, it was not true (as he had boasted) that a marvelous machine was running his company's paperwork. The cleverness hadn't been in the machine but in the brain of a man. With the man gone, the machine was just a pile of cold metal. The company nearly foundered in the ensuing year while struggling to unravel the mess.

Computers are that way: They absorb credit for human cleverness. Often a computerized operation will seem to go much more smoothly than it did in the old eyeshade-and-ledger days and the feeling will grow that the machine itself smoothed things out. What has really happened, however, is that the availability of the computer has forced human programmers to think logically about the operation and make it straightforward enough for the machine to handle. Professor David Johnson recalls a time when a company called him in to program an accounting operation for a computer. In previous years, this operation had taken two men ten months to perform by hand and brain. Johnson drew his flow charts, saw ways of simplifying, finally came up with an operation so organized that one man could do it in two days with a desk calculator. The company promptly abandoned its dreams of EDP—but if it had used a computer as planned, the machine rather than the programmer would doubtless have been showered with praise for the new simplicity.

Computers have been given credit for many things they haven't done. Even more, they've been given credit for things they were going to do in the future. The loudest crescendo of computer prognostications occurred in the late 1950s. Future-gazers went wild with enthusiasm. Soon, they said, computers would translate languages, write superb music, run libraries of information, become chess champions. Ah, those fantastic machines! Unfortunately, the whole history of computers—going all the way back to the pioneering Charles Babbage in the 19th Century—has been a series of manic-depressive cycles: early wild enthusiasm, followed by unexpected difficulties, followed by puzzled disappointment and silence.

Music? An amiable professor at the University of Illinois, Lejaren Hiller, Jr., has programed a machine to write music. One of the machine's compositions is the *Illiac Suite*. Says Hiller: "Critics have found it—er—interesting."

Chess? A computer in Russia is now engaged in a long-distance match with one at Stanford University in California. The match began awkwardly, with both machines making what for humans would be odd mistakes. Everybody concerned now seems somewhat embarrassed. Stanford's Professor of Computer Science John McCarthy, when asked recently how the game was going, said: "I have decided to put off any further interviews until the match is over."

Translate languages? There's something about human speech that computers just don't seem to get. It isn't rigid or formal enough; it's too subtle, too idiomatic. An IBM computer once translated "out of sight, out of mind" from English to Russian and back to English. The phrase returned to English as "blind, insane."

Libraries of information? "We don't know a good enough way to make a computer look up facts," says Honeywell programming researcher

Roger Bender. He leans forward abruptly and jabs a finger at you. "Who wrote *Ivanhoe?*" he asks. You say, "Walter Scott." Bender says, "How did you know? Did you laboriously sort through books in your memory until you came to *Ivanhoe*? No. And how did you even know it was a book? You made the connections instantly, *and we don't know how.*"

Superbrains? Dr. Hubert L. Dreyfus, professor of philosophy at the Massachusetts Institute of Technology, recently published a paper called "Alchemy and Artificial Intelligence." In it, he expresses amusement at the prognosticators' claim that today's computers are "first steps" toward an ultimate smarter-than-human brain. The claim, he says, makes him think of a man climbing a tree, shouting, "Hey, look at me, I'm taking the first steps toward reaching the moon!" In fact, says Professor Dreyfus, computers don't and can't approximate human intelligence. They aren't even in the same league.

Honeywell's Roger Bender agrees. "We once had a situation where we wanted a machine to take a long list of numbers and find the highest number," he recalls. "Now, wouldn't that seem to you like an easy problem? Kids in first grade do it. Nobody has to tell them how. You just hand them a list and they look at the numbers and pick the highest. Of all the simple-minded ——Well, it just shows what you have to go through with computers."

In this case, a programmer tried to figure out how he himself would tackle such a problem. He told the machine: "Start with the first number and go down the list until you come to a number that's higher. Store that number in memory. Continue until you find a still higher number," and so on. The last number stored would obviously (obviously to a man, that is) be the highest number on the list.

The machine imbibed its instructions, hummed for a while and stopped. It produced no answer.

"It was baffling," says Bender. "Nobody knew what the trouble was, until someone happened to glance down the list by eye. Then the problem became apparent. By great bad luck, it turned out, the highest number on the list was *the first number*. The computer simply couldn't figure out what to do about it."

Consultant John Diebold says: "Computers are enormously useful as long as you can predict in advance what the problems are going to be. But when something unexpected happens, the only computer in the world that's going to do you any good is the funny little one beneath your scalp."

V CONCEPTUAL FRAMES OF REFERENCE

Previous sections of this book have considered the nature of behavioral inquiry and its implementation in the social sciences. Evidence concerning the widespread acceptance of behaviorialism in political science is provided not only by the research methods employed, but also by the conceptual frames of reference used. These conceptual frames of reference serve as guides telling the researcher what to look for in his investigation while also providing him with an abstract scheme which integrates his findings into a coherent and meaningful explanation of political phenomena.

A useful distinction can be drawn between conceptual frames of reference that focus on political units and those that focus upon political processes. The former type seeks to explain political phenomena in terms of persons, groups, or societies. A focus of this type calls for the study of characteristics, behaviors, and organizations of units and for the explanation of political phenomena to be made in terms of these units. On the other hand, a process focus concentrates on sequences of events surrounding some phenomenon. Information about these events is used to explain the phenomenon. The unit focus is thus concerned with the nature and consequences of *action of units* while the process focus is concerned with the nature and consequences of *sequences of events*.

The most basic focal unit in political science is the individual political actor. Studies from this vantage point frequently use the concept "role" for explaining personal attitudes, opinions, and behaviors. Heinz Eulau points out in the first reading selection that much human activity is determined by one's position in the social context. Variations in this context produce differing roles which, in turn, produce diverse behavior. In addition to offering explanations about the individual, Eulau contends that role analysis can also be useful for understanding the behavior of conglomerates of individuals, that is, for groups and societies.

A second frame of reference focusing on the individual can be seen in those studies labeled socialization. In his essay, William C. Mitchell notes that socialization is a concern of all social sciences because it affects all aspects of human behavior—social, economic, and political. The socialization frame of reference stipulates that behavior is partially the result of internalized values, beliefs, and norms. In order to understand politics and political behavior, the researcher should direct inquiry toward understanding the nature of what is being internalized, the agents for this internalization, and the circumstances that surround this learning.

Just as the individual is considered an important element in the study of politics, so also are groups of individuals. This frame of reference seeks to explain politics by studying group activity. The application of this approach is clearly visible in studies of *interest* or *pressure* groups and their roles in politics which were quite common in the late 1940's and early 1950's. The dimensions of group analysis are discussed in the reading by Charles B. Hagan. In more recent years, political scientists have examined interpersonal behavior in group contexts, but not necessarily in the context of formally organized interest groups. The study of small-group dynamics has revived interest in the group frame of reference for political science.

Another type of group study focuses on entire societies. A useful model at this level is one developed in the natural sciences which conceptualizes a set of interrelated activities as a system. David Easton was one of the first political scientists to develop a system-oriented frame of reference for political science. The selection included here is one of Easton's early statements on this subject which he has subsequently expanded and more fully elaborated in two seminal books, *A Framework for Political Analysis* and *A Systems Analysis of Political Life*.[1] In systemic analysis, all political activity is viewed as taking place within the context of self-contained units called political systems. This framework integrates individual aspects of politics into one unified scheme by using the familiar notions of input, output, feedback, and processor. The system frame of reference has had considerable impact upon almost all areas of political inquiry.

[1] David Easton, *A Framework for Political Analysis* (Englewood Cliffs, N.J.: Prentice-Hall, Inc., 1965); and David Easton, *A Systems Analysis of Political Life* (New York: John Wiley & Sons, 1965).

A somewhat related approach which also uses society as the unit is the "political culture" frame of reference. As defined in the Sidney Verba selection, this scheme explains politics in terms of the context created by politically relevant values, attitudes, and beliefs. Political differences among as well as within societies can thus be explained by variations in political culture.

Some dissatisfaction has been voiced about approaches that use political units as the basic factors in explanatory schemes. One criticism is that these schemes are by nature static in that they consider political phenomena at a particular point in time. Another criticism is that a unit focus is weak in explaining cause and effect relationships because it deals only with those factors related to the unit under study. This dissatisfaction manifested itself in the development of frames of reference that specifically seek to deal with the dynamic elements of politics and that do not restrict the researcher's ability to consider a wide variety of seemingly unrelated factors. One example of such a frame of reference is the communications model. As depicted by Karl W. Deutsch, this approach stresses the importance of the flow of information. It considers, among other things, the volume and flow of information, the content of messages, the media of communication, and response to information stimuli. The emphasis in this frame of reference is upon change.

Another framework which also attempts to deal with dynamics and causal relationships is based upon the notion of power. In an earlier selection, Vernon Van Dyke noted the importance of power for defining the scope of politics. Robert A. Dahl also emphasizes this notion, but indicates that it can be useful for other purposes as well. Initially power must be distinguished from influence before meaningful analysis can be undertaken from this frame of reference. Once this distinction has been set out, Dahl discusses the potential of this frame of reference and concludes that the difficulties inherent in operationalizing power make it an elusive framework for viewing politics.

Perhaps the epitome of a process frame of reference is seen in what is labeled the decision-making approach. Richard C. Snyder in his essay explains decision making as a study of the sequence of events surrounding a decision. On the basis of information about political decision making, he argues, conclusions may be reached about political structures and about the behavior of individuals. Moreover, generalizations about one particular decision have explanatory significance for the phenomena of decision making in general. Of course, this explanatory capability is refined as more and more studies of single decisions are made.

The growing trend in political science for cross-national studies and the increasing desire for truly comparative research has resulted in specialized frames of reference. One of these, somewhat reminiscent of the political culture approach, purports to explain socio-political development in terms of the unique aspects of the society's cultural environment. This cultural relativistic approach actually tended to frustrate comparative analysis by emphasizing the unique rather than the shared aspects of cultural experience. As Lucian W. Pye points out, cultural relativism has been refined into an outlook called the developmental approach. Although several different interpretations have been given to this outlook, discussed by Pye, the developmental approach fostered cross-national comparisons by stressing the important similarities among phenomena.

Another frame of reference appropriate for comparative research, structural functionalism, is discussed by William Flanigan and Edwin Fogelman. Structural functionalism is closely related to and often considered a part of systems analysis. Underlying the structural functional framework is the assumption that all social or political systems must fulfill the same functional prerequisites in order to continue to exist as operating systems. Various systems perform these functions in different ways and by means of different structures. The purpose of inquiry, with this approach, is to identify these different structures and their functions and to explain why they are in fact different.

These reading selections are representative of the major frames of reference currently used by political scientists. A review of the literature of the discipline will indicate, however, that researchers tend to be eclectic in their choice of analytical schemes. The schemes outlined here are not always found in "pure" form. Researchers tend to select whichever conceptual tools they consider appropriate to their purpose.

Heinz Eulau

Role: A Basic Unit for Analyzing Political Behavior

In its simplest form, political behavior, like all social behavior, involves a relationship between at least two human beings. It is impossible to conceive of political behavior on the part of a person that does not have direct, indirect, or symbolic consequences for another person. The most suitable concept for analyzing a relationship between at least two actors and for determining the political relevance of the behavior characteristic of the relationship is "role," for we are not interested in all of a person's behavior but only in that aspect which is relevant to a political relationship.

The concept of role is familiar to most people. We speak of the father's role, the teacher's role, the minister's role, the judge's role, and so on. What we mean in all of these instances is that a person is identified by his role and that, in interpersonal relations activating the role, he behaves, will behave, or should behave in certain ways. In looking at man's social behavior or judging it, we do so in a frame of reference in which his role is critical. If we do not know a person's role, his behavior appears to be enigmatic. But a child ringing a doorbell is unlikely to be mistaken for a political "doorbell ringer." Political behavior, then, is always conduct in the performance of a political role.

Out of observations as simple as these, social scientists have built a variety of theories about the origins, structure, functions, and meanings of

© Copyright 1963 by Random House, Inc. Reprinted from *The Behavioral Persuasion in Politics*, by Heinz Eulau, by permission of Random House, Inc.

social roles. Whatever its uses in everyday language or scientific research, role seems to commend itself as a basic unit of social and political analysis.

Role can be used as a conceptual tool on all three levels of behavioral analysis: the social, the cultural, and the personal. It is a concept generic to all the social sciences. On the social level, it invites inquiry into the structure of the interaction, connection, or bond that constitutes a relationship. On the cultural level, it calls attention to the norms, expectations, rights, and duties that sanction the maintenance of the relationship and attendant behavioral patterns. And on the personal level, it alerts research to the idiosyncratic definitions of the role held by different actors in the relationship. Role is clearly a concept consistent with the analytic objective of the behavioral sciences. It lays bare the *inter*-relatedness and *inter*-dependence of people.

On the social level, many of the most immediate interactions can be analyzed in terms of polar roles: husband implies wife; student implies teacher; priest implies communicant; leader implies follower; representative implies constituent, and so on. The behavior of one actor in the relationship is meaningful only insofar as it affects the behavior of the other actor or is in response to the other's behavior. Whatever other acts a representative may perform, for instance, only those in the performance of his constituent relationships are of immediate interest in political behavior analysis. I say of immediate interest because, in actuality, no single relationship is isolated from other social relationships in which the partners to the focal relationship are likely to be involved.

Many relationships are not structured by unipolar roles alone. In most cases, a role is at the core of several other roles, making for a network of roles that can be very complex. A legislator is "colleague" to his fellow legislators, "representative" to his constituents, "friend" (or "enemy") to lobbyists, "follower" to his party leaders, "informant" to the press, and so on. Whatever role is taken, simultaneously or seriatim, what emerges is a very intricate structure of relations in which one role is implicated in several other roles.

A role may be implicated in several networks. For example, the mayor of a city is not only a chief executive, a role that implicates him in several other role relationships related to his position, but he is also involved in many other relationships of more or less direct relevance to his political roles. He may be a husband and father, an alumnus of the local college, a member of the Rotary Club, a lawyer, a churchgoer, an investor in local business, and so on. Depending on circumstances, these roles may complement each other, be mutually exclusive, or conflict. A network of roles reflects the complexity of social and political behavior patterns and warns against treating any one role as if it were exclusive.

Analytically, each network of roles can be thought of as a "role system." This has two corollaries. First, some roles are more directly related to each other than are other roles. For instance, the roles of husband and father or legislator and representative are intimately connected. Other roles may be less so. The existence and degree of their mutual implication is always subject to empirical determination. The legislator's role as a lawmaker is less likely to be related to his role as a parent than it is to his occupational role as, say, an insurance agent. This does not mean that the parent role is altogether irrelevant in his legislative behavior. A legislator with children attending public schools is probably more interested in school problems than a legislator who is a bachelor. The notion of role system directs attention to the totality of social behavior. At the same time, it points out the need of specifying the boundaries of the particular system under investigation.

The second corollary of role system implies that a change in one role may have consequences for the actor's other roles and, therefore, for the relationships in which he is involved by virtue of his roles. (This must not be confused with a change in position. When a Senator becomes President, his Senatorial role is terminated. His new position will make for new roles that greatly affect his other role relationships). As an example of role change, take the representative who finds it impossible to accept instructions from his constituents and increasingly relies on his own judgment. In the technical language of role analysis, this is a change from the "delegate" to the "trustee" role. It is likely that this change in the representational role will have consequences for the legislator's party-relevant roles. He might change from a partisan follower into an independent.

The structure of role relationships is not only patterned but also fluid. One source of change in role is a change in the expectations of others in the role system. Another source may be an actor's own redefinition of his role. These possibilities suggest the importance of treating role concepts from a cultural and personal standpoint as well as a social one.

On the cultural level, role refers to those expectations of a normative sort that actors in a relationship entertain concerning each other's behavior. These are the rights and duties that give both form and content to the relationship. A relationship can be maintained only as long as the participants are in agreement as to what each actor must or must not do in the performance of his role. If there is disagreement over what kind of behavior should be expected, the relationship is likely to disintegrate.

Expectations which define roles and give direction to the behavior of actors in a role relationship are cultural in two ways. People do not continuously define and redefine their mutual relations and expectations. If a relationship had to be defined anew with each interaction, or if expectations had to be elaborated with every new encounter, stable social life would be impossible. In fact, most of the crucial role relationships are well defined. They are well defined because expectations are widely shared and transmitted through time. There is, then, a broad cultural consensus as to what the rights and duties pertaining to social roles are, and there is consensus on the sanctions available to participants in a relationship if behavior should violate agreed-on norms.

There may be more or less agreement from one role to another, from one culture to another. In Western culture, there is a broad consensus as to what kind of behavior the role of parent vis-à-vis the child calls for, though there are differences in role conceptions from one subculture to another. But if a role is located at the center of a network of roles, consensus on expectations is more difficult to identify. Only some minimum agreement might exist. For instance, it is difficult to say without inquiry just what behavior is expected of the politician. For the politician is involved in a multitude of relationships, with other politicians, community opinion leaders, financial patrons, spokesmen of interest groups, friends and neighbors, government bureaucrats, and so on. Each set of these others, themselves role-takers in the relationship, may have its own particular expectations as to how the politician should conduct himself. Consensus cannot be taken for granted.

Precisely because role expectations may be widely shared and relatively permanent, they give stability to the relationship. Role relationships thus make for stable patterns of behavior and minimize what would otherwise have to be considered arbitrary behavior. Understanding a role means that we know how a person should behave and what he should do in the performance of a role. This includes knowledge of probable sanctions and thus makes accurate prediction in social relations possible. This ability to predict another's behavior, always, of course, within the limits set by expectations and on the assumption that behavior will agree with the role, permits the partners in a role relationship to shape their own conduct in anticipation of the other's reactions. In some respects, this is a kind of guessing game without beginning or end. The repetitiveness of the game makes for patterns of behavior that produce those uniformities of behavior whose cultural-normative source is not especially felt. . . .

Role analysis aids in discriminating between norms for behavior and actual performance of a role. It may be argued that the best way to identify a man's role is to see how he actually behaves. A role, it would seem, is best reconstructed from performance. But this procedure, apparently so objective, ignores an important aspect of behavior, its meaning. The same bit of action may have different meanings for different actors (and, of course, different observers). Meanings are important in politics because politics is eminently concerned with the consequences of behavior. These consequences require evaluation. Roles as normative expectations of an actor himself concerning his conduct or of others provide meaningful criteria of evaluation that would otherwise remain quite arbitrary. For this reason, the distinction between the normative and behavioral components of a role is analytically and empirically necessary.

Even if there is a wide consensus on roles, there is always a good deal of variation in their performance. This may simply be due to the fact that a role is defined not only by others' expectations but also by an actor's own conception of his role. Admittedly no self-conception of a role can be completely different from the conceptions of others in the role relationship. In spite of differences in behavior, most conduct is recognized for what it is because roles can be identified. But though we may see ourselves as others see us and take appropriate roles, roles are never taken in identically similar ways. The explanation may be that two actors taking the same role may have somewhat different self-conceptions of the role because the others to whom they react are different actors with different expectations. This interpretation remains on the level of social and cultural analysis. And if the deviation is minor, socio-cultural analysis is sufficient. If it is major, it is necessary to find more personal clues.

Actors do bring idiosyncratic perceptions of the interpersonal situation, attitudes, and motivations to a role. Role analysis does not preclude, but may require, investigation of role conceptions from the point of view of the actor's personality. An actor's capacity to take certain roles is predicated on the possession of certain personality characteristics. Just what these are is a subtle problem of theory and research. . . . But whatever hypotheses are formulated about personality in politics, they cannot ignore the wide range and the great variety of possible political roles. It would be quite erroneous to assume a one-to-one relationship between a political role and a given personality type, possibly treating personality as the independent and role as the dependent variable. This would deny the autonomy of analysis on the social level.

Role conflict may stem from various conditions, but two are noteworthy. These may actually be divergent expectations of a person's behavior. A city councilman may expect the city manager to guide and direct the council's legislative business, while another councilman may expect him to abstain from policy recommendations. Or there may be disagreement between others' expectations and an actor's own conceptions of his role. Moreover, the demands made from one role system to another may be so intense that behavior in the performance of various roles cannot satisfy role requirements.

For example, involvement in the life of the Senate may so absorb a member's time that he cannot meet his obligations as a representative of his state. In all of these cases, role conflict is likely to have dysfunctional consequences of either a social or a personal sort. On the social level, certain functionally necessary roles may not be taken. For instance, conflicting expectations concerning the democratic politician's role may deprive a group of strong leadership. On the personal level, role conflict may so disorganize behavior that it becomes highly erratic, irregular, and even irrational.

Study of how role conflict is avoided or resolved suggests a number of possibilities. I shall only list them. First, some roles are more pervasive than others and conflict is resolved in their favor. Second, some roles are more clearly defined than others, which again aids the resolution of conflict in their favor. Third, some roles are more institutionalized than others, leaving the actor relatively little choice. Finally, roles are more or less segmentalized so that, depending on circumstances, even potentially conflicting roles can be taken.

William C. Mitchell

Individual Socialization: Maintaining the Political System

Children are not born democratic nor American; they must be taught the tenets of both throughout their lives. Indeed, if Hobbes and Freud were correct, children are born with egotistical and aggressive drives that require either reduction or redirection and adaptation if society and government are to be realized. And as the demands of democracy are high, the moulding of these assumed hostile natures into effective citizens cannot be an easy one. As Reinhold Niebuhr put the matter, "Man's capacity for justice makes democracy possible; but man's inclination to injustice makes democracy necessary."[1] Thus, the task of making or converting man's "capacity for justice" into concrete motivations and behavior sustaining democratic rule is a major task of the socialization process.

The concept "socialization," an old one in sociology,[2] is of very recent vintage in political science.[3] The notions conveyed by the term, however, are as ancient as Greek philosophy. Indeed, both Plato and Aristotle were greatly concerned over the training or educating of youth so as to preserve the social and political systems of which they sought and defended. The phenomenon of political socialization, then, is an ancient one, but one that has not been adequately studied in contemporary social science.

Socialization, in short, has to do with the civilizing of the members of society. From the political scientist's point of view, the important aspect concerns the creation of citizens by instilling them with the desired ideals and practices of citizenship. Most often, the socialization process is thought of with relation to the young, for it is they who have so much to learn. Yet all members of society are in the process of learning all their lives. One way of getting at the socialization of citizens is to ask a series of logically related questions, which might read as follows:

> What is taught?
> To whom?
> By whom?
> How?
> Under what conditions?
> With what consequences?

---

[1] Reinhold Niebuhr, *The Children of Light and the Children of Darkness* (New York: Charles Scribner's Sons, 1944), p. xiii.

[2] For an excellent introduction to the study of socialization and bibliography, see Frederick Elkin, *The Child and Society* (New York: Random House, 1960).

[3] Charles E. Merriam was responsible for much of what has been done in political socialization. See the series he edited and for which he contributed a volume: *Studies in the Making of Citizens* (Chicago: University of Chicago Press, 1931); and his contribution to *Civic Education in the United States*, Report of the Commission on Social Studies: American Historical Association, (New York: Charles Scribner's Sons, 1934), part 6. Only recently have we seen a revival of interest in political socialization. Cf. David Easton, "The Perception of Authority and Political Change," in C. J. Friedrich, (ed.), *Authority* (Cambridge, Mass.: Harvard University Press, 1958), pp. 170–96; David Easton and Robert D. Hess, "Youth and the Political System," in Seymour M. Lipset and Leo Lowenthal (eds.), *Culture and Social Character* (New York: The Free Press of Glencoe, 1961), pp. 226–51; Robert D. Hess and David Easton, "The Child's Image of the President," *Public Opinion Quarterly*, 25 (Winter, 1961), 632–44; Fred I. Greenstein, "The Benevolent Leader: Children's Images of Political Authority," *American Political Science Review*, 54 (Dec., 1960), 934–43; Lewis A. Froman, Jr., "Political Socialization," *Journal of Politics*, 23 (May, 1961), pp. 341–352; Herman Hyman, *Political Socialization* (New York: The Free Press of Glencoe, 1959); and Robert A. LeVine, "The Internalization of Political Values in Stateless Societies," *Human Organization*, 19 (Summer, 1960), 51–58.

Reprinted with permission of the Macmillan Company from *The American Polity* by William C. Mitchell. Copyright The Free Press of Glencoe, a Division of The Macmillan Company, 1962.

When we have acquired answers to this set of queries, we will have characterized the socialization process.

Before we begin our analysis of political socialization in the United States, it might be wise to identify its functions or contributions to the maintenance of our society and polity. And while it is difficult to specify in exact terms just how significant socialization is, it is relatively easy to state that socialization must not be underestimated as an integrative and tension-management process.[4] In terms of our analytical scheme, political socialization has much to do with shaping the *inputs* of demands, expectations, resources, and support that enter the polity. In like fashion, socialization has a vital impact on the *processes* by which these inputs are converted into *outputs* of power and decisions.

In teaching the young for citizenship we are, in effect, providing them with political motivation, notions of participation, and conceptions of their own role and others' in the political process. In short, we are preparing Americans for democratic citizenship as that ideal is understood in our culture.

Our analysis begins with some observations on the content of political socialization, or, in terms of the above question, "what is taught?"

WHAT IS TAUGHT?

The simple answer to the question posed in the title of this section is *citizenship*. But, correct as that may be, it is really not very informative. What we need is an elaboration of what goes into citizenship in the United States as contrasted to other soceites.

Citizenship is first of all *membership*, meaning a set of actions and a state of mind. In the United States, we teach our people this state of mind, plus a set of appropriate behaviors or actions. They learn how to perform the various roles that constitute the polity and what to expect from others in different roles. On breaking down this state of mind and set of activities into its component parts, we note that the following types of things are taught as citizenship: political motivation; political values; political norms or roles; and political information. Let us discuss these elements of socialization in the order just stated.

Political Motivation

No social system can function successfully unless its members are motivated to perform the various roles that constitute the system. It is possible, of course . . . for a society and polity to coerce its members into action or obedience. But in the long run this is generally thought to be an ineffective and inefficient means of running the society. And in the case of democracies, coercion is expected to be, and is, minimized, for it is rightly considered to be morally inconsistent with the ethical foundation of the system. Democracies, therefore, devote considerable effort to inculcating the proper motivations into citizens so that they will voluntarily perform their functions and roles. Motivation, thus, is really a kind of "motive" force spurring citizens to action. We are not born with political motivation, but acquire it perhaps at great expense to both the individual and society.

. . . .

Political Values

No society is without a set of ideals or "things" valued. Certain objects, states of being, and actions are considered as more worthy to acquire, contemplate, and do than are other objects, states of being, and actions. In the United States, for example, high values are placed upon success, achievement, activity, work, efficiency, practicality, progress, material comfort, science, and, the individual.[5] Other societies, however, including both the advanced and pre-industrial, frequently place high value upon the very opposites

[4]Talcott Parsons, *The Social System* (New York: The Free Press of Glencoe, 1951), Chap. vi.

[5]Robin Williams, Jr., *American Society* (New York: Alfred A. Knopf, 1960), pp. 415–467.

of the American values. Efficiency, for example, is most highly esteemed in industrialized societies, while it is hardly ever heard of in the primitive and preindustrial communities. In India, mysticism is probably still more highly valued among many people than is science. There is, then, a relativity among societal values.

If we inquire into the values held by a society, we will discover that some of these values relate to the area of life we have designated as political. Political values are those having to do with the goals that the society collectively seeks and the means whereby they are sought. Accordingly, political values will be stated in terms of power and authority relationships among citizens and officials or governments, and in terms of goals that the society, as a society, would like to see achieved.

. . . .

Partisan Values

The rather abstract, although meaningful, values we have been discussing are those taught by all the agents of political socialization and which command widespread and deep respect among the American people. Yet, they are hardly the only values taught; we are also instructed in the values of political preferences or partisanship, in the sense of political parties. *We learn, in other words, to be Republicans or Democrats.* And each of the agencies of socialization has something to do with these identifications, and their reinforcement. To be sure, some of the agencies are more involved in this than others, but all participate in some manner in different ways and in different directions, with somewhat different results. In short, no agent of socialization can be neutral in the matter of party affiliation, although some try and others pretend to do so. The schools and the churches are usually among the latter category.

. . .

One learns not only to be a Republican or a Democrat, but also to prefer certain approaches to the operation of government, to prefer conservative or liberal public policies. A citizen's views of government's responsibilities, its role in public welfare, the level of its expenditures and taxes, and its regulation of business, agriculture, and labor—all are matters of preference that must be learned. They are not biologically inherited. And while they may be influenced profoundly by self-interest, self-interest is not something that registers itself automatically upon the mind. It has to be perceived, even if falsely.[6] We learn our party preferences and policies, in short, through experience and the conscious teaching and appealing of others. To this end, every agent of socialization participates and, in some cases, competes, as is the case with the parties and interest groups. Obviously, the agent who gets to the child first—the family and the father, in particular[7]—has the greatest advantage. Because of the family, as most voting studies confirm,[8] most of us become "little Republicans" or "little Democrats" at a very tender age. We have no awareness of why or how, of course, except that Dad always says nice things about one of the parties and not the other, or at least not as often. Consequently, every primary school teacher can predict party preferences of the parents of her students with considerable accuracy by the simple test of identifying the preferences of the children. Unlike their parents, the latter normally suffer no inhibitions about the public expression of preferences.

Political Norms

. . . With respect to the sources of the norms, we must distinguish between those norms which

---

[6]Talcott Parsons, "The Motivation of Economic Activities," in William C. Mitchell, *Essays in Sociological Theory* (New York: The Free Press of Glencoe, 1954), pp. 50–68.
[7]Herbert Hyman, *Political Socialization* (New York: The Free Press of Glencoe, 1959), p. 69.
[8]Paul F. Lazarsfeld, et al., *The People's Choice* (New York: Columbia University Press, 1944), pp. 140–45; Angus Campbell, et al., *The Voter Decides* (Evanston, Ill.: Row, Peterson and Co., 1954), pp. 97–107, 199–206; Bernard R. Berelson, et al., *Voting* (Chicago: University of Chicago Press, 1954), pp. 88–92; Angus Campbell, et al., *The American Voter* (New York: John Wiley and Sons, Inc., 1960), pp. 146–49.

are learned and have first application in the primary groups and private associations, and those which specify behavior in the nation-state. The norms of behavior that apply to the family, the neighborhood gang, the peer-group at school, the trade union, the corporation, and other associations may not be the same as the norms that apply in an adult's role as a citizen. Whether they do or not is, of course, an empirical matter; for the moment, we simply call attention to the distinction. Another distinction pertains to the functional consequences of these various sets of norms. For while all norms have consequences for both the allocative and integrative processes in a society, individual and certain other types of norms may have greater significance for one of these processes than the other. Thus, while the norms of "fair play" have their greatest consequences for the allocation of values, norms of loyalty are most relevant to the integrative problems of society or its solidarity.

Let us start at the beginning of the norm-learning process, in the primary groups and secondary associations. It is in the family and at school among the peer-groups that children learn their first norms of behavior, or what is considered proper and wrong.[9] They learn the rules by which values and costs are distributed, and the norms of loyalty to the group. In other words, the child at this time has his first experience as a "citizen" in a group, learning his roles, and learning about loyalty, conflict, leadership, power, and authority. He is participating, in fact, in a small-scale social system with a polity, an informal one to be sure, but a polity nevertheless. And, if Freud is correct as to the enormous importance of childhood to later adult behavior, we must attach considerable meaning to these early experiences as a "citizen" for later adult citizenship. But what, in the sense of political norms of behavior, is learned at this stage?

One of the first norms taught to every child, whether he learns them or not, specifies that he control his *demands* upon others, that is, that he not be selfish. The norm of self-discipline, of course, is a rather diffuse and generalized one, and not always easy to define. Indeed, most of the parent's problems stem from trying to get the child to be less self-oriented, urging him to think of others before making demands or holding expectations of them. The parent usually tries, then, to control the demands made by the child, just as does every other group with which the child becomes affiliated. The interests of the group, or the public interest, is thus experienced and learned.

Along with learning the need for self-discipline, the child is subjected to norms relating to the *allocation of resources* within the group. He learns that he is expected to contribute his time, energy, and whatever other resources children can produce to keep the group functioning. This may mean providing sports equipment, marbles, a play-space, and the like for the other children. In older groups it may mean holding office or doing any of the numerous tasks that all clubs require in order to survive. The type and quantities of resources are matters for each group or organization, but all require some kind, and each member must learn what they are as well as the expectations of others regarding resources.

The last set of norms learned concerns his *support* for the group: what to support, how, and how much. Generally, this set of norms defines patriotism or loyalty. Thus, the child learns in some families that loyalty to the family is the alpha and omega; that all internal squabbles end at the front-door; and that the family present a united front to the world. In the gang, the boy learns not to "rat" on his buddies and to defend them at all costs. Indeed, every primary group and most secondary associations have some definition of treason and minimal support, even though we seldom think of these more casual activities in such terms.

---

[9] Jean Piaget, *The Moral Judgment of the Child* (New York: Harcourt, Brace & World, 1952), pp. 19–49.

Political Information

Historically, one of America's most cherished ideals has been that of a highly informed nation and electorate. To this end, we have devoted a considerable amount, although not percentage, of our national resources. Over 40 million students are enrolled in our colleges and secondary schools. And, today, more than 97 per cent of the people can read and write.

To a large extent, the purpose of this educational faith has been to prepare citizens for a democratic way of life. In our conception of that way of life we have, traditionally, placed great faith in the efficacy of political information of the factual sort, that is, the learning of countless facts about our history and formal political institutions. There is great faith in the rational solution of social problems by people who must know what they are supposed to be doing. In America, this means reciting facts, not philosophy. Consequently, the required courses for our students are more often American history, American government, or a civics course, than they are courses in logic, ethics, or psychology. It is widely assumed that one can attain an understanding of our politics if he memorizes legal facts about the government. For generations, primary and secondary students, especially, have thus committed to memory such things as the number of states, the size of Congress, the Bill of Rights, the legal procedures of the legislatures, the sizes of city councils, the dates of Civil War battles, and the Gettysburg Address. Presumably, these facts equip one to be a useful and appreciative citizen.

According to the above conception, political information is the key to active mastery of the political world. Without it we cannot be good citizens. Nor must we terminate our studies at school graduation, for facts are constantly changing and the citizen must keep up with the news. Thus, we are exhorted to keep informed about both national and international events. To this end, literally dozens of organizations, programs, and events are sponsored daily. Thousands of radio and television stations, newspapers and magazines, books, and special publications are distributed or broadcast to keep a flow of news to citizens. In short, there is no paucity of a certain kind of political information.

STUDENTS OF CITIZENSHIP

We observed at the beginning . . . that everyone in society from the youngest to the oldest, from the poorest to the wealthiest, is being continuously socialized. Not all, however, are being equally socialized in politics. In general, it may be asserted that children, having the most to learn about everything, are engaged more continuously than their elders in learning how to be members of society. Yet the question is moot as to whether they acquire most of their *political* awareness, information, norms, and values during the younger years. No doubt they do learn a great deal. Nevertheless, much of what they learn in the family and schools is often in for some sharp changes as they mature into adult citizenship. Early beliefs about authority and power may well be influential throughout one's life, but practical experience as a citizen with the act of voting and dealing with political officials may also provide some new perspectives. In any case, a person learns about politics all his life, though certain early orientations may form the basis of all that is later selected from the environment.

Among those who learn most in adult life about politics are those who actually participate in politics, either as party workers or as politicians. Learning then frequently becomes an unlearning of much that was formerly uncritically accepted. Men who become politicians, for example, must discover how to campaign, run their offices, and deal with the public, press, and other politicians. For some men such learning is one of the more significant experiences in life. Moreover, for those who had highly unrealistic images of political life, a major reorientation in values and perspectives is a definite possibility.

THE AGENTS AND MEANS OF SOCIALIZATION

No one is ever solely self-taught; teachers are always present in the learning situation. This is no less true in the political system than elsewhere. Becoming a citizen, a partisan, a government official, or a member of an interest group, all requires instruction from others even though it be informal and hardly a conscious or deliberate activity on the part of teacher or student.

Schools and Formal Education

As might be expected in a complex society the size of the American, the agents of political socialization are many. For convenience we might approach the matter by first classifying the types of agents as either formal or informal and organizational or nonorganizational. Societies almost never allow the general socialization process to take place without having developed a set of formalized or highly institutionalized agencies of education. Thus, in all industrialized societies, a school system is a conspicuous element. And, from our point of view, the most important aspect and function of this school system is its role in making citizens out of the children entrusted to it. Indeed, the long history of public education in this country partially confirms the idea that schools are considered to or actually have a prime responsibility in citizenship training.[10] The elementary and secondary schools provide the most dramatic instance of political socialization, but hardly the only one. Universities and colleges also perform functions along these lines, and particularly state universities, which are required by state law to teach a certain minimum amount of political science or American history to the students. Another aspect of education in politics and the obligations of citizenship can be found in the requirement at many state universities that military training be a part of the male student's preparation for serving his country. Yet the colleges and universities of the country tend to treat the socialization process somewhat differently than do the primary and secondary institutions in that *knowledge* of politics is stressed rather than values and norms. But even this statement must be qualified, for the values and norms are more or less assumed to be democratic. Few if any professors of history, political science, or any of the social sciences, for example, ever deliberately educate students to believe that systems other than democracy are worthy. Fascism, communism, and varieties thereof are almost never taught as exemplary models for Americans. And while information about them may be conveyed, it is usually accompanied by condemnations. Indeed, the higher institutions of learning appear to operate on the belief that truth leads to greater faith in democracy. In any case, these formal institutions are important agents of political socialization in terms of developing citizenship. They are also important in another regard suggested above: that of equipping students with the knowledge and skills to become not merely citizens capable of performing the minimal duties of citizenship, but citizens capable of developing into future leaders for the polity. The fact that most of our leaders are college educated should come as no surprise.[11] Moreover, it may be assumed that many of their skills received their first sharpening during college years, either in the classroom or in campus politics. The law schools, too, seem to play a profound role in the preparation of political leaders.

. . . .

---

[10] V. O. Key, Jr., *Politics, Parties and Pressure Groups*, 2nd ed. (New York: Thomas Y. Crowell Co., 1948), especially chapter xxi; and V. O. Key, Jr., *Public Opinion and American Democracy* (New York: Alfred A. Knopf, 1961), p. 315.

[11] Cf. Donald Matthews, *The Social Backgrounds of Political Decision-Makers* (New York: Doubleday and Co., Inc., 1954), pp. 28-29.

The Family

No one would seriously disagree with the proposition that the family is the major agent of socialization in all societies. But while we are all familiar from personal experience, if not from academic study, with the role of the family in creating civilized beings, many of us, including political scientists, have forgotten how vital is the family in creating citizens. Our knowledge of the processes involved is not extensive, but there is reliable data on certain aspects of the process and particularly on the content of what is taught by the parents to their children. We will consider this data shortly. For the moment, we are simply concerned with the family as an agent of political socialization.

The family is apt to be crucial in the entire socialization process for the simple reason that the child spends most of his more formative years within its orbit. It is here that the child has his first experiences with power and authority. And authority, as we have indicated, is fundamental in the polity whether it be in a family or a nation-state.

One of the better-known facts about the United States, for example, and one which distinguishes it from many other societies, is the looseness of the kinship system and the permissiveness with which children are raised by their parents.[12] The father is not usually a powerful authoritative figure either toward the wife or the children, as is the case in, say, Germany. In many situations, the whims and wishes of the child are indulged to the extent that the child learns little if any form of self-discipline. All this is done in the interests of democracy and personality development in which expressiveness and adjustment are regarded as prime virtues. At one time, "children were to be seen, not heard"—a maxim that has disintegrated today with a vengeance.

The child growing up in such an environment sees power diffused among at least three persons: mother, father, and himself, and, perhaps, other children. Because he sees the mother on a more or less equal plane with the father as an authority, and his own wishes taken into account, he is likely to learn not only that power need not be respected, but that indeed it may even be manipulated.[13] Thus, he soon learns to play the parents off against one another and thereby to gain his own ends. He will also note that he may be used by one parent against the other, and appreciate the bargaining strength he has acquired. All this means that the child views power not in mystical terms and unquestioned obedience, but as a pragmatic thing with which one can bargain. Thus, rationality of action and skepticism about power-holders are encouraged. It should also be noted that these two approaches to power are further stimulated by the practice of justifying the exercise of power by appeal to reason, either to the child or in front of him. When the child is expected to do some chore, for example, he is generally *asked*, not ordered, and provided with reasons for so doing. American children soon learn, therefore, to demand "why" when commands are given; likewise, parents soon learn to ask or request, not order.

The child learns not only how power and authority operate within the kinship group, but also acquires certain attitudes toward power-holders outside of the family unit. Thus, while parents and other elders may not be politically-oriented, this attitude in itself, plus the comments on such authorities as the father's boss, the police, the politicians, and the bureaucrats, are apt to become internalized by the child as his own set of attitudes. Especially important in this regard are the stereotypes of such persons and the manner in which they are discussed before the child. The words "copper," "flatfoot," "politician," "boss," "political deal," and many others are not apt to convey favorable impressions of public authorities.

---

[12]Sister Francis Jerome Woods, *The American Family System* (New York: Harper and Brothers, 1960), pp. 108–110.

[13]Erik H. Erikson, *Childhood and Society* (New York: W. W. Norton & Co., Inc., 1950), pp. 273–77.

Likewise, the behavior of parents before authorities will impress the child in striking fashion. Should the father talk back to the policeman who has flagged him down for speeding or should he defer to the officer, the child may respond in similar ways himself as an adult. The respect shown by a parent for public property is also preparation for the adult civic life of the child. These, then, are illustrations of a basic orientation that children learn from the parents, an orientation concerning one's rights and responsibilities as a citizen. In short, whether a child grows up to emphasize rights and demands on the government or to fulfill his own obligations is surely strongly influenced, if not actually determined, by early family experiences.

. . . .

Youth Organizations

Although American youth organize themselves in both formal and informal groups, they are also organized by their elders into groups having considerable relevance for political socialization.[14] Usually these organizations are set up for more general socialization, but all inculcate political values, norms, and information. For some, moreover, it is their *raison d'être*. Organizations such as the Boy Scouts of America, Girl Scouts of America, Future Farmers of America, and 4-H Clubs of America are all generalized socializing groups, but each devotes much time to matters of political relevance. Other groups, including the Young Democrats and Young Republicans, are purely political in orientation. Still others, including the American Legion Baseball Leagues, are meant to provide recreation, attempting in the process to instill values and norms such as "competitiveness" and "good sportsmanship" that may serve to condition later political values and norms.

Each of these organizations inculcates ideal notions of citizenship and indoctrinates American values. Dominant values include: thrift, respect for private property, self-control, self-reliance, duty, good deeds, clean living, courtesy, productivity, and reverence for religion. Basically, these are the traditional values of the American middle class, and as such serve to rationalize and defend the status quo. In addition, the youth organizations also support the existing order by isolating youth from political conflict. American youth seldom learn or engage in the extreme forms of political behavior typical to the young of other countries. Nor are the youth of this country an active force in politics; their attentions are devoted rather to preparing for an occupation, and having a good time in the meanwhile—orientations that are supported by youth organizations.

. . . .

Political Parties

Most discussions of the functions of political parties suggest, if they do not prove, that parties serve to educate the citizenry.[15] How well they accomplish this function is debatable, but surely it is true that the parties do attempt to educate the citizens in their voter's role. Indeed, parties devote enormous amounts of time and energy, as well as other resources, in trying to communicate with the voters. And while their objective is obviously a self-oriented one—to win elections—they also convey information in the process about issues, candidates, and the parties. They also disseminate norms and values for the voter to use in his own process of evaluation. And they generally exhort the voter to vote regardless of the direction of his vote. Thus, by encouraging citizens to take advantage of one of their basic rights, voting, they thereby contribute to the polity.

Parties in the United States are agents of citizenship only in a partial sense in that they do not exist to socialize, nor do they "teach" continuously. Rather, their actions are likely to be peri-

---

[14]Bessie J. Pierce, *Citizens' Organizations and the Civic Training of Youth* (New York: Charles Scribner's Sons, 1933).

[15]For a representative statement, see Dayton D. McKean, *Party and Pressure Politics* (Boston, Mass.: Houghton Mifflin Co., 1949), p. 25.

odic or cyclical, mounting in frequency and intensity as elections approach, then suddenly diminishing and leveling off during the long interim periods. In short, they do not directly impinge on many of the citizens for any sustained period of time. Then, too, the parties are less concerned with the provision of factual data about political processes than with providing evaluative clues or perspectives. They train in partisanship.

But the inculcation of partisan values and information is not the only contribution made to political socialization. The American parties also teach and preach the superiority of our way of life. Both parties are nationalistic and both honor the political system that created and maintains them. Thus, even while the parties may each have their partisan emblems and other symbols, they also have national heroes, display the national flag, sing the national anthem, pray for America and divine guidance, and claim to speak for the United States. Their candidates do not claim to be just Democrats or Republicans, but Americans first and party members second or third in the hierarchy of loyalties. In such ways do the parties socialize both for partisanship and integration.

The ways in which the parties accomplish these functions is, as was said above, somewhat unsystematic and erratic. Some are accomplished through the formal methods of campaigning and the distribution of literature. Much, however, is done informally and symbolically by the actions of office-holders. Moreover, each party is organized at all levels of government, but not evenly throughout the nation. Services are provided by party officials ranging from ticket-fixing to legitimate help during crises. Many social events are utilized by parties, especially around election time, to educate the voters and maintain supporters. Clubs of all sorts are used to spread the word. To reach the young, each party has a Young Democrat or Young Republican affiliate, as the case may be. Thus, conscious teaching and learning from experience in internal party activities serve to socialize those who have contact with the parties.

Interest Groups

Interest groups, like political parties, are also primarily concerned with "educating" the adults of the community.[16] And, like the parties, they are mostly interested in providing evaluative cues to their members and the publics outside their domain. As interest groups want to realize specific goals, usually to augment their own positions in society, they too teach value positions. What information is provided is dealt with as it relates to the goals or objectives of the group. Thus, they maintain bulletins or newspapers with political sections or columns to keep members informed on the group's political action. Likewise, they publish and distribute special studies having to do with political affairs, and often engage speakers from public office and universities to inform them on political issues. Indeed, campaigns may well be run to encourage their members to participate in politics. To this end, books and lectures are provided to show how such activities can be accomplished. Apparently, the encouragement of political action by members is becoming increasingly popular. These groups, of course, have always engaged in politics, but have deceived themselves that only their opponents do so. Some even run schools as do the trade unions with COPE, or Committee for Political Education.

Interest groups cannot help but be agents of political socialization. They become so simply by being members of the political system. What distinguishes the modern interest group from the old is the self-conscious manner in which the former approach their activities. They have become highly specialized bureaucracies in socialization. In still other words, interest groups are rationalizing those aspects of political socialization in which they are interested. Note the programs they conduct to "educate" the citizenry or public opinion to aid their cause. To accomplish the task, highly skilled technicians in mass

---

[16]David Truman, *The Governmental Process* (New York: Alfred A. Knopf, 1951), pp. 213-61.

The Mass Media

In one sense, the mass media are simply a technical means of socialization. To treat them only as such, however, would be highly superficial. Those working with the mass media are, in fact, agents of socialization who have particular ideas of what should be conveyed to their consumers. In order to appreciate those in the mass media as agents of political socialization, we must then have some knowledge of who they are, of what their interests consist, and how they operate. Fortunately, students of public opinion have been quite diligent in collecting such data, and it is to their work that we will turn for most of what we want to know.[17]

By mass media we mean to include television, radio, newspapers, magazines, and books. These are the media that reach or are intended to reach large audiences, although considerable differences do exist in the sizes of the various audiences of each. Television, for example, influences many more Americans than do books. But whatever the differences, the numbers involved in even the smallest sphere are sizable.

. . . .

THE PROCESS OF SOCIALIZATION

One of the less known areas of political socialization has to do with the *processes* of teaching and learning. We can only convey impressionistic notions of the techniques or methods and informal processes.[18] If they should sound familiar to the reader, we shall have at least partial confirmation of their reliability.

How a person is taught and learns citizenship in a democracy is not likely to be a very clear picture, for many influences impinge upon the individual, and not all may be equal in their impact. Only one thing is certain: namely, that all of us are subject to both formalized teaching about politics and informal pressures to perceive and act in certain ways. No one group or institution employs a single procedure; all engage in many methods. Thus, the schools formally teach history and civics and engage in rituals of respect for the nation. The teachers themselves act informally as symbols of power and authority for the students and even their parents. And, in the schoolroom, the teachers—unknowingly in some instances—convey political preferences via verbal and nonverbal expressions. In other cases, they do so knowingly and openly. On the school grounds, the child is socialized by the peer-group with ideals of group behavior, and learns notions of authority, leadership, loyalty, bargaining, and conflict. All these "lessons in citizenship" are taught and learned informally without awareness of the process, as such, to its participants. Later in life, the adult citizen will have his political preferences constantly confirmed by the groups to which he belongs, even though he has no awareness of the acts of confirmation. And, he may confirm these same preferences by an unconscious selective choice of the mass media.

Although political socialization is considered of paramount importance in this society as in all societies, it is a highly confused and unsystematic process. Both formal and informal techniques are employed by all the agents and agencies, with greater reliance placed on one or the other. The schools, for example, tend to formalize their methods and become very self-conscious about education. The family, on the other hand, socializes by example and imitation, both primarily informal and even unconscious means.

---

[17] An excellent reader in public opinion is Daniel Katz et al. (eds.), *Public Opinion and Progranda* (New York: Holt, Rinehart & Winston, 1954). The primary professional publication is, of course, the *Public Opinion Quarterly*.

[18] Undoubtedly, we know more about political socialization in totalitarian than in democratic societies. Hardly a volume on the former is without chapters on schools, propaganda, and indoctrination, while very few books on American *politics* treat the same institutions and phenomena.

Charles B. Hagan

The Group in Political Science[1]

I

One of the first conditions for progress in a particular "discipline" of scientific inquiry is the ability of its practitioners to communicate with each other. Their ability to do that, in turn, depends upon at least these two things: an agreement among them upon what they are studying and an agreement among them upon certain categories of description, in accordance with which they can sector and investigate their problem.

For example, despite the many disagreements among economists on such questions as whether the progressive income tax is a good thing or whether we are presently in a "recession," they are all agreed that *the* problem of economics is "the allocation of scarce goods," and they are all agreed that the basic system by which those goods are allocated is the interplay of two forces, "supply" and "demand."

In his recent inquiry into the state of political science, entitled *The Political System*, David Easton raised this question: Do the extremely variegated and multiform writings that make up the literature of political science have anything in common that justifies their being placed under the same taxonomic tent? He concluded that they do, and that what they share is a concern with the same basic problem. That problem he defined as "the authoritative allocation of values for a society." He did not give a complete definition of the term "authoritative," and, indeed, it is probably impossible—and unnecessary—to do so.

Easton has come close to making articulate the common preoccupation of the polyglot "group" that marches under the banner of "political science." Some may, in research and writing, be working at the margins of the problem rather than at its center; but Easton has shown that all are working at the same problem. It is surprising that the core of concern should not have been made clear sooner, but Easton has come close enough to the mark to assist further analysis.

The most cursory survey of the literature that appears in the political science periodicals, however, reveals an immense *lack* of agreement on the matter of *how* to go about studying Politics. It is probably not too much to say that any essay that some political scientists regard as really significant and first-rate work hardly communicates at all to most political scientists, and an essay that most political scientists understand and approve is regarded by some as unimportant. This inability to communicate stems from a lack of fundamental units of description. Easton has shown that there is a common core of concern, just as there is in economics. But there is no agreement upon a system of description, like the economists' supply-and-demand system, to explain *how* values are authoritatively allocated for a society.

Some of the articles and essays can communicate only to fellow practitioners in some narrow segment of the total discipline, while other pieces are of such general character that they mainly pass on lore, to borrow a term from Charles Hyneman's memorandum on the subject matter of political science. The political science of a half-century ago undertook to spell out in considerable detail the legal structure of society with especial emphasis on those aspects that related to the election of officials and the operation of the various governmental agencies within themselves and in relation to each other. The development of the political party literature pushed the margins

[1] This essay is a modification of a paper prepared for a Conference of Political Scientists at Northwestern University, June 15–19, 1954.

Reprinted from Richard W. Taylor, ed., *Life, Language and Law: Essays in Honor of Arthur F. Bentley* (Yellow Springs, Ohio: The Antioch Press, 1957) pp. 109–24. By permission of the author and the publisher.

of inquiry out into new areas of investigation. Morality and ethics were proper questions of inquiry and logical analysis. As the probing continued the categories of explanation were widened to incorporate the data experienced in the research. On the whole the categories were able to swallow the data. Occasionally, on a second look from a different vantage point, it seems as if the researcher sometimes missed the point of his findings. That is merely one way of saying that a different frame of reference would have yielded different results.

The purpose of this paper is to suggest a descriptive system, which, properly understood and employed, can be as useful—*and* as uniting—for political scientists as the supply-and-demand categories are for the economists. This system of description rests on group activity and is most simply stated thus: values are authoritatively allocated in society through the process of the conflict of groups.

II

Contemporaneously with the elaboration of the traditional categories there appeared alternative suggestions for ordering the materials relevant to political life. The new materials came from a variety of sources. One of the books that seriously challenged the usual description of the ways of political life was Lincoln Steffens' *Shame of the Cities*, a famous muckraking study. For present purposes there are two comments to be made about that study: 1. the account that it gives of the operations of the political system differs greatly from other accounts to be found in the social sciences, and 2. the labels that social scientists attach to situations described exhibit a moral indignation pointing clearly to inconsistency between the expected patterns of behavior and the ones that are found. An observer of Steffens' type, on the other hand, found enough material to show that his account probably depicted more accurately the routine activities of political life than did the customary descriptions of the orators and genteel writers. At any rate political science has never recovered its confidence in the traditional descriptive categories.

The findings of the muckrakers were supplemented by studies of psychologists and sociologists. These people opened up new frames of reference for individual and social analysis. Their studies have influenced the categories of political science in the same way that "creeping socialism" is alleged to have influenced private enterprise. Often their techniques and results have been grafted on to the traditional categories. Some readers have found the results offensive to their sense of order and discipline, and there has been a search for new systems which provide more order and more system. It may be doubted that either of those results has been achieved. As evidence of that conclusion look at the categories that Stephen Bailey borrowed from Pendleton Herring and used in his study *Congress Passes a Law*.

These categories are to be found in Herring's *Politics of Democracy*[2]: "It is enough if we achieve a working union of interests, ideas, institutions, and individuals." There is little indication of how a working union of these four I's is to be achieved. Mr. Herring's prescription is tolerance, and that seems to accord more with an ethical prescription than with a scientific frame of reference. Neither Herring nor Bailey make any effort to show how you commingle those items; nor is there any effort to show how one of the I's differs from another. Most studies in political science make little or no effort to fix with any precision the range of meaning of the basic descriptive system.

The traditional categories as they play their role in contemporary descriptions of politics leave a lot to the imagination of the reader, and as a result the communication process is not always precise. The four I's of Herring and Bailey are close enough to the traditional categories to illustrate this thesis. How does an interest differ

[2] E. Pendleton Herring, *Politics of Democracy* (New York: W. W. Norton & Company, Inc., 1940), p. 421.

from an idea, and how do you tell when one is operating rather than the other? The same questions can be made about the other two categories: the institution and the individual. The answers will be just as ambiguous. These categories are usually mixed in with the notions of a legal system as formulated by John Austin or Westel W. Willoughby. Lord Bryce's introduction of the notion of Political sovereignty saved the descriptive system for a bit longer. At least that was the source of most descriptive apparatuses until recent years. It is difficult to fit the materials which pass for *facts* among Political Scientists into those categories without doing violence to the legal system or to the facts.

Now, leaving the communication process to the imagination of the reader has several "virtues." In the first place it allows the writer to avoid the spelling out of his meaning in more precise terms; in the second place it allows the reader to supply his own meaning; in the third place it allows both to agree on one another's brilliance; and finally it avoids controversy. These are "valuable" traits, but it may well be doubted that they have any place in a scientific descriptive system. The four I's comprise the same data viewed from slightly different angles of emphasis. *Idea* emphasizes the talk and writing facet, *Institution* emphasizes the customary modes of acting, *Individual* emphasizes the part that the physical entities play, and *Interest* usually emphasizes or explains the behavior not accounted for on the other principles. In every instance the phenomenon to be explained is a political decision of masses of people, and somehow the decision gets made. It is not clear how the *Idea* could "cause" anything to occur or exist apart from the activity of the participants in the decision. An *Institution* is more easily understood, perhaps, as expected activity of participants. And *Individuals* have meaning only as participants in the decision and only that part of them that operates in the decision-making is relevant to the analysis. *Interest* is likewise the activity of the decision-makers either pro or con the issue. Its meaning is found only in the activity. In fact the other I's get their meaning from the interest or activity that is to be explained.

While it is dangerous to reason by analogy, the danger may be risked briefly. The point has been made earlier that Easton sees the problem of political science as the authoritative allocation of values. To examine the equivalent problem in economics: the test of whether a study is an economic study or not is does it concern itself with the allocation of scarce resources. The device used by the economists to explain how such resources are allocated is the supply and demand situation. All economics, Professor Robbins has said in his *The Nature and Significance of Economic Science*, can be subsumed under those rubrics. The most recondite discussions in economics deal with these problems in more or less clear terms. The higher reaches of contemporary mathematical exercises in economics as well as the most traditional discussions are all concerned with the same issues. It may be difficult to recognize the relations but they are always there for the economists.

The analogical need is then for a process of allocating *authoritative* values. That is if supply and demand allocate scarce resources, there is a corresponding need for a process of allocating authoritative values. This process is the group struggle. In continuing the analogy the political scientist should search for the groups that are contesting one with the other for the values that are in controversy. The one side is seeking to gain its goals and the other is seeking to modify or prevent such goals. The notion of a struggle for conflicting goals is not new, but the implications that it carries are not always clearly envisioned. The next section undertakes to formulate some of the implications in a clearer manner than is to be found in most of the existing literature.

III

There are two distinct strands in the discussions which use the group as the explanatory principle in political science. The first strand still is found primarily in the writing of A. F. Bentley. He introduced the phraseology of group interest or group or interest (these are synonyms) in his volume *The Process of Government* which was published in 1908. He carried the *analysis* a little further in his volume *Relativity in Man and Society* in 1926. In all of his writing Bentley has been concerned with what might be called semantic or epistemological problems. He has tried especially to elucidate the ambiguities that are found in the discussions of social activity.

There have been few, if any, consistent followers of Bentley in political science. Two recent authors have admitted their great debt to him. David Truman in his volume entitled *The Governmental Process* and Bertram Gross in his *The Legislative Struggle* are clearly working in the direction suggested by the earlier Bentley volume. Truman has sometimes deviated from the requirements of Bentley's "tool," but this is done deliberately and with critical acumen. Gross adheres more closely to the framework suggested in *The Process of Government*.

A second strand of writing in political science has also utilized the group as an explanatory principle. This strand has many variations and many practitioners. In these studies the group is used to supplement the individual as an explanatory phenomenon, and both of these in turn are used to supplement ideas and institutions as explanatory principles. These categories take on a real existence or in other language they are reified. In more abstract expositions they may even have essences. These "objects" would seem to have careers of their own and are capable of producing results. And as has been stated earlier these "objects" or essences intermingle one with another to cause the results in politics. The intermingling can apparently be accomplished in varying degrees, so that a given political result may be ten per cent ideas, twenty-five per cent individual, thirty per cent institution, and thirty-five per cent interest or any other mathematical arrangement which adds to a unity of 100 per cent. The writer has never seen any study that undertook to establish the percentages, but it is the clear implication of such approaches that it can be done. Another variation of this approach is to allocate the explanatory principles to environment, technology, education and other such variables. The latter do not necessarily involve a group base but these variables are frequently interjected into studies allegedly proceeding on a group base.

As an aside to this general discussion attention may be called to the explorations of the sociologists and psychologists. These studies to which Homans' *The Human Group* is a recent addition are undertaking basic investigations into small face to face groups. These clearly have potential implications for a political system, but as yet the practitioners are extremely modest in their claims. It seems evident that political scientists will have to do just such detailed investigations before there can be much advance in theoretical sophistication. However, it should also be pointed out that some of these inquiries are operating on an epistemological basis similar to that of the second strand of political studies that are indicated above. In short there are implicit or explicit notions of the group as a real entity rather than as an hypothesis for organizing data.

To make clear the basis for these remarks it may be necessary to state a few premises. All verbal formulae are treated as principles in a system of description. The system may be implicit or explicit to the author. The task of the formulae is to organize and manage the data which the author finds it necessary to manage or organize. The system of description or theory may more or less "explain" the findings of an empirical character. If the explanation "works" it is useful, and if it doesn't it is inadequate. On the other hand those authors who reify the phenomena that they are studying expect their verbal formulae to have some kind of correspondence to a world external to themselves and independent of their mental constructs. In this framework "facts" exist and the function of a theory is to tie them together. If the theory does not possess this quality of "correspondence" it is wrong rather than, as the present approach would describe it, inadequate.

So far the effort has been to provide an intellectual setting to enable further analysis. One more warning and it will be possible to outline the political system of the first strand of writers about the group. It has been stated here that the second strand of writers reify the group. By this it is meant that the group is somehow imagined as having an existence independent of its surroundings. The group is external to other groups or to situations or to its environment. It, the group, is an aggregation of individuals who operate as a unit. There is implicit in this analysis or approach an acceptance of traditional notions of objects. A group, so to speak, is an enlarged individual. Sometimes the discussions proceed as if the group had a brain and physique like that of the individual, the physiological entity. The assumptions that are implicit are seldom elaborated or explored, but when they are, there are qualifications made to take care of the physical aspects of the analogy. The English language, it may be added, is so deeply committed to this reification process that it is almost impossible to escape from the consequences. If this discussion does so, it will be a major accomplishment. In order to move into an understanding of the alternative hypothesis about groups one needs a sort of brainwashing.

The group theory of the writers who follow Bentley starts from quite a different epistemological base. It is closer to Dewey's epistemological notions, and, of course, its record can be traced in philosophical literature. Its pedigree is long, but it is not as prestigious in contemporary society as the other. Perhaps the best way of conveying the basic supposition is to start with the mass of human activity out of which is to be abstracted that which is relevant to the authoritative allocation of values. As to how one knows what is relevant that is a matter for investigation and research. The method of study would be the same as now, but the emphasis of explanatory principles would be different. It will be necessary to return to this later.

Now what is this mass of activity? It includes all of the talk and writing and all of the public meetings as well as all of the physical violence that ever takes place. The relevance of particular phases of the activity and its meaning or significance in any political problem is a matter to be determined. It is determined by the use of principles or hypotheses to explain relations between the activities that are under study. The problem for political science is to arrange the activity on one side or the other of the question that is being investigated. For example, the proposal to alter the term of the members of the House of Representatives to four years can be explained in terms of the groups pro and con the proposal. There will be a wide variety of activity pro the proposal and a wide variety con the proposal. The activity is activity of human beings, and it may take the form of writing abstract treatises on the desirability of long or short terms for such offices. It may take the form of labor unions adopting resolutions, of political parties adopting platforms, or of conversations between friends enjoying a social evening. It may take the form of providing transportation to the polls on election day and in some instances stuffing the ballot box. It may be activity engaged in keeping the bars open or closed and fist fights or riots. In short all activity that can be tied to the proposal by one principle or another is political activity and it is activity by a group. The political meaning of the activity is to be found in its relation to the proposal to change the term of the Representative in this example. Other aspects of the activity may also be tied to other political goings on, but that would involve organizing the activity around another issue. The distribution of the population on one issue may be quite different from that on another. It is extremely unlikely that any two issues would divide the population exactly alike. Also, it is unlikely that at present there exist any methods of counting that can give more than a very approximate notion of the distribution.

The groups in the above illustration are those activities of individuals which support or oppose the matter at issue. It is important to grasp the notion that an individual in his role on the issue may play more than one part. The individual, the physiological one, may participate in a number of organizations or associations without formal organization. David Truman has called this phenomenon "overlapping memberships." Each organization or association may play a role in the process of making a governmental decision, and their roles may be contradictory. It is a matter of common observation that not all members of an organization support the program of the organization with equal zest. The phenomenon must be accounted for in any adequate theory of the political process. The variation in the amount of activity of any organization on different issues is a manifestation of the differences. In short, to use the terminology of Bentley, the organization is representative of the activity and is itself an activity. The activity *is* the interest. The interest is a short hand way of saying there is a mass of activity operating in a given direction. The point here is that the terms, individual and interest, which have been used above as illustrations in the reified type of analysis can be given a function in group analysis. The individual *is* his activity and his acts are the manifestation of the interest. This definition of individual includes all that is useful in the explanation of the political process and leaves out the physiological aspects which are of no importance in explaining the political process. If one wants to say that without the physiology there is no individual, there is no objection of course, but physiology does not aid in explaining politics.

Interest is often utilized to designate a kind of activity which follows as an inference from a premise. Thus one finds statements in the literature that it is to the interest of labor unions to oppose legislation prohibiting union contributions to political parties. If, on examination, it is discovered in the Congressional hearing that the unions were unopposed to such legislation, one could say that the unions acted contrary to their interest. That kind of explanation is often found in the literature. One wonders what kind of a science it is that finds activity contrary to its explanatory principles. The present assignment of meaning to interest is to find the activity and to call that the interest which the activity connotes. Interest is *a posteriori* not *a priori*, and it is consistent with the observed behavior and not contrary to it. In another way of speaking, the interest and the group are the same phenomenon observed from slightly different positions, and interest group is a tautologous expression. The interest is not a thing that exists apart from the activity or that controls activity. To make interest a phenomenon external to the behavior of the group is to reify interest and to direct attention away from the requirements of competent investigation.

The other two explanatory principles, institutions and ideas, gain meaning in this type of analysis. An institution is a way of action, and the action is clearly that of individuals. An institution is not a control outside of the persons who are acting, but *is* their acting. A constitution is often defined as an institution controlling the behavior of those who live under it. Such phraseology reifies the document and makes it possess an authority of some kind. However, if the constitution possesses any authority independent of the behavior under it, then its meaning and guidance should be the same at all times. No one believes that about a constitution, and so there is need for another explanation. In the approach here suggested a constitution is the behavior that the dominant groups manifest in the operation of the community. If the groups change then the constitution changes with them. The stability of the constitution is the stability of the underlying group support. There is nothing in the group hypothesis that requires groups to be evanescent or transitory in character. Neither on the other hand is there any requirement that the group be stable or long-lived. The changes in the group combinations allow a means of explaining the shifts in political behavior. To take an example from contemporary affairs. The Supreme Court has recently ruled that segregation of the young for educational purposes is a *per se* violation of the equal protection clause of the Fourteenth amendment. The decision marks a shift in the groups that are receiving representation in that Court. Previously that body had given its sanction to the view that separate but equal facilities would meet the requirements of that clause. Actually, in the years since that verbal formula was developed to meet the ideological quarrel, the facilities have been separate but hardly equal. Yet the factually unequal character of the educational facilities was not enough to bring the clause into operation in the period between 1868 and the present. The opinions that have been expressed since the recent decision and the Court's conclusion about what to order suggest, at the very least, that the vision of the Court as to the meaning of the Constitution is not enough to prevent immediately the "unconstitutional" segregated school. In short, the wording of the Constitution is not a precise guide to its meaning; a better guide to its meaning is to be found in the conflicting groups that seek to clothe themselves in one or another interpretation of the words and phrases. For example, if you look at the Constitution in 1868, in 1896 and in 1954, the words in the Fourteenth Amendment remain constant, yet the actual meaning has indeed changed. Some people have called this a revolution. How to explain this? Neither the words nor the governmental offices have changed. What has changed is the interest groups. The interest that secured representation in 1868 and in 1896 has been supplanted by another interest— that which gained representation in the recent decisions. There is no evidence that ideas played any significant or distinctive role in "causing" this change in the judicial interpretation of the Fourteenth Amendment, for the ideas, like the words, have been constant through this period.

This brings up the treatment of ideas. Discourses as to the meaning of the Constitution are ideas, but the word *idea* also embraces a wide range of literary productions. There are abstract and philosophical treatises, there are popular books and magazines, there are learned journals as well as the comics, there are speeches and art objects and all the other devices for conveying significance to others. The question is what is the meaning of these in the group theory of politics. The short answer is that all of these forms of communication get their political meaning from group conflicts. All or some of the media can be used in the group struggle, and they may be used in a wide variety of ways. Propaganda has become a commonplace explanation of the political process, and here it is asserted that all forms of communication may be treated as propaganda in the process. Obviously it would be foolish to say that all forms are equally important or that they convey the same meaning to all participants. Once one of these "products" starts on its career it is impossible to forecast the role that it will play. In short, at one stage it may be used by one group to flay another, and again the latter group or its remnants may use it to flay the former. That is, the idea or the communication has no fixed meaning, since its meaning derives from the context in which it is used. The arrangement of groups may be such that in one layer on one side the idea may play an opposite role to that which it plays in another layer on the same side. These are hypotheses about the idea and its operation and are not to be taken as having been demonstrated. They have been used to convey the sense in which ideas may be construed in the group theory. The importance of the idea is in its representative quality in the underlying group struggle. The idea must ultimately be traced into the activity, and its importance for this approach lies in the activity which is in turn related to all other activities.

Of course, the same observations apply to philosophy and to ideology. In some contemporary discussions of the role of ideas, distinctions are made between ideology and philosophy. Ideology is frequently given a role closer to action than to philosophy. Both are data in the political process and both have to be given their measurable importance and significance in that process. The influence of both in the political process is denoted in activity of those pro or con the group struggle.

Those who are seeking to develop a political

science which utilizes the group as the basic unit of analysis can explain all that the other writers explain with their reification. The group theory has the added capacity to cope with changes and shifts in the activity of the governmental process. The *individual* has many facets and these are reflections of the groups in which he is a participant. He is not a fixed point of reference but is an act in the totality of activity that makes up social action. *Interests* are the activity, and the association of acts of physiological beings in a common effort for or against another collocation of common activity is the manifestation of the group or group interest. An *institution* is a stable group of individual activities, and a crowd or mob is a temporary group. *Ideas* are expressions in verbal or other forms of activity, and in the governmental process they also take their meaning from the group association. To illustrate, a statue of Lincoln has a meaning quite as well as the Gettysburg address, but the meaning is to be found in the group activity associated with each. In the extreme case any connection between the group significance and the original existential record is coincidental. For example, Armistice Day is currently being made over into Veterans Day.

Nothing is lost that is significant in the political process by giving meaning through this method. The gain is in the freeing of the observer from the alleged meanings so that he may observe untrammeled by the clutterings of former meanings. The framework of analysis is flexible, and it is able to incorporate all social data that is relevant to a social theory. However, to state that this is a desirable method of approach is not to deny the usefulness of other frameworks. This approach is more desirable than the alternative frameworks because it is so much more capable of managing the data of political science.

IV

The three preceding parts of this essay have attempted to give first, a setting to the role which the group plays in political analysis, and second, to delineate the characteristics of the group upon which a science of politics may be based. Sizeable problems remain even though acceptance is given to the preceding comments.

In the first place it should be emphasized that the method of study indicated here is dialectical, that is, one side of the proposed topic of investigation always has its other side. One side is always to be treated as related to the other. Movement may be in either direction or both. This rejects the notion of interaction of separate, independent and discrete factors that is found in most political studies. Interaction in such studies postulates factors as the basic units of inquiry. Furthermore, each factor exists independently of the other. A combination of factors may cause a third phenomenon. All of these are real entities rather than intellectual inventions.

An alternative method of studying such activity is the "transaction" which is beginning to find its way into contemporary discussions of social phenomena. The transaction regards both sides of the group struggles as a single process. Each side gets its meaning from that struggle. One does not exist without the other. The relation is not one of cause in the sense that one side exists before the other. The meaning of one side *is* its relation to the other side.

An illustration may help to clarify the position. A loan is a transaction. It involves a borrower and a lender. Neither has meaning apart from the relation to the other. In order to get the full significance of the loan it would be necessary to describe all the activities of those who provide the monetary system and its accompanying credit system plus all of the potential activity involved in collecting of the loan. The loan is embedded in a pattern of habits of activity which are widely understood and acted out easily and with dispatch. (This illustration has been taken from John R. Commons' stimulating study, "Legal Foundations of Capitalism").

For the moment the concern is only with the notion of the transaction which makes the loan a dialectical process, and it is a glimpse of that feature of the political process that this essay wishes to convey. The group struggle is datum of this sort.

A group is a segment of human activity focused upon by the analyst or investigator for the purposes of his particular inquiry. The membership of the group is to be determined by the purposes of the inquiry rather than by "inherent" or "essential" criteria. For example if one were to study the groups supporting and opposing classical neutrality, both groups would certainly include persons labelled for other purposes as Englishmen, Frenchmen, Brazilians, Americans, etc. For other purposes these categories of Englishmen, Frenchmen, Brazilians, and Americans may be useful, but they would not assist in the study of the neutrality contest. National groups, in other words, are not the only or even the most important groups to describe some political struggles.

One aspect of research in the group would involve the study of the size of the group, that is the number of persons who participate in it. This would not always be an easy task. The difficulties would vary with different kinds of issues.

On some matters there would be vast quantities of data, and on other matters there would be little or none. Crude measuring devices are used in almost all political studies, frequently without the recognition that they are measures. For example, words like important, great, significant, etc. are such measures. There is need for a great deal of work on techniques for counting and on developing new methods of solving problems as they are detected and delineated. The materials compiled by polling organizations illustrate my point. These organizations started out with simple questions to determine public opinion and soon found that the answers were poor guides to political behavior. As a result they are constantly sharpening their tools. It is desirable that similar development should take place in the comparable study of the means of determining the numbers on one side or the other of the group struggle.

Political speculation has always placed a great emphasis on numbers. But the group theorist's concept of numbers or size as a facet of activity is a different concept from the traditional. In traditional investigation attention is focused on physiological entities. Each individual is counted. But in research based on the group theory, attention is focused on the activity not on the individual.

Numbers, or a better way to state it, the size of the group, is obviously not the only factor to be considered. It is a matter of common observation that the largest group's activity does not always dominate in the group struggle. Another aspect of the activity which needs investigation is what Bentley called "intensity." This does not refer to any "mental force" but to the observable variations in the nature of the activity under investigation. David Truman has called our attention to the importance of "overlapping membership." Another way of phrasing the phenomenon could be that the activity of an individual (and a number of individuals) is multi-directional. The activity of a relatively small group may dominate that of the larger group because the activity of the smaller number is in one direction and constant. "Hard core Communists" or "hard core Republicans" are examples of behavioral patterns of intense activity.

A third aspect of group activity has to do with technique. Technique would include such things as organization, cohesion, leadership. These are different ways of talking about the activity. It is true that these group techniques have been studied in the traditional literature. But most of these studies are set within the framework of formal and informal categories. It is consistent with group theory to consider the formally organized groups as a manifestation of the group struggle representing "underlying" groups so long as it is clearly understood that the "underlying" groups are merely that part of the group which is not within the formal organization. The detection and delineation of the "underlying" groups is a matter for investigation. But what is needed is a broader view of the techniques. Traditional studies center attention on the logic of the arguments and on writing and talking activity as means of influencing others. Although they have recognized that there are other techniques of influence, these have been frowned upon and left for study by sociologists and revolutionaries. But these other kinds of activity and the relations between them and the writing and talking activity cannot be left out of account if the purpose is adequate description.

V

A political science must reduce its problems to their simplest terms and the smallest number of explanatory principles. Under that rule only one of the above conceptions of the group can hope to become the basis for a science of politics. The conception of the group that meets the requirements of the above rule is the one of the group as activity of human beings. It has been shown that all the important qualities of ideas and institutions and individuals and interests meet on common ground in that conception. It has also been shown that whatever qualities those words have that are not incorporated into the group activity are irrelevant for a political science. The same can be done for all the other explanatory principles.

The reasoning that leads to the rejection of the alternative conception of the group as an explanatory device in the study of government may be briefly stated. This alternative conception expands the number of factors that are needed to explain behavior, and, moreover, it offers no means by which one factor can be related to the others. The highest level to which generalizations can rise on this basis is common sense, and by that is meant the kind of explanations that any person can give. Science seeks order and not disorder, and the factorial method is more likely to attain the latter than the former. That is to say, the constant addition of new and different factors leads to disorder.

It is not meant a scientific theory will always be the one preferred. However, it should be the basis until a better one is found. In order to be better the new theory will have to manage the data relevant to an understanding of the governmental process on simpler and more comprehensive principles.

David Easton

An Approach to the Study of Politics: The Analysis of Political Systems[1]

I. Some Attributes of Political Systems

. . . .

The study of politics is concerned with understanding how authoritative decisions are made and executed for a society. We can try to understand political life by viewing each of its aspects piecemeal. We can examine the operation of such institutions as political parties, interest groups, government, and voting; we can study the nature and consequences of such political practices as manipulation, propaganda, and violence; we can seek to reveal the structure within which these practices occur. By combining the results we can obtain a rough picture of what happens in any self-contained political unit.

In combining these results, however, there is already implicit the notion that each part of the larger political canvas does not stand alone but is related to each other part; or, to put it positively, that the operation of no one part can be fully understood without reference to the way in which the whole itself operates. I have suggested in my book, *The Political System*,[2] that it is valuable to adopt this implicit assumption as an articulate premise for research and to view political life as a system of interrelated activities. These activities derive their relatedness or systemic ties from the fact that they all more or less influence the way in which authoritative decisions are formulated and executed for a society.

Once we begin to speak of political life as a system of activity, certain consequences follow for the way in which we can undertake to analyze the working of a system. The very idea of a system suggests that we can separate political life from the rest of social activity, at least for analytical purposes, and examine it as though for the moment it were a self-contained entity surrounded by, but clearly distinguishable from, the environment or setting in which it operates. In much the same way, astronomers consider the solar system a complex of events isolated for certain purposes from the rest of the universe.

Furthermore, if we hold the system of political actions as a unit before our mind's eye, as it were, we can see that what keeps the system going are inputs of various kinds. These inputs are converted by the processes of the system into outputs and these, in turn, have consequences both for the system and for the environment in which the system exists. The formula here is very simple but, as I hope to show, also very illuminating: inputs—political system or processes—outputs. These relationships are shown diagrammatically in Figure 1. This diagram represents a very primitive "model"—to dignify it with a fashionable name—for approaching the study of political life.

Political systems have certain properties because they are systems.[3] To present an over-all view of

---

[1] In modified form, the substance of this article was presented to a meeting of the New England Political Science Association in May, 1956, and to a special conference of the International Political Science Association held in Switzerland in September, 1956.

Reprinted from *World Politics*, Vol. 9, No. 3 (April, 1957), 383-400, where it appeared under the title "An Approach to the Analysis of Political Systems." By permission of the author and the publisher.

[2] David Easton, *The Political System* (New York: Alfred A. Knopf, Inc., 1953).

[3] My conceptions relating to system theory have been enriched through my participation in the Staff Theory Seminar of the Mental Health Research Institute at the University of Michigan. There has been such thorough mingling of ideas in this Seminar that rather than try to trace paternity, I shall simply indicate my obligation to the collective efforts of the Seminar.

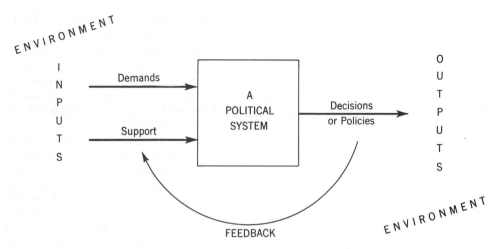

Figure 1

the whole approach, let me identify the major attributes, say a little about each, and then treat one of these properties at somewhat greater length, even though still inadequately.

(1) Properties of identification. To distinguish a political system from other social systems, we must be able to identify it by describing its fundamental units and establishing the boundaries that demarcate it from units outside the system.

    (a) Units of a political system. The units are the elements of which we say a system is composed. In the case of a political system, they are political actions. Normally it is useful to look at these as they structure themselves in political roles and political groups.

    (b) Boundaries. Some of the most significant questions with regard to the operation of political systems can be answered only if we bear in mind the obvious fact that a system does not exist in a vacuum. It is always immersed in a specific setting or environment. The way in which a system works will be in part a function of its response to the total social, biological, and physical environment.

The special problem with which we are confronted is how to distinguish systematically between a political system and its setting. Does it even make sense to say that a political system has a boundary dividing it from its setting? If so, how are we to identify the line of demarcation?

Without pausing to argue the matter, I would suggest that it is useful to conceive of a political system as having a boundary in the same sense as a physical system. The boundary of a political system is defined by all those actions more or less directly related to the making of binding decisions for a society; every social action that does not partake of this characteristic will be excluded from the system and thereby will automatically be viewed as an external variable in the environment.

(2) Inputs and outputs. Presumably, if we select political systems for special study, we do so because we believe that they have characteristically important consequences for society, namely, authoritative decisions. These consequences I shall call the outputs. If we judged that political systems did not have important outputs for society, we would probably not be interested in them.

Unless a system is approaching a state of entropy—and we can assume that this is not true of most political systems—it must have continuing inputs to keep it going. Without inputs the system can do no work; without outputs we cannot identify the work done by the system. The specific research tasks in this connection would be to identify the inputs and the forces that shape and change them, to trace the processes through which they are transformed into outputs, to describe the general conditions under which such processes can be maintained, and to establish the relationship between outputs and succeeding inputs of the system.

From this point of view, much light can be shed on the working of a political system if we take into account the fact that much of what happens within a system has its birth in the efforts of the members of the system to cope with the changing environment. We can appreciate this point if we consider a familiar biological system such as the human organism. It is subject to constant stress from its surroundings to which it must adapt in one way or another if it is not to be completely destroyed. In part, of course, the way in which the body works represents responses to needs that are generated by the very organization of its anatomy and functions; but in large part, in order to understand both the structure and the working of the body, we must also be very sensitive to the inputs from the environment.

In the same way, the behavior of every political system is to some degree imposed upon it by the kind of system it is, that is, by its own structure and internal needs. But its behavior also reflects the strains occasioned by the specific setting within which the system operates. It may be argued that most of the significant changes within a political system have their origin in shifts among the external variables. . . .

(3) Differentiation within a system. As we shall see in a moment, from the environment come both energy to activate a system and information with regard to which the system uses this energy. In this way a system is able to do work. It has some sort of output that is different from the input that enters from the environment. We can take it as a useful hypothesis that if a political system is to perform some work for anything but a limited interval of time, a minimal amount of differentiation in its structure must occur. In fact, empirically it is impossible to find a significant political system in which the same units all perform the same activities at the same time. The members of a system engage in at least some minimal division of labor that provides a structure within which action takes place.

(4) Integration of a system. This fact of differentiation opens up a major area of inquiry with regard to political systems. Structural differentiation sets in motion forces that are potentially disintegrative in their results for the system. If two or more units are performing different kinds of activity at the same time, how are these activities to be brought into the minimal degree of articulation necessary if the members of the system are not to end up in utter disorganization with regard to the production of the outputs of interest to us? We can hypothesize that if a structured system is to maintain itself, it must provide mechanisms whereby its members are integrated or induced to cooperate in some minimal degree so that they can make authoritative decisions.

II. INPUTS: DEMANDS

Now that I have mentioned some major attributes of political systems that I suggest require special attention if we are to develop a generalized approach, I want to consider in greater detail the way in which an examination of inputs and outputs will shed some light on the working of these systems.

Among inputs of a political system there are two basic kinds: demands and support. These inputs give a political system its dynamic character. They furnish it both with the raw material or information that the system is called upon to process and with the energy to keep it going.

The reason why a political system emerges in a society at all—that is, why men engage in political activity—is that demands are being made by persons or groups in the society that cannot all be fully satisfied. In all societies one fact dominates political life: scarcity prevails with regard to most of the valued things. Some of the claims for these relatively scarce things never find their way into the political system but are satisfied through the private negotiations of or settlements by the persons involved. Demands for prestige may find satisfaction through the status relations of society; claims for wealth are met in part through the economic system; aspirations for power find expression in educational, fraternal, labor, and similar private organizations. Only where wants require some special organized effort on the part of society to settle them authoritatively may we say that they have become inputs of the political system.

Systematic research would require us to address ourselves to several key questions with regard to these demands.

(1) How do demands arise and assume their particular character in a society? In answer to this question, we can point out that demands have their birth in two sectors of experience: either in the environment of a system or within the system itself. We shall call these the external and internal demands, respectively.

Let us look at the external demands first. I find it useful to see the environment not as an undifferentiated mass of events but rather as systems clearly distinguishable from one another and from the political system. In the environment we have such systems as the ecology, economy, culture, personality, social structure, and demography. Each of these constitutes a major set of variables in the setting that helps to shape the kind of demands entering a political system. For purposes of illustrating what I mean, I shall say a few words about culture.

The members of every society act within the framework of an ongoing culture that shapes their general goals, specific objectives, and the procedures that the members feel ought to be used. Every culture derives part of its unique quality from the fact that it emphasizes one or more special aspects of behavior and this strategic emphasis serves to differentiate it from other cultures with respect to the demands that it generates. . . . The culture embodies the standards of value in a society and thereby marks out areas of potential conflict, if the valued things are in short supply relative to demand. The typical demands that will find their way into the political process will concern the matters in conflict that are labeled important by the culture. For this reason we cannot hope to understand the nature of the demands presenting themselves for political settlement unless we are ready to explore systematically and intensively their connection with the culture. And what I have said about culture applies, with suitable modifications, to other parts of the setting of a political system.

But not all demands originate or have their major locus in the environment. Important types stem from situations occurring within a political system itself. Typically, in every on-going system, demands may emerge for alterations in the political relationships of the members themselves, as the result of dissatisfaction stemming from these relationships. For example, in a political system based upon representation, in which equal representation is an important political norm, demands may arise for equalizing representation between urban and rural voting districts. Similarly, demands for changes in the process of recruitment of formal political leaders, for modifications of the way in which constitutions are amended, and the like may all be internally inspired demands.

I find it useful and necessary to distinguish these from external demands because they are, strictly speaking, not inputs of the system but something that we can call "withinputs," if we can tolerate a cumbersome neologism, and because their consequences for the character of a political system are more direct than in the case of external demands. Furthermore, if we were not aware of this difference in classes of demands, we might search in vain for an explanation of the emergence of a given set of internal demands if we turned only to the environment.

(2) How are demands transformed into issues?

What determines whether a demand becomes a matter for serious political discussion or remains something to be resolved privately among the members of society? The occurrence of a demand, whether internal or external, does not thereby automatically convert it into a political *issue*. Many demands die at birth or linger on with the support of an insignificant fraction of the society and are never raised to the level of possible political decision. Others become issues, an issue being a demand that the members of a political system are prepared to deal with as a significant item for discussion through the recognized channels in the system.

The distinction between demands and issues raises a number of questions about which we need data if we are to understand the processes through which claims typically become transformed into issues. For example, we would need to know something about the relationship between a demand and the location of its initiators or supporters in the power structures of the society, the importance of secrecy as compared with publicity in presenting demands, the matter of timing of demands, the possession of political skills or know-how, access to channels of communication, the attitudes and states of mind of possible publics, and the images held by the initiators of demands with regard to the way in which things get done in the particular political system. Answers to matters such as these would possibly yield a conversion index reflecting the probability of a set of demands being converted into live political issues.

If we assume that political science is primarily concerned with the way in which authoritative decisions are made for a society, demands require special attention as a major type of input of political systems. I have suggested that demands influence the behavior of a system in a number of ways. They constitute a significant part of the material upon which the system operates. They are also one of the sources of change in political systems, since as the environment fluctuates it generates new types of demand-inputs for the system. Accordingly, without this attention to the origin and determinants of demands we would be at a loss to be able to treat rigorously not only the operation of a system at a moment of time but also its change over a specified interval. Both the statics and historical dynamics of a political system depend upon a detailed understanding of demands, particularly of the impact of the setting on them.

III. INPUTS: SUPPORT

Inputs of demands alone are not enough to keep a political system operating. They are only the raw material out of which finished products called decisions are manufactured. Energy in the form of actions or orientations promoting and resisting a political system, the demands arising in it, and the decisions issuing from it must also be put into the system to keep it running. This input I shall call support.[4] Without support, demands could not be satisfied or conflicts in goals composed. If demands are to be acted upon, the members of a system undertaking to pilot the demands through to their transformation into binding decisions and those who seek to influence the relevant processes in any way must be able to count on support from others in the system. Just how much support, from how many and which members of a political system, are separate and important questions that I shall touch on shortly.

What do we mean by support? We can say that A supports B either when A acts on behalf of or when he orients himself favorably toward B's goals, interests, and actions. Supportive behavior may thus be of two kinds. It may consist of actions promoting the goals, interests, and actions of another person. We may vote for a political candidate, or defend a decision by the

[4] The concept support has been used by Talcott Parsons in an unpublished paper entitled "Reflections on the Two-Party System." I am pleased to note that in this article Professor Parsons also seems to be moving in the direction of input-output analysis of political problems, although the extent to which he uses other aspects of system theory is not clear to me.

highest court of the land. In these cases, support manifests itself through overt action.

On the other hand, supportive behavior may involve not external observable acts, but those internal forms of behavior we call orientations or states of mind. As I use the phrase, a supportive state of mind is a deep-seated set of attitudes or predispositions, or a readiness to act on behalf of some other person. It exists when we say that a man is loyal to his party, attached to democracy, or infused with patriotism. What such phrases as these have in common is the fact that they refer to a state of feelings on the part of a person. No overt action is involved at this level of description, although the implication is that the individual will pursue a course of action consistent with his attitudes. Where the anticipated action does not flow from our perception of the state of mind, we assume that we have not penetrated deeply enough into the true feelings of the person but have merely skimmed off his surface attitudes.

Supportive states of mind are vital inputs for the operation and maintenance of a political system. For example, it is often said that the struggle in the international sphere concerns mastery over men's minds. To a certain extent this is true. If the members of a political system are deeply attached to a system or its ideals, the likelihood of their participating in either domestic or foreign politics in such a way as to undermine the system is reduced by a large factor. Presumably, even in the face of considerable provocation, ingrained supportive feelings of loyalty may be expected to prevail.

We shall need to identify the typical mechanisms through which supportive attitudes are inculcated and continuously reinforced within a political system. But our prior task is to specify and examine the political objects in relation to which support is extended.

(1) *The Domain of Support*

Support is fed into the political system in relation to three objects: the community, the regime, and the government. There must be convergence of attitude and opinion as well as some willingness to act with regard to each of these objects. Let us examine each in turn.

(a) The political community. No political system can continue to operate unless its members are willing to support the existence of a group that seeks to settle differences or promote decisions through peaceful action in common. The point is so obvious—being dealt with usually under the heading of the growth of national unity—that it may well be overlooked; and yet it is a premise upon which the continuation of any political system depends. To refer to this phenomenon we can speak of the political community. At this level of support we are not concerned with whether a government exists or whether there is loyalty to a constitutional order. For the moment we only ask whether the members of the group that we are examining are sufficiently oriented toward each other to want to contribute their collective energies toward pacific settlement of their varying demands.

The American Civil War is a concrete illustration of the cessation of input of support for the political community. The war itself was definitive evidence that the members of the American political system could no longer contribute to the existence of a state of affairs in which peaceful solution of conflicting demands was the rule. Matters had come to the point where it was no longer a question of whether the South would support one or another alternative government, or whether it could envision its demands being satisfied through the normal constitutional procedures. The issue turned on whether there was sufficient mutual identification among the members of the system for them to be able to work together as a political community. Thus in any political system, to the extent that there is an in-group or we-group feeling and to the extent that the members of the system identify one another as part of this unit and exclude others according to some commonly accepted criteria, such as territoriality, kinship, or citizenship, we shall say that they are putting in support for the political community.

(b) The regime. Support for a second major

part of a political system helps to supply the energy to keep the system running. This aspect of the system I shall call the regime. It consists of all those arrangements that regulate the way in which the demands put into the system are settled and the way in which decisions are put into effect. They are the so-called rules of the game, in the light of which actions by members of the system are legitimated and accepted by the bulk of the members as authoritative. Unless there is a minimum convergence of attitudes in support of these fundamental rules—the constitutional principles, as we call them in Western society—there would be insufficient harmony in the actions of the members of a system to meet the problems generated by their support of a political community. The fact of trying to settle demands in common means that there must be known principles governing the way in which resolutions of differences of claims are to take place.

(c) The government. If a political system is going to be able to handle the conflicting demands put into it, not only must the members of the system be prepared to support the settlement of these conflicts in common and possess some consensus with regard to the rules governing the mode of settlement; they must also be ready to support a government as it undertakes the concrete tasks involved in negotiating such settlements. When we come to the outputs of a system, we shall see the rewards that are available to a government for mobilizing support. At this point, I just wish to draw attention to this need on the part of a government for support if it is going to be able to make decisions with regard to demands. Of course, a government may elicit support in many ways: through persuasion, consent, or manipulation. It may also impose unsupported settlements of demands through threats of force. But it is a familiar axiom of political science that a government based upon force alone is not long for this world; it must buttress its position by inducing a favorable state of mind in its subjects through fair or foul means.

The fact that support directed to a political system can be broken down conceptually into three elements—support for the community, regime, and government—does not mean, of course, that in the concrete case support for each of these three objects is independent. In fact we might and normally do find all three kinds of support very closely intertwined, so that the presence of one is a function of the presence of one or both of the other types. . . .

This very brief discussion of support points up one major fact. If a system is to absorb a variety of demands and negotiate some sort of settlement among them, it is not enough for the members of the system to support only their own demands and the particular government that will undertake to promote these demands. For the demands to be processed into outputs it is equally essential that the members of the system stand ready to support the existence of a political community and some stable rules of common action that we call the regime.

(2) *Quantity and Scope of Support*

How much support needs to be put into a system and how many of its members need to contribute such support if the system is to be able to do the job of converting demands to decisions? No ready answer can be offered. The actual situation in each case would determine the amount and scope required. We can, however, visualize a number of situations that will be helpful in directing our attention to possible generalizations.

Under certain circumstances very few members need to support a system at any level. The members might be dull and apathetic, indifferent to the general operations of the system, its progress or decisions. In a loosely connected system such as India has had, this might well be the state of mind of by far the largest segment of the membership. Either in fact they have not been affected by national decisions or they have not perceived that they were so affected. They may have little sense of identification with the present regime and government and yet, with regard to the input of demands, the system may be able to act on the basis of the support offered by the known 3 per cent of the Western-oriented politicians and intellectuals who are politically active. In other words, we can have a small minority putting in quantitatively sufficient supportive energy to keep the system going. However, we can venture the hypothesis that where members of a system are putting in numerous demands, there is a strong probability that they will actively offer support or hostility at one of the three levels of the system, depending upon the degree to which these demands are being met through appropriate decisions.

Alternatively, we may find that all the members of a system are putting in support, but the amount may be so low as to place one or all aspects of the system in jeopardy. Modern France is perhaps a classic illustration. The input of support at the level of the political community is probably adequate for the maintenance of France as a national political unit. But for a variety of historical and contemporary reasons, there is considerable doubt as to whether the members of the French political system are putting in anything but a low order of support to the regime or any particular government. This low amount of support, even though spread over a relatively large segment of the population, leaves the French political system on somewhat less secure foundations than is the case with India. There support is less widespread but more active—that is, quantitatively greater—on the part of a minority. As this illustration indicates, the amount of support is not necessarily proportional to its scope.

It may seem from the above discussion as though the members of a political system either put in support or withhold it—that is, demonstrate hostility or apathy. In fact, members may and normally do simultaneously engage in supportive and hostile behavior. What we must be interested in is the net balance of support.

IV. Mechanisms of Support

To this point I have suggested that no political system can yield the important outputs we call authoritative decisions unless, in addition to demands, support finds its way into the system. I have discussed the possible object to which support may be directed, and some problems with regard to the domain, quantity, and scope of support. We are now ready to turn to the main question raised by our attention to support as a crucial input: how do systems typically manage to maintain a steady flow of support? Without it a system will not absorb sufficient energy from its members to be able to convert demands to decisions.

In theory, there might be an infinite variety of means through which members could be induced to support a system; in practice, certain well-established classes of mechanisms are used. Research in this area needs to be directed to exploring the precise way in which a particular system utilizes these mechanisms and to refining our understanding of the way in which they contribute to the making of authoritative policy.

A society generates support for a political system in two ways: through outputs that meet the demands of the members of society; and through the processes of politicization. Let us look at outputs first.

(1) *Outputs as a Mechanism of Support*

An output of a political system, it will be recalled, is a political decision or policy. One of the major ways of strengthening the ties of the members to their system is through providing decisions that tend to satisfy the day-to-day demands of these members. Fundamentally this is the truth that lies in the aphorism that one can fool some of the people some of the time but not all of them all of the time. Without some minimal satisfaction of demands, the ardor of all but the most fanatical patriot is sure to cool. The outputs, consisting of political decisions, constitute a body of specific inducements for the members of a system to support that system.

Inducements of this kind may be positive or negative. Where negative, they threaten the members of the system with various kinds of sanctions ranging from a small monetary fine to physical detention, ostracism, or loss of life, as in our own system with regard to the case of legally defined treason. In every system support stems in part from fear of sanctions or compulsion; in autocratic systems the proportion of coerced support is at a maximum. For want of space I shall confine myself to those cases where positive incentives loom largest.

Since the specific outputs of a system are policy decisions, it is upon the government that the final responsibility falls for matching or balancing outputs of decisions against input of demand. But it is clear that to obtain the support of the members of a system through positive incentives, a government need not meet all the demands of even its most influential and ardent supporters. Most governments, or groups such as political parties that seek to control governments, succeed in building up a reserve of support. This reserve will carry the government along even though it offends its followers, so long as over the extended short run these followers perceive the particular government as one that is in general favorable to their interests. One form that this reserve support takes in Western society is that of party loyalty, since the party is the typical instrument in a mass industrialized society for mobilizing and maintaining support for a government. However, continuous lack of specific rewards through policy decisions ultimately leads to the danger that even the deepest party loyalty may be shaken.

For example, labor has continued to support the Democratic Party even though much of the legislation promoted by members of that party has not served to meet labor's demands. In some measure, large sections of labor may continue to vote and campaign vigorously on behalf of the Democratic Party because they have no realistic alternative other than to support this party; but in addition the Democrats have built up in recent years, especially during the Roosevelt era, a considerable body of good will. It would take repeated neglect of labor's demands on the part of the Democratic Party to undermine the strong urban working-class support directed toward it and the government that the party dominates from time to time.

Thus a system need not meet *all the demands* of its members so long as it has stored up a reserve of support over the years. Nor need it satisfy even *some of the demands* of all its members. Just whose demands a system must seek to meet, how much of their demands, at what time, and under what conditions are questions for special research. We can say in advance that at least the demands of the most influential members require satisfaction. But this tells us little unless we know how to discover the influentials in a political system and how new sets of members rise to positions of influence.[5]

The critical significance of the decisions of governments for the support of the other two aspects of a system—namely, the political community and the regime—is clear from what I have said above. Not all withdrawal of support from a government has consequences for the success or failure of a regime or community. But persistent inability of a government to produce satisfactory outputs for the members of a system may well lead to demands for changing of the regime or for dissolution of the political community. It is for this reason that the input-output balance is a vital mechanism in the life of a political system.

(2) *Politicization as a Mechanism of Support*

It would be wrong to consider that the level of support available to a system is a function exclusively of the outputs in the form of either sanctions or rewards. If we did so conclude, we could scarcely account for the maintenance of numerous political systems in which satisfaction of demands has been manifestly low, in which public coercion is limited, and yet which have endured for epochs. Alternately, it might be difficult to explain how political systems could endure and yet manage to flout or thwart urgent demands, failing thereby to render sufficient *quid pro quo* for the input of support. The fact is that whatever reserve of support has been accumulated through past decisions is increased and reinforced by a complicated method for steadily manufacturing support through what I shall call the process of

[5] See C. W. Mills, *The Power Elite* (New York: Oxford University Press, 1956).

politicization. It is an awkward term, but nevertheless an appropriately descriptive one.

As each person grows up in a society, through a network of rewards and punishments the other members of society communicate to and instill in him the various institutionalized goals and norms of that society. This is well known in social research as the process of socialization. Through its operation a person learns to play his various social roles. Part of these goals and norms relate to what the society considers desirable in political life. The ways in which these political patterns are learned by the members of society constitute what I call the process of politicization. Through it a person learns to play his political roles, which include the absorption of the proper political attitudes.

Let us examine a little more closely something of what happens during the process of politicization. As members of a society mature, they must absorb the various orientations toward political matters that one is expected to have in that society. If the expectations of the members of society with regard to the way each should behave in specific political situations diverged beyond a certain range, it would be impossible to get common action with regard to the making of binding decisions. It is essential for the viability of an orderly political system that the members of the system have some common basic expectations with regard to the standards that are to be used in making political evaluations, to the way people will feel about various political matters, and to the way members of the system will perceive and interpret political phenomena.

The mechanism through which this learning takes place is of considerable significance in understanding how a political system generates and accumulates a strong reserve of support. Although we cannot pursue the details, we can mention a few of the relevant dimensions. In the first place, of course, the learning or politicization process does not stop at any particular period for the individual; it starts with the child and, in the light of our knowledge of learning, may have its deepest impact through the teen age. The study of the political experiences of and the influences operating on the child and the adolescent emerges as an important and neglected area of research.[6]

In the second place, the actual process of politicization at its most general level brings into operation a complex network of rewards and punishments. For adopting the correct political attitudes and performing the right political acts, for conforming to the generally accepted interpretations of political goals, and for undertaking the institutionalized obligations of a member of the given system, we are variously rewarded or punished. For conforming we are made to feel worthy, wanted, and respected and often obtain material advantages such as wealth, influence, improved opportunities. For deviating beyond the permissible range, we are made to feel unworthy, rejected, dishonored, and often suffer material losses.

This does not mean that the pattern of rewards and punishments is by any means always effective; if it were, we would never have changed from the Stone Age. A measure of non-conformity may at certain stages in the life history of a political system itself become a respected norm. Even where this is not the case, the most seductive rewards and the severest punishments will never succeed in preventing some of the members of a system from pursuing what they consider to be their inextinguishable interests and from seeking, with varying degrees of success, to change the goals and norms of the system. This is one of the important sources of political change closely associated with changes in the inputs of demands

---

[6] I am happy to say that, since I wrote this statement, the neglect has begun to be remedied. My colleagues at the University of Chicago, Robert Hess of the Committee of Human Development and Peter Rossi of the Department of Sociology, and I have undertaken a questionnaire-interview study of the development of the political attitudes, opinions, and images held by children and adolescents. This research is an attempt to develop some useful generalizations about major aspects of the processes of politicization in the American political system and to formulate a design that, for comparative purposes, could be applied in other political systems as well.

that are due to a changing environment. But we cannot pursue this crucial matter of the nature of political change, as it would lead us off in a new direction.

In the third place, the means used for communicating the goals and norms to others tend to be repetitive in all societies. The various political myths, doctrines, and philosophies transmit to each generation a particular interpretation of the goals and norms. The decisive links in this chain of transmission are parents, siblings, peers, teachers, organizations, and social leaders, as well as physical symbols such as flags or totems, ceremonies, and rituals freighted with political meaning.

These processes through which attachments to a political system become built into the maturing member of a society I have lumped together under the rubric of politicization. They illustrate the way in which members of a system learn what is expected of them in political life and how they ought to do what is expected of them. In this way they acquire knowledge about their political roles and a desire to perform them. In stable systems the support that accrues through these means adds to the reservoir of support being accumulated on a day-to-day basis through the outputs of decisions.[7] The support obtained through politicization tends to be relatively—although, as we have seen, not wholly—independent of the vagaries of day-to-day outputs.

When the basic political attachments become deeply rooted or institutionalized, we say that the system has become accepted as legitimate. Politicization therefore effectively sums up the way in which legitimacy is created and transmitted in a political system. And it is an empirical observation that in those instances where political systems have survived the longest, support has been nourished by an ingrained belief in the legitimacy of the relevant governments and regimes.

What I am suggesting here is that support resting on a sense of the legitimacy of a government and regime provides a necessary reserve if the system is to weather those frequent storms when the more obvious outputs of the system seem to impose greater hardships than rewards. Answers to questions concerning the formation, maintenance, transmission, and change of standards of legitimacy will contribute generously to an understanding of the way in which support is sufficiently institutionalized so that a system may regularly and without excessive expenditure of effort transform inputs of demand into outputs of decisions.

That there is a need for general theory in the study of political life is apparent. The only question is how best to proceed. There is no one royal road that can be said to be either the correct one or the best. It is only a matter of what appears at the given level of available knowledge to be the most useful. At this stage it appears that system theory, with its sensitivity to the input-output exchange between a system and its setting offers a fruitful approach. It is an economical way of organizing presently disconnected political data and promises interesting dividends.

---

[7] In primitive systems, politicization, not outputs of decisions, is normally the chief mechanism.

Sidney Verba

The Political Culture Approach

... The political culture[1] of a society consists of the system of empirical beliefs, expressive symbols, and values which defines the situation in which political action takes place. It provides the subjective orientation to politics. The political culture is of course but one aspect of politics. If one wanted a full picture of the process of politics in a nation one would have to consider many other aspects as well. We have focused on the cultural aspect for several reasons. For one thing, though political systems represent complex intertwinings of political culture with other aspects of the political system both formal and informal, it is difficult with the tools currently at hand to deal with the totality of political systems all at once. One is almost forced to look at one aspect or another. Second, we believe that the political culture of a society is a highly significant aspect of the political system.

There are many other aspects of the political system that could have been selected for close analysis. For a long time students of politics dealt with formal political institutions. If one wanted to explain why a political system survived or failed, why it was successful or unsuccessful, one looked at its constitution. How were laws passed? Was the system a presidential or a parliamentary one? Was there an independent judiciary? Of particular interest to those studying democratic systems were the structures of electoral systems. Did a nation have a single-member district electoral system or some form of proportional representation? Political scientists have since turned to other questions as well, particularly questions about what has been called the "infrastructure" of politics—those institutions not directly within the government that play a major role in political decisions. ...

This concentration upon political culture should not be taken to imply that other aspects of the political system are not important for the functioning of that system. They are both important and intimately related to the political culture. The political culture of a nation, for instance, derives from, among other things, the experiences that individuals have with the political process. One way to learn about political beliefs is to observe the ways in which political structures operate. These beliefs affect and are affected by the way in which the structures operate and there is a close circle of relationships between culture and structure. The justification for separating out the cultural aspect for attention is that it

---

[1] The term "political culture" is beginning to find currency in the political science literature. The first use of the term in roughly the way it is used in this volume (at least that I am aware of) is in Gabriel Almond, "Comparative Political Systems," *Journal of Politics*, 18 (1956). Another early use of the term is in Samuel Beer and Adam Ulam (eds.), *Patterns of Government* (New York: Random House, 1958).

The use of the word "culture" is perhaps a bit unfortunate in this context. Surely few other words in the social sciences have a greater variety of uses. It is used here because it has some currency in the literature, and any substitute word would just introduce more confusion. Furthermore, as used in this essay, the term refers to a rather general approach to politics and some imprecision in its definition is probably not too crucial. This point will be discussed further below.

Reprinted from Lucian W. Pye and Sidney Verba (eds.), *Political Culture and Political Development*, (Princeton, N.J.: Princeton University Press, 1965), pp. 513–26, where it appeared under the title, "Comparative Political Culture." Reprinted by permission of Princeton University Press.

(*Editor's note*) Footnote 1 of the original (located in a portion which was omitted) is included here because it is relevant for understanding the concept "political culture." However, footnote 9 of the original has been omitted.

may facilitate the analysis of these relationships.

The study of political culture is not new. It would be both presumptuous and historically inaccurate to argue that . . . the works of . . . political scientists concerned with what we have called political culture deal with aspects of the political system that have previously been ignored. Writers on politics, particularly those who have tried to explain why the political system of a nation operated the way it did, have been aware of and commented upon the belief systems of the members of the nations about which they have written. Surely the works of Montesquieu, Tocqueville, and Bagehot represent contributions to the study of political culture, and one finds concern with such problems at least as far back as the Greeks. Nor is the interest in political culture a subject that we can claim to rediscover for contemporary political science. Many recent students of politics have focused upon the factors we consider to be important aspects of political culture.

These disclaimers about the political culture approach are needed because of an unfortunate tendency in the social sciences to oversell new concepts and to assume that the mere labeling of an old phenomenon with a new term represents a breakthrough in our understanding. The term "political culture" refers . . . to a very general phenomenon which can be approached from many points of view. The concept of political culture serves to focus our attention on an aspect of political life, and such a focus of attention is useful. The concept makes it easier for us to separate the cultural aspect of politics from other aspects (as well as the political culture from other forms of culture) and to subject it to more detailed and systematic analysis. The process of separating out the cultural aspects of politics puts us in a position to see more clearly the place they have within the political system. The term "political culture" ties our study of political beliefs to sociological and anthropological works on culture and focuses our attention on basic values, cognitions, and emotional commitments. It also focuses our attention on a subject that has been of major concern to the students of culture—the process by which such values, cognitions, and emotional commitments are learned. The study of political culture leads invariably to the study of political socialization, to the learning experiences by which a political culture is passed on from generation to generation and to the situations under which political cultures change. The study of political culture may also lead to a new perspective on the political history of a nation . . . by which one focuses on the ways in which basic political beliefs are affected by the memories of political events.

To focus one's attention on a significant aspect of political life is useful, but it is only the beginning of the analysis and explanation of political phenomena. What really is important is not that one deals with political culture, but how one deals with it and how it is used to further our understanding of politics. To say that political culture is important is not very informative; to say what aspects of political culture are determinants of what phenomena—what the significant political beliefs are, and how they are related to other aspects of politics—may be very important. . . .

What exactly is it that we focus upon when we use the term "political culture"? Political culture does not refer to the formal or informal structures of political interaction: to governments, political parties, pressure groups, or cliques. Nor does it refer to the pattern of interaction among political actors—who speaks to whom, who influences whom, who votes for whom. As we use the term "political culture" it refers to the system of beliefs about patterns of political interaction and political institutions. It refers not to what is happening in the world of politics, but what people believe about those happenings. And these beliefs can be of several kinds: they can be empirical beliefs about what the actual state of political life is; they can be beliefs as to the goals or values that ought to be pursued in political life; and these beliefs may have an important expressive or emotional dimension.[2]

Political culture forms an important link between the events of politics and the behavior of individuals in reaction to those events; for, although the political behavior of individuals and groups is of course affected by acts of government officials, wars, election campaigns, and the like, it is even more affected by the meanings that are assigned those events by observers. This is to say no more than that people respond to what they perceive of politics and how they interpret what they see.[3] From the cultural point of view, for instance, we would look at political history not so much as a series of objective events but as a series of events that may be interpreted quite differently by different people and whose effects on future events depend upon this interpretation. The terms "meaning" and "interpretation," it should be stressed, are relational terms. They do not refer to what exists in the mind of the individual or to what happens in the outside world, but to the interaction between the two. And in this interaction it would be wrong to assume that either the previously held beliefs or the external events necessarily are dominant. An event will be interpreted in terms of previously held beliefs; but preconceptions can only go so far in affecting interpretation. (How far, indeed, may be a very important question and a very important source of differentiation among political cultures.)

Of course basic patterns of political belief affect not merely how individuals respond to external events. Since these basic belief systems consist of existential beliefs, general values that set the goals of behavior, norms that regulate the means used to achieve goals, as well as emotional attachments, these belief systems also affect when and in what ways individuals become involved in political life. Looked at this way, it can be seen that political culture represents a system of control vis-à-vis the system of political interactions.[4] Political culture regulates who talks to whom and who influences whom. It also regulates what is said in political contacts and the effects of these contacts. It regulates the ways in which formal institutions operate as well. A new constitution, for instance, will be perceived and evaluated in terms of the political culture of a people. When put into practice in one society it may look quite different from the same constitution instituted in another nation with another political culture. Similarly, political ideologies are affected by the cultural environment into which they are introduced. History is full of examples of constitutions that did not "take" as the constitution writers had hoped because their application was mediated through a particular political culture, and history is full of examples of the ways in which political ideologies have been adapted to fit the pre-existing culture of the nation into which they were introduced.

POLITICAL BELIEFS

Political culture is very broadly defined in this essay. The term refers to all politically relevant

---

[2] The term "belief" is one that can cause almost as much trouble as "culture." It is used in this essay to refer not only to the cognitive aspects of thought—which will be referred to as "empirical beliefs"—but to the evaluative and expressive aspects as well. The specific thoughts that people have about politics involve no clear differentiation into their cognitive, evaluative, and expressive components, but usually involve a combination of all three. Furthermore, I use the term "belief" rather than "attitude" or "opinion" because I am interested in patterns of thought more deeply rooted and more general than the latter two terms imply.

[3] Though politics is often defined in terms of its relationship to violence and coercion (the state defined as the possessor of the monopoly of legitimate coercion, for instance), it is interesting how little of political importance involves the direct application of violence. One can of course affect a person's political behavior through the direct use of violence or coercion, in which case the beliefs about or interpretations of the violent act by the victim may be irrelevant. But violence plays a larger political role as a threat. Or, if violence is actually applied, what is of political relevance is more often the effect of the violent act on those who learn of it. And the two latter uses of violence involve interpretations and beliefs.

[4] On the role of cultural systems as controllers of social systems see Talcott Parsons, "General Introduction" in Parsons et al., *Theories of Society*, Vol. 1 (New York: The Free Press of Glencoe, 1961), p. 35.

orientations whether of a cognitive, evaluative, or expressive sort. It refers to the orientations of all the members of a political system; and it refers to orientations to all aspects of politics. This broad and rather loose definition is useful for this essay, one purpose of which it to direct attention to a general area of concern. And it is useful so long as political culture in the general sense is not used as an explanatory term in propositions about political systems. If political culture is so generally defined, it is of little use to say that the political culture of nation X explains why it has political structures of the form Y. Rather one must specify what aspects of political culture—what beliefs about what subjects —are the important elements for explaining the operation of political systems. . . .

Thus the empirical beliefs we are interested in in this approach to political culture are the fundamental beliefs about the nature of political systems and about the nature of *other* political actors. In particular it is quite important to discover what political beliefs are—to borrow a term from Milton Rokeach—primitive beliefs. Primitive political beliefs are those so implicit and generally taken for granted that each individual holds them and believes all other individuals hold them. They are the fundamental and usually unstated assumptions or postulates about politics. In this sense they are unchallengeable since no opportunity exists to call them into question.[5] It may be, for instance, that one of the major characteristics of the transitional political cultures within developing nations and one of the major sources of their instability is the fact that there are few such unchallenged primitive political beliefs.

Similarly, in terms of the evaluative mode of orientation, we shall be concerned with the most general level of values—the guiding principles that set the general goals of a political system— rather then with the preferences for specific kinds of policies. And in terms of expressive commitment we shall try to deal with the fundamental symbols of political integration and with fundamental patterns of loyalty rather than with the specific satisfactions and dissatisfactions concerning politics. The selection of such a set of political orientations involves the risk that they may be so general as to be of little use in explaining behavior. But in the perspective of cross-national comparisons of political cultures this is not so, for one can find rather sharp differences among different political cultures in terms of these most general beliefs, a fact that makes them a useful explanatory tool.

Fundamental political beliefs are, furthermore, particularly relevant to the study of change. They play a major role in guiding the ways in which institutions develop and change. To a large extent these beliefs may represent stabilizing elements in a system; they may motivate the actors in a political system to resist change in the name of traditional beliefs or they may lead to fundamental modifications of innovative institu-

---

[5] See Milton Rokeach, *The Open and the Closed Mind* (New York: Basic Books, 1960), pp. 40–42. I am using the concept of primitive beliefs in connection with political beliefs somewhat differently from the way Rokeach uses the term but in a way that is not, I believe, inconsistent with his meaning. By primitive beliefs he refers to such basic beliefs as that in one's own identity and in the nature of physical reality. These are so taken for granted that one assumes all others agree with one. Therefore the question is never opened. The primitive political beliefs I refer to are not quite as primitive as the ones Rokeach cites, for the political beliefs held in one society are not necessarily shared by those in other political systems. Or indeed the assumptions of one sub-group of a society may not be shared by the other members of that society. (It is perhaps only when one has a sub-group of society whose fundamental assumptions about politics differ from the assumptions predominant in that society that one can properly talk of a political "sub-culture.") Thus, political beliefs may be primitive and unquestioned only for the group within which the individual lives. He may, for instance, be aware that there are other patterns of political belief but not consider them as relevant. And of course the political beliefs of the ordinary man are not structured according to the logic of comparative political analysis. The fact that the universality of the belief would be called into question if he considered political systems other than his own may offer no challenge to the depth of his belief in them. Furthermore one basic political belief—that in one's political identity—is the sort Rokeach has in mind and, as we shall see, one of the most crucial of political beliefs.

tions so that they fit the traditional culture. . . . But though fundamental political beliefs may be closely connected with the maintenance of existing patterns of politics, this is not necessarily the case. In a number of ways the cultural patterns in a society may generate change. Not all political cultures are well integrated and consistent. There may be many sources of strain within such cultures: sets of beliefs that are incompatible with other beliefs, sets of beliefs held by one segment of society and not another, or unmanageable incongruities between belief and reality. Under such circumstances the culture may accelerate change as part of a search for a new and more integrative set of beliefs. Furthermore it is possible for a culture to incorporate change as part of its fundamental belief system. One of the hallmarks of modernity within the realm of culture may be the acceptance and indeed the positive value of continuing change and innovation—to put it somewhat paradoxically—the institutionalization of innovation. And lastly, it must be pointed out that the fundamental general beliefs we have been talking about are themselves not immutable. The basic political values of a group may not be easily changed, yet under certain types of pressure and over time they can change rather drastically.[6] . . . The political cultures in the nations studied impinge upon attempted changes in patterns of political interaction—in many cases to impede these changes, in some cases to facilitate them. But the political cultures in these nations are themselves in flux.

The changeability of basic political beliefs is indeed a crucial question to the elites of the developing nations. It is customary to think that cultural dimensions are unchanging factors that form the setting within which politics is carried on; that culture conditions politics, but not vice versa. Certainly this was the main argument of much of the national character literature. But the situation is sharply different today. Basic beliefs have now become the object of direct concern and attempted manipulation by the political elites in many nations. This is especially true in the new nations, but it may also be true in more established nations when a political elite is trying to found a new type of political system. Thus in Germany as well as in Egypt or India basic political beliefs have become the object of direct governmental concern.[7] With a great deal of sophistication and with a great deal of self-consciousness, elites in those countries have taken upon themslves the task of remaking the basic belief systems in their nations as part of their over-all task of nation building. Basic political attitudes have of course always been in part the objects of conscious manipulation—as when adults teach children what are considered to be the basic virtues—but the new cultural policies involve attempts to create new patterns of beliefs, not merely to transmit established patterns to new generations. And these attempts are being made at a time when technological changes in the realm of communications and symbol manipulation may make such policies particularly effective. In any case, the student of political culture would be wise to accept the fact that what seem today to be fundamental sets of political beliefs may be quickly cast aside. . . .

Culture and Political Culture

The distinction between political culture and the more general cultural system of a society is an analytical one. Political culture is an integral aspect of more general culture, the set of political beliefs an individual holds, being of course part of the totality of beliefs he holds. Furthermore the basic belief and value patterns of a culture—

---

[6]On the stability but changeability of basic values see Florence R. Kluckhohn and Fred R. Strodtbeck, *Variations in Value Orientation* (New York: Harper & Row, Publishers, 1961), pp. 9–10.

[7]References in this essay to cultural patterns of the nations covered in this book are, unless otherwise specified, to the respective essays in this volume. See also on this point McKim Marriott, "Cultural Policies in the New States," in Clifford Geertz (ed.), *Old Societies and New States: The Quest for Modernity in Asia and Africa* (New York: The Free Press of Glencoe, 1963), pp. 27–56.

those general values that have no reference to specific political objects—usually play a major role in the structuring of political culture. Such basic belief dimensions as the view of man's relation to nature, as time perspective, as the view of human nature and of the proper way to orient toward one's fellow man, as well as orientation toward activity and activism in general would be clearly interdependent with specifically political attitudes.[8] In a culture in which men's orientation toward nature is essentially one of fatalism and resignation their orientation toward government is likely to be much the same. Political cultures in which the activities of the government are considered in the same class with such natural calamities as earthquakes and storms—to be suffered but outside of the individual's control—are by no means rare, and one would assume that such an attitude would be closely related to a fatalistic attitude toward man's role in relation to nature. . . .

The focus on the relationship between basic belief structure and political beliefs is of great use in determining what political attitudes are important to consider in describing a political culture. Too often, as Robert Lane has pointed out forcefully,[9] students of politics have asked questions about those political attitudes which political scientists consider important—about attitudes toward political issues or toward partisan affiliation. When the individual does not respond in ways that fit the researcher's view as to what a consistent political ideology is (for instance, he does not take consistent and well-reasoned liberal or conservative positions), he is considered to have no political ideology. But by focusing on basic value orientations—often implicit assumptions about the nature of man and the nature of physical reality—we may find a set of political attitudes that, though not structured as the political philosopher might structure them, nevertheless have a definite and significant structure. Since an individual's involvement in society is likely to be only peripherally political—since he is likely to invest more concern and affect in his personal relations or economic relations than in his political ones—it is quite likely that he will structure his political attitudes in ways that derive from his structuring of attitudes toward these more salient areas of activity rather than in terms of the ways in which political scientists or political theorists structure the political world.[10]

Though political culture is closely connected with other aspects of the cultural system, the analytical separation from general culture of those values, cognitions, and expressive states with political objects is useful. It is useful because it allows us to concentrate on those areas of attitudes that are most relevant for politics. It is useful also because the connection between general culture and political culture is not one of complete identity. Under some circumstances, for instance, there may be discontinuities between the values associated with political interactions and those associated with interactions of other

---

[8] This set of basic values is taken from Kluckhohn and Strodtbeck, *Variations in Value Orientation*, . . . pp. 11–20.

[9] Robert Lane, *Political Ideology* (New York: The Free Press of Glencoe, 1962), chap. 26.

[10] On this general point see Gabriel A. Almond and Sidney Verba, *The Civic Culture* (Princeton, N.J.: Princeton University Press, 1963), chaps. 6 and 10. The mode of structuring political beliefs by the students of the subject suggests in turn that much of what, in the absence of direct evidence on political attitudes, we have assumed to be the political culture of a society may in fact be the political ideology of political elites or the political theory of political scientists. The attitudes of the ordinary man may not be structured around those aspects of politics that concern political elites or political scientists. Some recent work by Philip Converse and Georges Dupeux ("Politicization of the Electorate in France and the United States," *Public Opinion Quarterly*, 26 (1962), pp. 1–23) suggests that the familiar characterization of French political culture as involving a number of highly ideological and principled political sub-cultures may be more a characterization of political elites than of the political mass. The French Electorate appears to be quite a bit less ideologically oriented than one would expect from a consideration of the nature of political debates in the National Assembly of the 4th Republic.

Of course the political culture of the elites may be more important than that of the mass, but the example suggests that one might have to be careful in describing the elite culture also. The categories of belief of the analyst may be different from those of the elites studied.

sorts—personal or economic interactions, for instance. By separating political beliefs from other beliefs one can explore the nature of their interrelationship.

The relationship among the various beliefs that individuals have—both political and non-political—represents one of the most important topics of discussion for the student of political culture. Though in dealing with culture we tend to think of patterned relationships among beliefs, this ought not to imply that all sets of beliefs are perfectly integrated. It may be that political beliefs are sharply discontinuous from or in some way inconsistent with other beliefs. One can conceive of a society in which cynicism is reserved for politics but does not pervade other social interactions. A more usual situation would be one in which the formal values stressed in the political realm were not consistent with those stressed in other areas of social life. One of the predominant features of the cultural patterns in most of the transitional nations . . . is that belief systems stressing modernity in politics are sharply different from the more traditional beliefs associated with other aspects of life, and this may cause severe strain for those who are forced to act within the political culture and the more general culture at the same time.[11] Similarly, the relationship between basic political beliefs and political behavior is not unambiguous. The same belief can be converted into action in a number of ways, just as the same action can have its roots in many alternate beliefs. But though the relationship between political beliefs and general beliefs on the one hand and political behavior on the other may vary, it is clear that they are never irrelevant to each other. Inconsistencies between belief and belief or between belief and action have significant implications for a political system.[12]

Furthermore, when a relationship is found between political beliefs and general social beliefs, one cannot assume that general social beliefs affect political beliefs with no reciprocal effects. Though it is probably true that individuals are more likely to generalize from basic social values to political values, political values may very well have effects on values in other spheres of life. In democratic political systems the belief that one ought to participate in political decisions is often used as a justification for participation in non-political decisions. . . .

THE HOMOGENEITY OF CULTURE

Lastly a word must be said about the problem of cultural homogeneity in a political system. The focus on political *culture* rather than political *attitudes* implies a concentration upon the attitudes held by all the members of a political system rather than upon the attitudes held by individuals or particular categories of individuals. As anthropologists have used the term "culture," it has frequently referred to those aspects of belief systems that are shared by members of a society and that are distinctive to that group.[13] Our approach is somewhat different. To concentrate only on shared beliefs might lead one to overlook

---

[11] In this connection see Almond and Verba, *The Civic Culture*, . . . chaps. 10 and 15. We argue there that the extent to which the political value system interpenetrates with the system of general social values is a major dimension of political culture with significant implications for the operation of the political system. See also the discussion below of discontinuities in the socialization process when the values in one sphere of life are not congruent with those in some other.

Harry Eckstein's discussion of Weimar Germany, where authority patterns in non-political relationships were sharply different from those formally institutionalized in politics, is relevant here. The attendant strains from such discontinuity resemble those in new nations where new political norms are laid upon a traditional base—see Eckstein's *A Theory of Stable Democracy* (Princeton, N.J.: Center of International Studies, 1961).

[12] Not all such inconsistencies are destabilizing. See the discussion of the role of inconsistencies between behavior and belief within democratic political systems in Almond and Verba, *The Civic Culture*, . . . chap. 15, which gives examples of the way in which inconsistencies play a positive role in democratic stability.

[13] "By 'culture' we mean that historically created definition of the situation which individuals acquire by virtue of participation in or contact with groups that tend to share ways of life that are in particular respects and in their total configuration distinctive." Clyde Kluckhohn, "The Concept of Culture," in Richard Kluckhohn (ed.), *Culture and Behavior* (New York: The Free Press of Glencoe, 1962).

situations where significant political beliefs were held only by certain groups, and where the very fact that these attitudes were not shared by most members of the system was of crucial importance. This is particularly a problem as one begins to deal with societies as large and complicated as the nation-state. Our approach is to begin with a set of belief dimensions that seem particularly crucial for the understanding of the operation—in particular the development and adaptability—of a political system, and then ask whether or not members of a political system share attitudes on these dimensions.

The degree to which basic political attitudes are shared within a political system becomes thus a crucial but open question. A major pole of differentiation among political cultures is the number of basic political attitudes that are widely shared and the patterning of the differences in political belief among the various groups in society. . . .

Of course for the purposes of predicting the political future of a nation the beliefs of certain groups are more crucial than others—those in actual political power, members of organized groups, those living near the centers of communications, and the like. But in an era of rapid political change when the mobilization of mass support is so eagerly pursued, few political subcultures can be ignored.

The degree to which political beliefs are shared may be a good indicator of the cohesiveness of a society. But this is probably more true of certain types of beliefs than of others. . . . Some basic political values may indeed lead to conflict if shared on the level of generality we are discussing. Thus all members of a system . . . may share a belief that output of the political system ought to benefit a fairly parochially defined group—their own family or perhaps their local region. But though this implies a sharing of political values on one level—a shared set of criteria for evaluating governmental output—the fact that on a different level the specific group used as a criterion differs among the members of the system could lead to conflict. Similarly, the existence of an "ideological" political style through a shared cultural norm may institutionalize divisiveness. Though the commitment to ideology is shared throughout the system, the ideologies to which individuals are committed vary from group to group.

Karl W. Deutsch

Political Communications and the Political System

The recent models of communication and control may make us more sensitive to some aspects of politics that have often been overlooked or slighted in the past. This, as we know, is a major function of models in their early stages. Well before they permit quantitative inferences, they may already aid in adding new criteria of relevance: What kinds of facts are now interesting for us, since we have acquired a new intellectual context for them?

. . . .

The Concept of Information

. . . [L]et us remember the distinction made by theorists like Norbert Wiener between communication engineering and power engineering. Power engineering . . . transfers energy which then may produce gross changes at its place of arrival. In the case of power engineering, these changes are in some sense roughly proportionate to the amount of energy delivered. Communication engineering transfers extremely small amounts of energy in relatively intricate patterns. It can produce sometimes very large changes at the point of arrival, or in the "receiv-

Reprinted with permission of the Macmillan Company from *The Nerves of Government* by Karl W. Deutsch. Copyright © 1966 by The Free Press, a Division of The Macmillan Company. Copyright © 1963 by The Free Press of Glencoe, a Division of The Macmillan Company.

er" of the "message," but these changes need in no way be proportionate to the amount of energy that carried the signal, much as the force of a gun shot need not be proportionate to the amount of pressure needed to set off the trigger.

Power, we might say, produces changes; information triggers them off in a suitable receiver. In the example just given, the most important thing was not the amount of pressure on the trigger, once it had reached the required threshold, but rather the fact that it was delivered at the trigger, that is, at one particular point of the gun. Similarly, the information required for turning the gun to a particular target need not be carried by any amount of energy proportionate to the energy delivered to the target by the gun. The important thing about information is thus not the amount of energy needed to carry the signal, but the *pattern* carried by the signal, and its relationship to the set of patterns stored in the receiver.

Generally, *information* can be defined as a patterned distribution, or a *patterned relationship between events*. Thus the distribution of lights and shadows in a landscape may be matched by the distribution of a set of electric impulses in a television cable, by the distribution of light and dark spots on a photographic plate, or on a television set, or by the distribution of a set of numbers if a mathematician had chosen to assign coordinates to each image point. In the case of photography or television the processes carrying this information are quite different from each other: sunlight, the emulsion on the photographic plate, the electric impulses in the cable, the television waves, the surface of the receiving screen. Yet each of these processes is brought into a state that is similar in significant respects to the state of the other physical processes that carried the image to it.

A sequence of such processes forms a *channel of communication*, and information is that aspect of the state description of each stage of the channel that has remained invariant from one stage to another. That part of the state description of the first stage of the channel that reappears invariant at the last stage is then the information that has been transmitted through the channel as a whole.[1]

The Reception of Information

The effectiveness of information at the receiver depends on two classes of conditions. First of all, *at least some parts of the receiving system must be in highly unstable equilibrium*, so that the very small amount of energy carrying the signal will be sufficient to start off a much larger process of change. Without such disequilibrium already existing in the receiver, information would produce no significant effects.

This obvious technical relationship might have some parallels in politics. The extent of the effect of the introduction of new information into a political or economic system might well be related, among other things, to the extent of the instabilities that already exist there. A crude empirical expression of this problem is found in the perennial debate concerning the relative share of "domestic instabilities" versus "foreign agitators" in strikes or political disturbances. On a somewhat more sophisticated level, the problem reappears as the question of the role of ideas in inducing or prompting social change, and it has relevance for studies of the conditions favoring political reform or technological innovations in different countries. In all such cases a search for "promising instabilities," that is, instabilities relevant for possible innovation, should be rewarding.

Richness of Information and Selectivity of Reception

The second class of conditions involves the *selectivity* of the receiver. What patterns are already stored in the receiver, and how specific must be the pattern of the incoming signal in

---

[1] George A. Miller states the same point in somewhat different language: "The 'amount of information' is exactly the same concept that we have talked about for years under the name of 'variance.' The equations are different, but if we hold tight to the idea that anything that increases the variance also increases the amount of information we cannot go far astray.

"The advantages of this new way of talking about variance are simple enough. Variance is always stated in terms of the unit of measurement—inches, pounds, volts, etc.—whereas the amount of information is a dimensionless quantity. Since the information in a discrete statistical distribution does not depend upon the unit of measurement, we can extend the concept to situations where we have no metric and we would not ordinarily think of using the variance. And it also enables us to compare results obtained in quite different experimental situations where it would be meaningless to compare variances based on different metrics. So there are some good reasons for adopting the newer concept.

"The similarity of variance and amount of information might be explained this way: When we have a large variance, we are very ignorant about what is going to happen. If we are very ignorant, then when we make the observation it gives us a lot of information. On the other hand, if the variance is very small, we know in advance how our observation must come out, so we get little information from making the observation.

"If you will now imagine a communication system, you will realize that there is a great deal of variability about what goes into the system and also a great deal of variability about what comes out. The input and the output can therefore be described in terms of their variance (or their information). If it is a good communication system, however, there must be some systematic relation between what goes in and what comes out. That is to say, the output will depend upon the input, or will be correlated with the input. If we measure this correlation, then we can say how much of the output variance is attributable to the input and how much is due to random fluctuations or 'noise' introduced by the system during transmission. So we see that the measure of transmitted information is simply a measure of the input-output correlation.

"There are two simple rules to follow. Whenever I refer to 'amount of information,' you will understand 'variance.' And whenever I refer to 'amount of transmitted information,' you will understand 'covariance' or 'correlation.'

"The situation can be described graphically by two partially overlapping circles. Then the left circle can be taken to represent the variance of the input, the right circle the variance of the output, and the overlap the covariance of input and output. I shall speak of the left circle as the 'amount of transmitted information,' you will understand 'covariance' or formation, and the overlap as the amount of transmitted information."

George A. Miller, "The Magical Number Seven, Plus or Minus Two: Some Limits on Our Capacity for Processing Information," *Psychological Review*, 63 (March, 1956), 81–82.

order to produce results? A simple example of this problem is furnished by the relationship of lock and key. How many tumblers and notches have been built, let us say, into a particular Yale lock, and what restrictions do they impose upon the distribution of notches on any key that is to turn it? Clearly, the effectiveness of any key in turning a particular lock depends only slightly on the energy with which it is turned (beyond a minimum threshold), and far more on the correspondence of the configuration of its notches with the configuration of the tumblers in the lock.

This crude example shows that there is a measurable difference between locks that are simple and those that are elaborate. Simple locks may have few tumblers in them, and may be turned by a wide variety of differently patterned keys, as long as each of these keys corresponds to the others and to the lock at the few relevant points determined by the distribution of the tumblers. A more elaborate lock will have more tumblers and thus is likely to impose more restrictions on the patterns of keys able to turn it. The selectivity of receivers, then, is related to, among other things, the richness and specificity of information already stored in them.

Similarly, there is a measurable distinction between the richness of information contained in different images. The amount of detail that a photographic film can record is limited by, among other things, the fineness of the grain. Reproductions of photographs in ordinary newspapers are made with the help of screens with only a few hundred lines to the inch, and are thus much poorer in detail and cruder in appearance than photographs. The same is true of pictures in television, and of details of the human voice in telephoning or recording. In all these processes details can be lost and the amount of lost information can be measured. Altogether a large amount of thought and experience has gone into the measurement of information, of the possible losses of information under certain conditions, and of the carrying capacity of certain communication channels in terms of quantities of information.

THE MEASUREMENT OF INFORMATION AND THE FIDELITY OF CHANNELS

The upshot of all this work has been the emergence of information as a quantitative concept. Information can be measured and counted, and the performance of communication channels in transmitting or distorting information can be evaluated in quantitative terms. Some of these measurements in electrical engineering have reached high levels of mathematical sophistication.[2]

Other methods of measuring information may be simpler. Information could conceivably be measured in an extremely crude way in terms of the percentage of image points transmitted or lost on a line screen of a given fineness, or in terms of the number of outstanding details lost as against the number of outstanding details transmitted; or perhaps in a slightly more refined way, information could be measured by the number of such details lost or transmitted in terms of their probability in the context of the set of details already stored in the receiver.

The fact that social scientists may have to use some of the cruder rather than the more refined methods for measuring the amounts of stored or transmitted information should not obscure the importance of being able to measure it at all. In the investigations of Gordon Allport and L. J. Postman on the psychology of rumor, quantitative measurements of information were used to good effect: a subject was shown a picture for a short time and then told to describe it to a second person who had not seen it. The second person then had to tell a third, and so on through a chain of ten, and the amount of details lost or distorted at each stage was recorded. When each successive stage of retelling was plotted

[2] Y. W. Lee, *Statistical Theory of Communication* (New York: Wiley, 1960).

along a horizontal axis, and the number of details retained correctly were plotted vertically, the result was a curve of the loss of details that paralleled strikingly a well-known curve of the forgetting of details by individuals in the course of several weeks. In both cases the details were flattened and sharpened, that is, simplified and exaggerated, and they were assimilated by distortion to the prevailing opinions and cultural biases of the individuals carrying the memories of rumors.[3]

INFORMATION AND SOCIAL COHESION

If we can measure information, no matter how crudely, then we can also measure the cohesion of organizations or societies in terms of their ability to transmit information with smaller or larger losses or distortions in transmission. The smaller the losses or distortions, and the less the admixture of irrelevant information (or "noise"), the more efficient is a given communications channel, or a given chain of command.

If we think of an ethnic or cultural community as a network of communication channels, and of a state or a political system as a network of such channels and of chains of command, we can measure the "integration" of individuals in a people by their ability to receive and transmit information on wide ranges of different topics with relatively little delay or loss of relevant detail.[4]

Similarly, we can measure the speed and accuracy with which political information or commands are transmitted, and the extent to which the patterns contained in the command are still recognizable in the patterns of the action that are supposed to form its execution.

The difference between a cohesive community or a cohesive political system, on the one hand, and a specialized professional group—such as a congress of mathematicians—on the other hand, consists in the multiplicity of topics about which efficient communication is possible. The wider this range of topics, the more broadly integrated, in terms of communications, is the community, or the "body politic." In traditional societies this range of topics may be broad, but limited to topics and problems well within the traditional culture; the ability to communicate widely and effectively on nontraditional topics may be relevant for the cohesion and learning capacity of peoples and political systems in countries undergoing rapid industrialization.

All this is not to say that the measurement of losses in the transmission of information on different ranges of topics is the only way in which the predisposition for political or social cohesion can be measured. Approaches in terms of interlocking roles and expectations might be another way. It is suggested, however, that the information approach offers an independent way of measuring basic cohesion, however crudely, and that it can do so independently from the current political sympathies of the participants. Such sympathies or conflicts might show up sharply in the execution of controversial commands, such as, let us say, between Northerners and Southerners in the United States in the 1850's and again during the Reconstruction period, or between nationalists and Social Democrats in Germany before 1914. Measurements of the

---

[3] G. W. Allport and L. J. Postman, "The Basic Psychology of Rumor," *Transactions of the New York Academy of Science*, ser. II, VIII (1945), 61–81, reprinted in Wilbur Schramm (ed.), *The Process and Effects of Mass Communication* (Urbana, Ill.: University of Illinois Press, 1954), pp. 141–55. Cf. also F. C. Bartlett, "Social Factors in Recall," in T. M. Newcomb and E. L. Hartley (eds.), *Readings in Social Psychology* (New York: Holt, 1947), pp. 69–76 especially on "The Method of Serial Reproduction," Newcomb and Hartley, *Readings in Social Psychology* . . . , p. 72; C. I. Hovland, I. L. Janis and H. H. Kelley, *Communication and Persuasion* (New Haven, Conn.: Yale University Press, 1953), pp. 245–49; and C. I. Hovland, "Human Learning and Retention" in S. S. Stevens (ed.), *Handbook of Experimental Psychology* (New York: Wiley, 1951), pp. 613–89.

[4] K. W. Deutsch, *Nationalism and Social Communication* (Cambridge-New York: M.I.T. Press-Wiley, 1953), pp. 70–74.

accuracy and range of topics of information transmitted in a state or a political or social group would also show the extent and depth of the remaining area of effective mutual communication and understanding among its members. In this manner we might gain important data for estimating the chances for strongly unified behavior of the political system, as well as of the underlying population, in later emergencies.

Face-to-Face Communication Networks and Legitimacy Symbols

If many studies of politics have stressed *power*, or enforcement, it should now be added that information precedes compulsion. It is impossible to enforce any command unless the enforcing agency knows against whom the enforcement is to be directed—a truism that has given much delight to readers of detective stories. The problem becomes more serious where enforcement is to be directed against a significant number of personally unknown members of an uncooperative population, as in situations of conspiracy, political "underground activities," resistance to military occupation, or guerrilla warfare.

Similarly, information must precede compliance. It is impossible for anyone to comply with a command unless he knows what the command is. In this sense, a "legitimacy myth," discussed by some writers, is an effective set of interrelated memories that identify more or less clearly those classes of commands, and sources of commands, that are to be given preferential attention, compliance, and support, and that are to be so treated on grounds connecting them with some of the general value patterns prevailing in the culture of the society, and with important aspects of the personality structures of its members. Yet, even where such legitimacy beliefs are effectively held in the minds of a large part of the population, its members must have ways of receiving the commands rapidly and accurately if they are to act on them. Governments-in-exile or leaders of underground movements during World War II not only had the task of maintaining their status as legitimate but also the task of maintaining an actual network of communication channels to carry the essential two-way flow of information.

In evaluating the political significance of this fact, two mistakes may easily be made. The first mistake consists in overestimating the importance of impersonal media of communication, such as radio broadcasts and newspapers, and underestimating the incomparably greater significance of face-to-face contacts. The essence of a political party, or of an underground organization, consists in its functioning as a network of such face-to-face contacts. These face-to-face contacts determine to a large degree what in fact will be transmitted most effectively and who will be the "insiders" in the organization, that is, those persons who receive both information and attention on highly preferred terms.

The second mistake might consist in considering legitimacy myths or symbols in isolation from the actual communications networks, and from the human networks—often called "organizations," "machines," "apparatus," or "bureaucracy"—by which they are carried and selectively disseminated. During World War II, several governments-in-exile continued to be considered legitimate by most of the population of their respective Nazi-occupied countries. The decisive failure of Nazi legitimacy beliefs to gain wide acceptance in those countries was followed by the growth of underground resistance organizations staffed to a significant extent by Communists or Communist sympathizers. The initial opportunities for Communist participation in this underground depended in part on the legitimacy beliefs that permitted it; without these beliefs, which transferred some of the prestige of the French or Czechoslovak republic or the royal government of Norway to all participants in the underground, the Communists would have had to carry on separate and weaker underground activities. While the governments-in-exile often retained control of radio broadcasts from London, the political outcome of the underground period depended in considerable degree upon the actual position of Communists in the underground network of face-to-face contacts and decisions. Where that position had been strong, the governments of National Committee in exile had to deal with a far stronger Communist organization and influence than the legitimacy beliefs of the country in themselves would have led one to expect.

Without widespread and favorable legitimacy beliefs, a face-to-face communications network is exceedingly hard to build, as, for example, the failure of the Quisling group in Norway has demonstrated. Without effective control of the bulk of the actual face-to-face communication networks, on the other hand, the nominal holders of the legitimacy symbols may become relatively helpless vis-à-vis those groups that do have this control. The Polish government-in-exile, as well as the group of President Beneš of Czechoslovakia, found themselves with far less power at the end of the war than their symbolic status of legitimacy would have suggested.

Perhaps we may suspect, accordingly, that it is rather in the more or less far-reaching *coincidence* between legitimacy beliefs and social communication channels that political power can be found. Thus, when we speak loosely of the "manipulation of political symbols" we might do well to distinguish sharply between their manipulation in a speech or book, and the manipulation of those human and institutional chains of communication that must carry and disseminate these symbols and all other information and that are crucial for the functioning of political power.

The frequent superiority of networks of face-to-face contacts over either isolated legitimacy symbols or even impersonal media of mass communication can be illustrated by two examples. The Democratic party in the big cities in the United States has shown persistent electoral strength despite the fact that it is notoriously weak in newspaper support; however, it is relatively strong in face-to-face contacts on the ward level. The second example was demonstrated to television viewers in the United States during the presidential compaign of 1952: the discrepancy between the amount of publicity and symbolic reputation attracted by Senator Estes Kefauver in his campaign for the Democratic presidential nomination, and his inability either to overcome the coldness or hostility of the "insiders" and "machines" of the Democratic party or to attract really substantial support without their aid.

As these examples indicate, the discrepancy between the "newspaper strength" of a leader or candidate and his real strength, not at the "grass roots" but *at the decisive middle level of communication and decision*, may be a promising field for comparative political research.

Robert A. Dahl

Power and Influence

Power and influence are . . . concepts of key importance in political analysis. Curiously enough, however, systematic analysis of these concepts it rather recent, partly because serious attempts have not been made until quite recently to formulate them rigorously enough for systematic study.

. . . .

INFLUENCE

Suppose you were to stand on a street corner and say to yourself, "I command all automobile drivers on this street to drive on the right-hand side of the road"; suppose also that all the drivers actually did as you "commanded" them to do. Still, most people would regard you as mentally ill if you were to insist that you had just shown enough influence over automobile drivers to compel them to use the right-hand side of the road. On the other hand, suppose a policeman is standing in the middle of an intersection at which most traffic ordinarily moves ahead; he orders all traffic to turn right or left; the traffic moves right or left as he orders it to do. Then common sense suggests that the policeman acting in this particular role evidently influences automobile drivers to turn right or left rather than go ahead.

Our common-sense notion, then, goes something like this: A influences B to the extent that

Robert A. Dahl, *Modern Political Analysis*, © 1963. Reprinted by permission of Prentice-Hall, Inc., Englewood Cliffs, New Jersey.

he gets B to do something that B would not otherwise do. An influence relation between two actors might be illustrated in this fashion:

A	doesn't influence	B	does → x
A	influences →	B	doesn't do x
			does → y

Influence, then, is a *relation*—among individuals, groups, associations, organizations, states. We can use a convenient bit of jargon and say that influence is a *relation among actors* in which one actor induces other actors to act in some way they would not otherwise act. Of course this definition also includes instances in which actor A induces B to go on doing something he is now doing, though B would stop doing it except for A's inducements.

In principle, then, we can determine the *existence* of influence, and also the *direction* of influence: who influences whom. In practice it is often difficult to find out who influences whom, but we shall postpone considering the practical obstacles in order to deal with a preliminary conceptual problem. Often it is not enough to know simply that some actors influence others. Frequently one also wants to know what the *relative* influence is among different actors. Is Thompson more influential with the mayor than Green is? Who are the most influential people in town on school appropriations? What senators have the most influence in Congress on matters of foreign policy? What countries are the most influential on disarmament questions in the United Nations? Which actors are the least influential in these situations? In the United States are poor people generally less influential on questions of taxation than rich people? In short, one wants to know *how much* influence actor A has over actor B, and one wants to *compare* the amount of influence different actors have over others.

One would find it almost impossible to discuss political life without comparing the influence of different actors. Even to distinguish a democracy from a dictatorship requires one to estimate the relative influence of citizens and leaders. Aristotle's famous classifications scheme obviously assumes that one can measure—and thus compare—relative amounts of influence; for in order to classify any particular political system one must first determine who has the most influence in the system—one person, a few, or a majority?

However, attempts to compare the relative influence of different actors in a political system are the source of great confusion in political analysis. Probably the most important cause of confusion is the fact that many different measures of influence are used, and they are almost always used implicitly rather than explicitly. It is as if two neighbors, one very tall and the other very short, each paced off the length of the boundary between their properties and fell to arguing over the results without ever noticing that one of them took long steps while the other took short steps. To reduce this kind of fruitless controversy, let us consider some of the underlying measures of influence that are used—usually implicitly—to compare the relative influence of different actors.

Five Ways of Comparing Influence

It will help if we pause to take another look at our common-sense definition of influence: A influences B to the extent that A gets B to do something that B would not otherwise do. Somehow, then, A changes B's behavior from what it would have been. If so, why not measure A's influence by the *extent or amount of the change* in B's behavior from what it would have been? The first three underlying measures of influence we are going to examine all rest on this simple idea: The greater the change in some aspect of B's inner or overt behavior that A induces, the greater A's influence over B.

This way of thinking about influence is analogous to the concept of force in mechanics. In mechanics object A exerts a force on object B if A produces a change in the velocity of B. Galileo's famous law of inertia states that a body left to itself will move with uniform velocity in one and the same direction. Any change in the velocity of a body, then, indicates the presence of a force. And the size of the force is proportional to the size of the change in velocity. Thus one object exerts more force than another on a third if the first produces a greater change in velocity.

Although we ought not to push such analogies very far, there is little doubt that our ideas about underlying measures of influence rest on intuitive notions very similar to those on which the idea of force rests in mechanics. The underlying idea in both cases can be represented in the following way:

$$A \text{ influences } B \begin{array}{l} \text{doesn't do } x \\ \text{does} \longrightarrow y_1 \\ \text{does} \longrightarrow y_2 \end{array} \left.\begin{array}{l}\\ \\ \end{array}\right\} \begin{array}{l}\text{Amount of A's}\\ \text{influence over B}\end{array}$$

Since the change represented by the vector from x to y_1 is less than the change represented by the vector from x to y_2, A's influence over B is greater in the second case than in the first.

Let us now turn to some underlying measures of influence that draw on this idea.

1. *The amount of change in the position of the actor influenced.* Here is a typical statement illustrating this way of comparing influence:

Considering what our foreign policy was when Truman came into office and what it was when he left, it is fair to conclude that no other President has ever exerted so much influence over Congress on foreign policy actions in times of peace. Joining the UN, Greek-Turkish Aid, the Marshall Plan, creating and joining NATO—these all represented profound innovations and great breaks with the past.

Now consider the following example:

Members of a school board are debating the question of raising teachers' salaries. One member, A, wants to keep salaries as they are. B would like to raise salaries by $250 a year, C favors an increase of $500, and D wants to add $1,000. The four members of the board can be visualized standing along a chalk line according to the amount of salary increase each one favors:

A	B	C	D
$0	$250	$500	$1,000

Two other citizens, Thompson and Green, would like the board to adopt D's position:

> *Thompson:* I know I can't budge either A or C, but I can get B to join C and accept a $500 increase. B gets a lot of business from me. I know from past experience with him that he'll be willing to compromise on that figure if I ask him to; but I'm sure if I try to push him beyond $500 he'll get stubborn and I won't get anywhere.
>
> *Green:* Well, I know from past experience that I can't budge C either. I just don't have any influence with C at all. But I can get both A and B to agree to $500.
>
> *Thompson:* Well, Green, you evidently have more influence with A than I do. We both seem to have the same amount of influence with B, and neither of us has any influence over C.

If Green and Thompson were accurate in their initial estimates, one would be inclined to agree with Thompson's conclusion. In this case as in many others when we say that one actor has "more" influence than another over some third actor, we mean that the one is capable of producing a greater change in the position held by the third.

However, it is not always satisfactory or even possible to measure the relative influence of different political actors in this way. For one thing, we do not always know what the initial positions of the different actors were, nor how much they have actually changed. In politics, individuals sometimes take extreme positions for bargaining purposes, hoping in this way to end up with a compromise that is very close or even identical to their real but concealed initial position. In fact, one of the most important skills in politics—one of the reasons why the practice of politics is an art—is the ability to detect a bargaining position and to guess correctly how much it diverges from the adversary's "real" position. Another skill is just the converse: the capacity to conceal one's real position and create belief in one's bargaining position. In international politics and labor-management disputes, parties often try to mask their real positions by threats, bluffs, and displays of strength—troop movements, angry words, strike-votes. The need to uncover an adversary's true position in turn leads to intelligence operations, research, espionage, and attempts to test his intentions by counter-threats and counter-moves.

Moreover, it is not always possible to decide whether a change in position is larger or smaller than another change. In the case of the school board, if A changes to B's position and C to D's, which change would be the larger? It is true that A's change involves less money, but perhaps A places a much higher value on being economical than C does; he may be much more reluctant to shift to B's position than C is to shift to D's position. In this sense it is reasonable to say that it takes more influence to induce A to vote for a $250 increase than B to vote for a $500 increase. What we need, in short, is a measure of what might be called the psychological distance involved in a change. This suggests a second underlying measure of influence.

2. *The subjective psychological costs of compliance.* Here are some typical statements embodying this measure of influence:

> The influence of the old-fashioned political boss declined as immigrant groups became assimilated into American life and as social security took over many of the functions of the boss, for the boss's followers had less and less need, or desire, for his services—the basket of food at Thanksgiving and Christmas, intervening with the police, the stray job with the city, and so on.
>
> It doesn't take a President with much

influence to get support for welfare measures among congressmen from urban areas, because they are already in favor of it. It takes a lot more influence to get congressmen from rural districts in the Midwest to go along.

These statements point up the fact that a seemingly equal change in the "objective" positions of two different actors may actually require quite different amounts of subjective change. The "costs" of complying with the wishes of someone else can be very different for different people, depending on their values and their situation.

For example, consider different members of a labor union faced with the prospect of a strike. Some members, let us say, have managed to put aside savings in anticipation of the strike; others have not. A union leader who has enough influence to persuade the first group to vote for a strike may very well not have enough influence to persuade the second group. To the members who have no savings, the "costs" of a strike are likely to seem much higher than to the members who do. Hence it takes correspondingly greater influence to persuade the second group to strike.

In the same way, it takes more influence to induce pacifists to support the draft than militarists; isolationists to support the UN than internationalists; southern Democrats rather than northern Democrats to support civil rights legislation.

In practice, to be sure, this underlying measure cannot always be applied, since we do not always know what the differences are in the real psychological costs of compliance among different individuals. Nevertheless, one often makes guesses based on known difference in values and situation; and there is no doubt that a good deal of confusion about relative influence might be avoided if the assumptions concerning subjective costs were perfectly clear.

3. *The amount of difference in the probability of compliance.* Consider again the hypothetical case in which you stand on the edge of a busy street in an American city and silently "command" all the automobile drivers to drive on the right. To use the fact that the drivers did indeed drive on the right as evidence for your influence would, it was suggested, make you a likely candidate for a mental institution. For everyone knows that whether or not you "command" them to do so, in this country drivers are compelled by law to drive on the right and would do so without your "command." In other words, the chances that any given driver will drive on the right are already very high—say 999 out of 1,000, allowing for the rare deviant case; and your silent "command" does not alter these chances in the slightest. The policeman, however, is in a different position. Perhaps if he were not standing there only 1 car out of 10 would turn right at that particular intersection. But on his signal, every car turns right. In your case the difference in the probability that they will turn right when you "command" them to do so, and when you don't "command" them to do so, is exactly zero. With the policeman, the change is from 1 chance out of 10 to 10 out of 10—a difference of nine chances out of ten.

Why not measure relative influence, then, by the size of the difference in probabilities of compliance? Notice that the *difference* in probabilities is what matters. If 59 out of 60 Democratic senators vote for a bill proposed by a Democratic President, it would be premature to conclude that the President has great influence with his Congressional party. What we must know first is how many Democratic senators would have voted for the measure anyway, even if the President took no position on it at all.

Here is a statement that implies this underlying measure:

> With Congressman X, the Farm Bureau Federation has enormous influence, the Chamber of Commerce considerably less, and the AFL–CIO almost none. Over the years, the Farm Bureau has learned that whenever Congressman X is not going to vote for a bill the Bureau wants to pass, all

they have to do is to get on the phone and indicate how they stand. The chances are 9 out of 10 that he will vote the way they want him to. In similar circumstances involving business questions the Chamber of Commerce can get him to change about one time out of three. The AFL–CIO, on the other hand, wins him over less than one time out of ten.

The example also illustrates some of the difficulties with this measure. First, well-grounded estimates of "chances" or probabilities require either random events, as with a coin or a die, or else a large number of past occurrences of equivalent events. Political decisions are ordinarily neither random nor equivalent. Hence it is usually difficult to estimate probabilities except in a very loose way. Second, it is often difficult to know the initial likelihood of a particular event because, as has already been indicated, the initial positions of various participants are often unknown. Finally, the mere difference in probabilities is not a wholly satisfactory measure precisely because it does not take into account either of the two previous dimensions—the extent of change in position and the costs of compliance.

4. *Differences in the scope of the responses.* Consider a statement like the following:

> Taken all in all, the Majority Leader is the most influential man in the Senate. A committee chairman is usually one of the most influential leaders on matters falling within the jurisdiction of his committee, but on other questions he usually wields little more influence than the average member. The Majority Leader, on the other hand, tends to be highly influential on practically any matter that comes before the Senate.

So far in our analysis the responses of those subject to influence have been considered roughly similar and therefore comparable. All actions were assumed to be of approximately the same nature. The decisions of different school-board members on teachers' salaries, for example, are comparable because they all deal with the same subject, teachers' salaries. But can we compare the relative influence of different actors when they are influencing other actors on different kinds of questions? For example, if by any of the three measures already discussed it were found that President Truman was more influential than President Eisenhower on matters of foreign policy, but President Eisenhower was more influential on matters of domestic policy, could we merge these two different kinds of issue into a single common "scope" and thus decide which of the two presidents was the more influential with Congress over the whole scope of congressional action?

This is a troublesome problem. No wholly satisfactory solution to it has yet been found. Meanwhile, however, it is possible to avoid some snares that observers frequently fall into when they have tried to analyze influence.

First, a statement about influence that does not clearly indicate the scope it refers to verges on the meaningless. When one hears or reads that X is highly influential, the proper question is: "Influential with respect to what?" The failure to insist on this simple question often leads political observers astray. For example, some studies of American cities report the existence of a "dominant" elite that "runs" the city. Thus in "Regional City," a large southern city, 14 well-informed persons were given four lists totaling more than 175 names and were asked to give "their opinions on who were the top leaders on each of the lists." The 10 names on each list most frequently named by the judges were assumed to be "the top leaders." The 40 persons chosen in this way were then described as the "top leaders" in Regional City.[1] Unfortunately, however, these conclusions are all vitiated by the fact that the 14 judges were never asked to specify in what areas of activity different leaders were influential. Are the men who decide on the candidates for mayor the same as those who determine school appropria-

[1] Floyd Hunter, *Community Power Structure* (Chapel Hill, N. Ca.: University of North Carolina Press, 1953).

tions? City policies on redevelopment? On segregation and desegregation?

On the evidence presented, it is impossible to know. We simply cannot tell whether men who are influential in Regional City on some questions are or are not influential on other questions. A concern for the scope within which different leaders exert their influence might have established the existence of a single, homogeneous ruling elite. On the other hand, conceivably it might have led to the discovery of leaders drawn from other social or economic groups. In fact, studies of American communities in which different issue-areas are examined often reveal a highly pluralistic structure.

Second, even when different sectors of influence are clearly designated, it may be difficult or impossible to say which of several actors is the most influential over the whole scope of their influence. An analogous problem arises if we try to compare two athletes who compete in different sports. Was Babe Ruth a better athlete than Jack Dempsey? The question seems unanswerable. We might measure Babe Ruth against other hitters by using the number of home runs as a measure, and we might measure Dempsey against other fighters by using the number of knockouts, but how to compare home runs with knockouts? If two athletes were to compete in two or more of the same sports, perhaps we could say that one was a better athlete than the other if the one was as good as or better than the other in all fields of competition. For example, if A was as good a hitter as B in baseball, an equally good boxer, but better in squash, we might then say that A was a better athlete than B.

In exactly the same way it seems reasonable to say that one actor is more influential than another over the whole scope of their influence only if A's influence is not less than B's in any particular issue-area and is greater than B's in at least one issue-area. If Green and Thompson are highly influential on issues having to do with the public schools and urban redevelopment, and if Green is more influential than Thompson on political nominations, then Green is more influential than Thompson over the whole scope of their influence.

Real life, however, does not always produce such neatly tailored situations. Green may be more influential than Thompson on school questions, while Thompson may be more influential than Green on political nominations. What do we say in this case? We might try to assign weights to different issue-areas, but weights are bound to be arbitrary. If schools are given a weight of 1, what weight should be given to political nominations—2, 5, $\frac{1}{2}$? How can we justify the weights we assign?

There is at present no single best way of solving the problem of comparability when different actors have different levels of influence in several different issue-areas. Perhaps the most important lesson the student of politics can gain from this is the need for caution and clarity in making comparisons of influence over several issue-areas. As in many other cases, it is wise in political analysis to specify whether we are adding oranges, or apples, or oranges *and* apples.

5. *The number of persons who respond.* If Green controls 5,000 votes and Thompson 10,000, it seems reasonable to say that Thompson has more influence in elections than Green.

To use the number of persons who respond in a certain way as a measure of influence is so obvious and clear that one might suppose it to be flawless. Alas, like other underlying measures this one also has its limitations. These are suggested by the other measures we have already discussed. Green may be able to induce his voters to make a greater change in their positions than Thompson can. Or Green may be able to sway his captive voters even when the disadvantages of compliance seem to them very great, whereas Thompson can only get his voters to do things that do not matter much to them anyway. Or Green's control may be much more definite: The likelihood that his captive voters will do what he asks is virtually certain, whereas Thompson's control is much more uncertain. Finally, Green may be able to control his voters over a broader range of issues than Thompson can; Green, perhaps, can deliver his 5,000 votes on any kind of election, local, state, or national, or on any referendum, whereas Thompson can deliver his 10,000 votes merely in local elections.

Summary. All five of the underlying measures of influence we have examined help to illuminate some important aspects of influence. All five have limitations. The analysis suggests the following injunctions to the student of politics:

1. Look for information that will enable you to use as many measures as possible in trying to determine relative influence.

2. Adapt your comparison to the kind of information available.

3. Be specific. A paradigm for any comparison of power might be: "—— is more influential than —— with respect to —— as measured by —— and ——."

4. Be constantly aware of what you have had to leave out of your appraisal. (In what follows, we shall frequently violate each of these injunctions except this one!)

Potential Influence Versus Actual Influence

It is one thing to describe or measure differences in influence, but it is quite another to explain these differences. . . . [T]he reasons why some individuals or groups acquire more influence than others over some scope of decisions are reducible to three:

1. Some actors have more political resources at their disposal than others. Or

2. Given the resources at their disposal, some actors use more of them to gain political influence. Or

3. Given the resources at their disposal, some actors use them more skillfully or effectively than others do.

Thus it is important to distinguish between the past or current influence of a particular actor within some scope of decisions, his probable future influence, and his *maximum potential influence* if he were to use *all* his existing political resources with optimum skill to acquire influence within that scope of decisions. An actor's current influence in any given scope always (or nearly always) falls short of his maximum potential influence. Consider the following imaginary case:

Rodney Brown is the richest man in town. In fact, he has made a small fortune on the stock market. He owns a controlling share in the most important local newspaper, he completely owns one radio station, he has a large interest in the biggest bank in town, and he owns dozens of down-town commercial properties. In addition, he has enough investments in gilt-edge securities to insure, as some of his envious fellow citizens put it, that Rodney will never have to put in an honest day's work as long as he lives.

Rodney happens to be deeply interested in only two things: elegant living and the ballet. He lives expensively and he pours his money prodigiously into support for eternally bankrupt ballet companies. He is considerably less interested in politics than in the coddled egg he eats for breakfast every morning. Rodney has few views on local and national affairs, except as these may touch on the ballet, and little influence.

"What with his wealth and all, Rodney has more potential influence on local affairs than anyone in this town," his friends sometimes say dolefully. "And he has a good deal less actual influence than his chauffeur, Woodrow, who happens to be the Democratic precinct leader in Rodney's neighborhood—though Rodney probably doesn't even know it."

Thus Rodney's *potential* influence on local affairs is, at a maximum, obviously high; his *actual* influence is negligible.

Rodney's brother Walter has never had Rodney's flair for outguessing the stock market. He works hard at a job in one of Rodney's radio stations, earns a modest living, and between his job and his family has little time for his great passion—local affairs. Rodney dreams of ballet, Walter dreams of what he could do in local politics if he only had half a chance.

One fine day, Rodney dies suddenly. To

everyone's surprise (including Walter's) it turns out that his whole fortune does not, after all, go to that experimental ballet company in New York. It goes to Walter.

As this example illustrates, we cannot infer what the differences in political influence will be among different actors simply from the differences in their access to political resources. In our example:

1. Given his resources, Rodney's potential influence on local affairs was, at a maximum, very much higher than Walter's or Woodrow's influence. But
2. Rodney's actual influence was probably less than his chauffeur's. And
3. With the same resources as Rodney had, Walter's probable future influence is far greater than Rodney's influence was ever likely to be.

Thus an attempt to rank the influence of Rodney, Walter, and Woodrow might look something like this:

Maximum potential influence on local affairs:	Actual influence up to the present:	Expected future influence on local affairs:
1. Rodney	1. Woodrow	1. Walter
2. Walter	2. Walter	2. Woodrow
3. Woodrow	3. Rodney	

. . . .

Thus the influence an actor exerts in some sphere of decisions may fluctuate, *but it rarely approaches his maximum potential influence in that sector*. Why is this so? There seem to be two main reasons. First, only a few actors ever acquire a high degree of political skill. Second, only a few actors ever feel that it is worthwhile to use their resources to the full in order to maximize their political influence in a given sector. . . .

Enough has been said, I hope, to indicate how important it is to maintain a clear distinction between an actor's past or present influence, the influence it is reasonable to expect he will exert in the future under certain specified conditions, and his maximum potential influence.

POWER

It is now easy to see why some individuals seek to increase their influence by gaining control over the State. For when an actor controls the State, he can enforce his decisions with the help of the State. More concretely, he can use the State's monopoly over physical coercion to try to secure compliance with his policies. But one can seek compliance not only through punishments but also through rewards; and control over the State generally provides resources that can be used to create large benefits as well as severe punishment. In short, the State is a peculiarly important source of *power*. Let us examine this point more closely.

Two kinds of influence are sometimes singled out for particular attention:

1. *Coercive influence*: influence based on the threat or expectation of extremely severe penalties or great losses, particularly physical punishment, torture, imprisonment, and death.

2. *Reliable influence*: influence in which the probability of compliance is very high.

Coercive influence can be illustrated graphically as follows: Imagine a continuum representing various degrees of some value that A regards as important—honor, wealth, prestige, popularity. A's present position is at A_0 on the continuum. That is, a gain in, say, honor, for A is represented by a move to the right; a loss in honor is represented by a move to the left.[2] Hence one might

[2]Many other terms used to describe this kind of situation can be used inter-changeably with gains and losses. Gains are equivalent to rewards, benefits, advantages, inducements, positive incentives, indulgences. Losses are equivalent to penalties, disadvantages, negative incentives, deprivations, sanctions.

try to influence A by offering him rewards or threatening him with

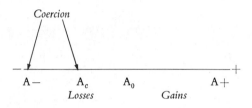

penalties, or some combination of the two. In other words, one might induce A to do something he would not otherwise do by promising to make him better off than he is now ($100 reward if he does), threatening to make him worse off than he is now ($100 fine if he doesn't), or both ($100 reward if he does, $100 fine if he doesn't). The domain of *influence*, one might say, runs all the way from one extreme to the other, from A− to A+ and includes all possible combinations. The extreme left portion of the continuum, let us say from A− to A_c, represents the most severe penalties. This is the domain of *coercive influence*, which is sometimes called *power*.

Exactly what constitutes a "severe" penalty or deprivation is, to be sure, somewhat arbitrary. No doubt what one regards as a severe penalty varies with his experiences, culture, bodily condition, and so on. Nonetheless, whoever can visit severe penalties on others, including imprisonment and death, is bound to be unusually important in any society. Indeed, the State is distinguishable from other political systems only to the extent that it successfully upholds its claim to the exclusive right to determine the conditions under which certain kinds of severe penalties, those involving physical coercion, may be legitimately employed.

But very substantial rewards can be made to operate rather like coercion. For if A is offered a very large reward for compliance, then once his expectations are adjusted to this large reward, he suffers a prospective loss if he does not comply. At the extreme ends of the continuum, near A− and near A+, the disadvantages of non-compliance are exceedingly high. In this sense the manipulation of substantial rewards can also be considered "coercive." Negative coercion is based on the threat of extreme punishment, whereas positive coercion is based on the prospect of very large gains. Both negative coercion and positive coercion are sometimes included in the term "power."[3]

The State is, then, a pawn of key importance in struggles over power, for the relatively great resources of the State and its exclusive claim to regulate severe physical coercion means that those who control the State inevitably enjoy great power.

How to Detect and Weigh Power

It is one thing to define power and quite another to observe it. How can we actually observe power relations, and particularly how can we determine the relative power of different actors in the real world? In short, with respect to any particular

[3] The existence of both negative and positive coercion is sometimes a source of confusion in political analysis, since writers often either confound the two or ignore positive coercion. A close reading of Harold D. Lasswell and Abraham Kaplan, *Power and Society* (New Haven, Conn.: Yale University Press, 1950), indicates that they include both negative and positive coercion in their definition of power, though the inclusion of positive coercion is not obvious. They write:

"A *decision* is a policy involving severe sanctions (deprivations). . . . *Power* is participation in the making of decisions. . . . It is the threat of sanctions which differentiates power from influence in general. Power is a special case of the exercise of influence: it is the process of affecting policies of others with the help of (actual or threatened) severe deprivations for nonconformity with the policies intended."

Lasswell cites as comparable Locke's use of the term in the *Two Treatises of Government*: "Political Power, then, I take to be a right of making laws, with penalties of death, and consequently all less penalties" (pp. 74–76). These passages appear to suggest that they included only negative sanctions. However, they define "sanctions" to include "reward or punishment by way of any value whatever. The sanction is *positive* when it enhances values for the actor to whom it is applied, *negative* when it deprives him of values." (pp. 48–49).

political system how can we answer the question: "Who rules"?

Efforts have been made to study power relations in experimental situations, in laboratories, or in other circumstances in which the observer can pretty well control the situation. Although these efforts are sometimes interesting and useful, obviously they are a long way from real life. We want to know who rules in Jonesville, or Washington, or Moscow, not in a contrived situation. We want to know how power is distributed among the various members of a concrete political system. We need this knowledge before we can go on to other important questions: What kind of a political system is it? How are we to classify it? Do the actors with the most power or the least power come from a particular socio-economic stratum of the community? If so, what stratum? If not, from what strata are they drawn? What are the goals and values of the most powerful actors? The less powerful? How much do these diverge? What changes are taking place in the sources of leadership? In the distribution of power?

Outside the laboratory there are four main ways to observe power relations in order to discern the way in which power is distributed among the members of a political system. Each has advantages and disadvantages. None is perfect. Skillfully used, each can be a highly valuable instrument for inquiry.

First, you can proceed on the assumption that an actor's power is closely correlated with his position in an official or semi-official hierarchy. Who occupies the offices in a political system? Who holds the "major" offices? The minor offices? No offices? What groups are over-represented or under-represented among the office-holders?

The great advantage of this method is its simplicity. Official hierarchies are usually well-defined, the incumbents are clearly designated, records are extensive, information is easily obtainable. The method is particularly useful for detecting large-scale historical changes and gross differences among broad political systems, such as the decline of the landed gentry in English politics and the rise of other socio-economic strata. This method is also helpful in discovering patterns of relationship among different elite groups—how much overlap is there, for example, among occupants of high social positions, high positions in business, high positions in the military, and high positions in politics?

The great weakness of the method, of course, is the shaky assumption on which it rests, for formal position is not necessarily correlated with power. This method would not necessarily uncover the *eminence grise*, the king-maker, the political boss, the confidante; nor would it record the power of a class or stratum that rules indirectly by allotting formal offices to others.

To overcome these defects you can turn to a second method. You can rely on well-placed judges. This was the method used in the study of Regional City and it has been widely imitated. Corrected for the neglect of scope that marred the original study and many others that have followed, this method can be highly useful. It is relatively simple, quick, and economical. The judgments of one set of observers can easily be checked against the judgments of others. The method is useful even in historical studies. It is the recorded judgments of well-placed observers that lead us to Father Joseph, the *Eminence Grise* of Cardinal Richelieu, or to President Wilson's confidante, Colonel House.

The big disadvantage of the method is that it puts us at the mercy of the judges—and how are we to determine who are the best judges? Even seemingly well-placed observers can be misled by false reputations; they may attribute great power where little or none exists. Do we need more judges, then, to judge the judges? Or can we somehow go behind the judgments of the judges and arrive at our own independent appraisal?

The third method is to pierce the facade of formal position and reputation by studying participation in decisions. Which actors actually participate most often in making decisions within this or that scope of activity? Which actors participate in several or many different kinds of activity, and which ones are highly specialized? This method focusses on what people do (or at least what they and others report they do), not on formal office or reputation. Moreover, it can help us trace out various patterns of power; we can distinguish power in one issue-area from power in another, general power over many issue-areas from the specialized power of an actor who participates in only one issue-area.

The great disadvantage of this method, however, is that participation or activity is not equivalent to power. A president and his confidential secretary may both participate in all the major governmental decisions of a country, but it would be folly to conclude that they had anything like equal power.

In order to get around this disadvantage, you might try a fourth method. You could weigh the activities of different participants in decisions. You could assign weights by means of some kind of operational definition of one or more of the underlying measures discussed earlier.

One study, for example, reconstructed in considerable detail a set of decisions made over a period of time in New Haven, Conn. From the reconstructed record of various decisions in different issue-areas, the author sought to determine which of the participants had most frequently initiated proposals that were later adopted as actual policy, or had successfully opposed proposals initiated by others. It was assumed that the actors who not only participated in the decisions but were most frequently successful according to these criteria were the most influential.[4] The great advantage of the method is that it uses an operational test, however crude, for appraising the relative power of different participants in decisions, and thus it enables the observer to go behind mere office, reputation, and activity. One disadvantage of the method is the time required to reconstruct decisions in sufficient detail; another is that operational definitions must sometimes be so crude as to lend themselves to serious criticism. For example, which man is the more powerful in a given issue-area: a man who initiates two proposals that are subsequently adopted without objection, a second man who carries through one proposal over very strong initial objections, or a third who initiates 20 policy proposals but succeeds in securing the adoption of only one-third of them?

Every method of detecting and weighing power, then, has decided advantages and disadvantages. None is foolproof. Without commonsense and judgment, all can produce ludicrous results. Used skillfully, however, each is highly useful.

Some Common Errors in the Analysis of Power

It is easy to see why the analysis of power is so full of pitfalls. Here are some common errors that the preceding discussion should help one to detect and avoid:

1. Failing to distinguish clearly between participating in a decision, influencing a decision, and being affected by the consequences of a decision.

2. Failing to identify the scope or scopes within which an actor is said to be powerful.

3. Failing to distinguish different degrees of power, for example, by equating the proposition that power is distributed unequally in a political system with the proposition that the system is ruled by a ruling class.

4. Confusing an actor's past or present power with his potential power, particularly by assuming that the greater the political resources an actor has access to, the greater his power must be.

5. Equating an actor's expected future power with his potential power, particularly by ignoring differences in incentives and skills.

---

[4] Robert A. Dahl, *Who Governs?* (New Haven, Conn.: Yale University Press, 1961).

Richard C. Snyder

A Decision-Making Approach to the Study of Political Phenomena

THE LOCATION OF THE DECISION-MAKING APPROACH

. . . The decision-making approach to the study of politics clearly belongs in the category of *dynamic*, as distinct from static, analysis. I hesitate to introduce this distinction because the words are ambiguous and because the line is much fuzzier than the words suggest. Relatively speaking, dynamic analysis is *process* analysis. By process is meant here, briefly, *time* plus *change*—change in relationships and conditions. Process analysis concerns a *sequence of events*,[1] *i.e.*, behavioral events. In general, static analysis is a snapshot at one point in time. One basic difference between the two types is in the way (or ways) the time factor is handled. An important brand of static analysis[2] (namely structural-functional analysis), can yield information on the nature of change between two periods in time and on the conditions under which change took place but not on the reasons for change or how it actually unfolded.

In turn, there are two kinds of process analysis: *interaction* and *decision-making*. So far as I can see, there are only two ways of scientifically studying process in the sense employed here: the making and executing of decisions and the patterns of interaction between individuals, states, organizations, groups, jurisdictions, and so on. Interaction analysis does not and cannot yield answers to "why" questions. Thus interactions can be described and measured but the explanation of the patterns—why they evolved as they did—must rest on decision-making analysis.[3]

These distinctions are neither intended to prejudice the case for or against decision-making analysis, nor are they intended to reflect favorably or unfavorably on static and dynamic analysis. As a matter of fact, I believe these types can and should supplement each other. I do not believe they are rivals (except for the energy and attention of scholars) because there are certain things each can and cannot do.[4] . . .

C. Wright Mills[5] has made a related and also useful distinction between what he calls macroscopic and molecular social research. This distinction, too, is one of degree and emphasis. The former embraces such things as the total social structure, global forces, great sweeps of history, and gross patterns of relationship. The latter embraces the actions and reactions of social beings in particular situations and under particular conditions. Obviously, decision-making belongs more in the latter category than in the former. It is worth noting that, as Mills says, we have as yet found no satisfactory way of relating these two types of research.

Thus decision-making is one phase or form of social action analysis. The term action has a technical meaning, not a commonsense meaning. Analytically, action depends on the empirical

[1] Event here is used as an analytic term, not as meaning a discrete occurrence.
[2] Other forms of static analysis are: requisite analysis (a modification of structural-functional analysis), equilibrium analysis, head-counting, and description of structure in the formal sense.

Reprinted from Roland Young (ed.), *Approaches to the Study of Politics* (Evanston, Ill.: Northwestern University Press, Copyright©1958), pp. 10–24, where it appeared under the title, "A Decision-Making Approach to the Study of Political Phenomena." By permission of the publisher.

[3] It may be true that analysis and investigation will show that social change can take place without conscious choice—if so, it will be necessary to take this into account.
[4] For example, see David Easton, "Limits of the Equilibrium Model in Social Research," *Chicago Behavioral Science Publications*, Number 1, 26ff.
[5] "Two Types of Social Analysis," *Philosophy of Science* (October, 1953), 266–75.

existence of the following components: actor (or actors), goals, means, and situation. While this formulation is borrowed from Parsons and Shils,[6] the conceptualization outlined below owes more of an intellectual debt to the writings of Alfred Schuetz.[7] Although the two schools overlap and agree in many particulars, there are fundamental differences in the observer's relationship to the actor. One such difference is that under the Parsonian scheme, a rational model of action is assumed: the observer's criteria of rationality are imposed on the actor. My own feeling is that, on balance, decision-making needs a phenomenological approach.[8] Hence no rational actor[9] is assumed in the present scheme and the observer's criteria are not imposed on the actor.

. . . .

The Decision-Making Approach[10]

There are two fundamental purposes of the decision-making approach: to help identify and isolate the "crucial structures" in the political realm where change takes place—where action is initiated and carried out, where decisions must be made; and to help analyze systematically the decision-making behavior which leads to action and which sustains action.

Some Postulates. (1) The decision-making approach herein formulated focuses inquiry on a class of actors called decision-makers. On the assumption that *authoritative* (i.e., binding on the whole society viewed as a political association or on some segment thereof also viewed as a political association, such as states, counties, and cities) action can be decided upon and initiated by *public officials*, who are formally or actually responsible for decisions and who engage in the making of decisions, our actors are official actors. These officials comprise a *reservoir* of decision-makers from which particular groups are drawn for particular decision-making purposes. We are concerned, then, primarily with the behavior of members of the total governmental organization in any society. And we are concerned therefore only with decisions made *within* that structure.[11] From past experience, I am aware that many will gag on this assumption. Without attempting to argue the case fully, let me anticipate some of the difficulty. An insistence upon a clear distinction between the governmental and nongovernmental realms for purposes of decision-making analysis *appears* to imply a narrowing of the definition of *political*.[12] This runs counter to the prevailing doctrine among behavioral political scientists and those who have been promoting "political process" research.[13] Also it seems to be a retreat

[6]Talcott Parsons et al., "Some Fundamental Categories of the Theory of Action: A general Statement," in Talcott Parsons and Edward A. Shils (eds.), *Toward a General Theory of Action*, (Cambridge, Mass.: Harvard University Press, 1951), pp. 3–30.
[7]See especially, "Choosing Among Projects of Action," *Philosophy and Phenomenological Research*, 12 (1951), 161–84.
[8]I also hesitate to introduce this troublesome term but it points to an important quality of my analysis. See Rohrer and Sherif, *Social Psychology at the Crossroads* (New York: Harper, 1951), pp. 215–42.
[9]This problem of rationality is more troublesome than we have recognized. I cannot go into the matter here. However, I do not deny that rationality may be a useful concept for some purposes. Cf. Schuetz, "The Problems of Rationality in the Social World," *Economics*, 10 (N.S., 1943), 130–49.
[10]I owe a great debt for many ideas and formulations to the Organizational Behavior Project at Princeton under the direction of Wilbert E. Moore and to my colleagues Henry Bruck and Burton Sapin of the Foreign Policy Analysis Project at Princeton.

[11]Contrast this with the position taken by Harold Lasswell, "The Selective Effort of Personality on Political Participation," in Richard Christie and Marie Jahoda (eds.), *Studies in the Scope and Method of "The Authoritarian Personality,"* (Glencoe, Illi.: Free Press, 1954), pp. 197ff.
[12]Lasswell, "The Selective Effort . . ." Lasswell makes a distinction between a *conventional* definition of political decision and a functional one which makes *all* "important" decisions political. He goes on to say that in order to locate functional elite, it is necessary to locate those making "actual" decisions (pp. 203–204). This implies that there is a difference between actual and nominal decisions.
[13]For example: Garceau, "Research in the Political Process," *American Political Science Review*, 45 (1951), 69–85.

from the discovery that noninstitutional social factors are basic to an understanding of political life. To focus on the behavior of official decision-makers seems to omit those powerful nongovernmental (but political by the broad definition) figures who (allegedly) *really* make the decisions.

To clarify, let it be said that this postulate does *not* imply that *all* politically important decisions are made *within* the governmental structure. I do insist that only decisions actually made by public officials are politically *authoritative*. A decision by a corporation or an organized group may be very significant politically and it may affect or be binding on certain persons, but it is not binding on the community politically organized. Furthermore, I know of no way that such nongovernmental decisions can be shown to have consequences for governmental decisions without accounting for the behavior of official decision-makers.

Earlier I made a distinction between two kinds of process analysis: *interaction* and *decision-making*. This may save some misunderstanding on the present point. Interaction process analysis does not require—and indeed would be handicapped by—a separation of decision-makers into official and nonofficial groups or a boundary line between governmental and nongovernmental decision-making. But the limitation here is that interaction analysis per se cannot answer "why" questions of decision-making activity. To reiterate, if one wants to analyze the "why" of governmental decisions, some other conceptual scheme is required.

I have become convinced that when one shifts to decision-making analysis, it is far less troublesome methodologically to *account* only for the behavior of official decision-makers and to relate them to decision-makers outside of government by some other scheme than one which requires that *both* groups be regarded as actors *in the same social system*—which means accounting for the behavior of both according to formal rules of action analysis.

2) The behavior of official decision-makers should be described and explained in terms of action analysis. This means treating the decision-maker as an "actor in a situation." In turn, this means we make a basic choice to take as our prime analytical objective the recreation of the "social world" of the decision-makers as *they* view it. Our task is to devise a conceptual scheme which will help us to reconstruct the situation as defined by the decision-makers. The key to political action lies in the way decision-makers as actors define their situation. Definition of the situation is built around the projected action as well as the reasons for the action. Therefore, it is necessary to analyze the decision-makers in the following terms:

a) their *discrimination* and *relating* of objects, conditions, and other actors—various things are perceived or expected in a relational context;

b) the existence, establishment, or definition of *goals*—various things are wanted from the situation;

c) attachment of significance to various courses of action suggested by the situation according to some criteria of estimation;

d) application of "standards of acceptability" which (1) narrow the range of perceptions; (2) narrow the range of objects wanted; and (3) narrow the number of alternatives.

Three features of all orientations emerge: *perception*, *choice*, and *expectation*.

Perhaps a translation of the vocabulary of action theory will be useful. We are saying that the actors' orientations to the action are reconstructed when the following kinds of questions are answered: What did the decision-makers think was relevant in a particular situation? How did they determine this? How were the relevant factors related to each other—what connections did the decision-makers see between diverse elements in the situation? How did they establish the connections? What wants and needs were deemed involved in or affected by the situation? What were the sources of these wants and needs?

How were they related to the situation? What specific or general goals were considered and selected? What courses of action were deemed fitting and effective? How were fitness and effectiveness decided?

In other words, the actor-situation approach to social analysis alerts the observer to the *discrimination of relevancies*—to the *selection and valuation* of objects, events, symbols, conditions, and other actors. These relevancies are, so to speak, carved from a total number of phenomena present in the over-all setting.[14] Of the phenomena which might have been relevant, the actors (decision-makers) finally endow only some with *significance*. Relevancies may be "given" for the actors (*i.e.*, not open to their independent judgment, and among the "givens" will be certain cues to the determination of other relevancies). The situation—as defined—arises from selective perception: it is abstracted from a larger setting.

3) "Situation" is an analytical concept pointing to a pattern of relationship among events, objects, conditions, and other actors organized around a focus (objective, problem, course of action) which is the center of interest for the decision-makers.[15] As noted above, typologies are important to unifying concepts. A decision-making frame of reference will require several, among them a typology of kinds of situations. Only a crude formulation is possible here:

> a) *Structured* vs. *unstructured situations*—pointing to the relative degree of ambiguity and stability; a situation for which the decision-makers find it difficult to establish meaning may be characterized by change as well as intrinsic obscurity.
>
> b) Situations having different degrees of *requiredness*, *i.e.*, the amount of pressure to act and its source (from within the decisional system or from the setting).
>
> c) The *cruciality* of situations—their relatedness to, and importance for, the basic purposes and values of the decision-makers.
>
> d) *Kinds* of affect with which the situation is endowed by the decision-makers—threatening, hostile, avoidance-inducing, favorable, unfavorable, and so on.
>
> e) How the problem is interpreted and how its *major functional characteristic* is assigned—political, moral, economic, military, or a combination of these.
>
> f) The *time* dimension—the degree of permanence attributed to various situations.
>
> g) The degree *to which objective factors impose* themselves on the decision-makers—the number of uncontrollable factors and imponderables.

Perhaps the chief advantage of such a breakdown is to remind us of the fact that certain objective properties of a situation will be partly responsible for the reactions and orientations of the decision-makers and that the assignment of properties to a situation by the decision-makers is indicative of clues to the rules which may have governed their particular responses.

The Organizational Context

All political decisions (as defined), on whatever level of government or wherever in the total structure of government, are formulated and executed in an organizational context. Having said that we will concentrate on decision-makers and how they orient to action, it is necessary to consider them as participants in a system of action. The concept of system is essentially an ordering device implying certain defined types of relationships among the decision-makers and patterns of activities which they engage in. Major characteristics of the system determine to a considerable extent the manner in which the decision-makers relate themselves to the setting. The type of social system with which we are primarily concerned is an organization. Many studies of politics ignore or merely assume the fact that decision-makers

---

[14] To be explained below.

[15] Situational analysis is discussed in Easton, *The Political System* (1953), pp. 149–170; Carr, *Situational Analysis* (1948), pp. 1–38, 45–61, 90–100; Cole, *Human Behavior* (1953), pp. 357–388; Cartwright (ed.), *Lewin's Field Theory in Social Science* (1951), pp. 30–60, 238–304. Compare the first chapter of Arthur Macmahon's excellent little book, *Administration in Foreign Affairs* (1953). This chapter is entitled the "Concept of Judgment" and should be compared with my concept of definition of the situation.

operate in a highly particular and specific context. To ignore this context omits a range of factors which significantly influence the behavior of decision-makers and omit not only the critical problem of how choices are made but also the conditions under which choices are made. I am convinced that many of the difficulties surrounding the attempt to apply personality theory, culture theory, and small-group theory have been due to a failure to consider the peculiar social system in which decision-makers function. Emphasis on personality and so-called informal factors[16] has tended to minimize the importance of formal factors. Combined with some of the consequences of the "political process" approach, the individual policy-maker has been regarded as operating in a vacuum.

Since we are interested in process analysis we shall take for granted many of the commonly recognized structural features of organization. In other words, such factors as personnel, internal specialization, authority and control, routinized relationships, professionalized positions and careers, and so on will be considered as given prerequisites.

Organizational Decision-Making[17]

Here is a tentative definition plus a commentary: *Decision-making results in the selection from a socially defined, limited number of problematical, alternative projects (i.e., courses of action) of one project to bring about the particular future state of affairs envisaged by the decision-makers.*

EXPLANATION AND ASSUMPTIONS. (1) Decision-making leads to a *course of action* based on the project. *Project* is employed here to include both objectives and techniques. The course of action moves along a *path* toward the outcome envisaged. Adoption of the project signifies that the decision-makers were motivated by an intention to accomplish something. The means included in the project are also socially defined.

2) Organizational decision-making is a *sequence of activities*. The particular sequence is an *event* which for purposes of analysis may be isolated. The event chosen determines in good part what is or is not relevant. To illustrate: if the event in which the observer is interested is the making of the Japanese Peace Treaty, then the focus of attention is the system that produced the treaty and the various factors influencing the decision-making in that system. NATO, EDC, ERP, the Technical Assistance Program, etc., are not relevant. If, on the other hand, the overall cluster of policy decisions with respect to the policy of containment is the focus, the Japanese Peace Treaty and NATO, EDC, ERP, the Technical Assistance Program and a number of other factors all become a part of the strategies of implementation.

3) The event can be considered a unified whole or it can be separated into its constituent elements. A suggested breakdown might be in terms of the sequence of activities: (*a*) pre-decisional activities, (*b*) choice, and (*c*) implementation.

4) Some choices are made at every stage of the decision-making process. The *point of decision* is that stage in the sequence at which decision-makers having the authority choose a specific course of action to be implemented and assume responsibility for it. The weeding out of information, condensation of memoranda, etc., all involve decisions which must be recognized as such by the observer.

5) Choice involves *evaluation* in terms of a *frame of reference*. *Weights* and *priorities* are then assigned to alternative projects.

6) The *occasion for decision* arises from uncertainty. Some aspect of the situation is no longer taken for granted and becomes problematical in terms of the decision-maker's frame of reference.

7) The problem requiring decision may originate within the decisional system or it may originate in a change in the internal or external setting.

---

[16] I have reservations on the formal-informal dichotomy, but I shall here let conventional meaning prevail.

[17] The lack of a commonly accepted, general concept of decision-making or decision-making process has already been commented on. However, that there are theories of decision-making is clear from the previous section.

8) The *range of alternative projects* which the decision-makers consider is limited. Limitations exist both as to means and ends. Limitations on the range of alternative projects are due in large part to the following factors: The individual decision-maker's past experience and values, the amount of available and utilized information, situational elements, the characteristics of the organizational system, and the known available resources.

Definition of the Decisional Unit and of the Decision-Makers

It is necessary to establish boundaries which encompass the actors and activities to be observed and explained. Here we specify that the organizational system within which a decision-making event takes place is the decisional *unit* which becomes the focal point of observation. The unit embraces, analytically, the actors and the system of activities which results in decision.

By what criteria is the decisional unit to be isolated and differentiated? The single criterion which seems at the moment to be most useful is the objective or mission. Objective or mission is taken to mean a particular desired future state of affairs having a specific referent. Specificity is most crucial because it is only possible to speak of the unit (or organization or system) with respect to a specified objective. In other words, regardless of the level of government or the size of the unit, it is constituted by the observer in terms of the decision-makers responsible for, and activities geared to, a particular policy, problem, or other specific assignment. With respect to any objective or mission, there is an organizational unit so constituted as to be able to select a course of action for that objective.

In passing, it might be noted that as the concept of decision-making is refined, two other typologies will be useful: a typology of kinds of political objectives and a typology of decisional units.[18]

Immediately, a two-headed question will be asked: how can the observer be sure he has all the actors in the unit who were involved in a decision and how is "involved" to be interpreted? This is mostly a matter for empirical investigation in the particular case. Very often there are established, well defined units. In some cases it may be necessary to do some detective work to reconstruct the unit. Undoubtedly, the observer will have awkward choices to make occasionally as to whether an actor or a function is to be included or excluded. When this is true, the observer will have to choose on the basis of his analytical purposes. The one great advantage of establishing the unit on the basis of the purposes of its activities is that we can avoid having to be content only with high level abstractions such as the State Department or the city government or the company, and when several agencies or other concrete structures are engaged in policy-making, only the relevant actors and functions need be considered.

The Unit as Organization

The constituent elements of *any* decisional unit will be suggested below. Here we shall only indicate that all units will be organizational in the sense that activities and relationships will be the outcome of the operation of formal rules governing the allocation of power and responsibility, motivation, communication, performance of function, problem-solving, and so on. Each unit will have its own organization in this sense. Obviously, the particular organizational form which a unit takes will depend on how and why the unit was established, who the members are, and what its specific task is. A unit may be a one-shot affair—as in the case of the Japanese Peace Treaty or an ad hoc investigating committee in Congress. Or, a unit may represent a typical decisional system for dealing with typical objectives as in the case of an interdepartmental committee at the federal level or the city council.

The Origins of Units

I have argued that the unit is an analytical tool—a device to aid the observer in reconstituting the

---

[18]Typologies would not be as necessary for historical studies as they would be for prediction.

decision-making universe and in establishing boundaries. However, as hinted above, the empirical question underlying the concept of the unit is: who becomes involved in a decision, how, and why? How does the group of officials (actors, decision-makers) whose deliberations result in decision become assembled? Often, of course, the answer to this question is essential to an explanation of why the decision-makers decided the way they did. Two methods of unit construction may be suggested: *automatic assignment* and *negotiation*. Sometimes the selection of decision-makers from the total number who might in any substructure of government become involved is based on a simple classification of problems or decisions. The formal roles of the actors provide the clue as to whether they will be part of the unit. Also, as already noted, there are standing units (*i.e.*, committees or groups) who are expected to act on given matters. A quite different method of selection is negotiation in cases where no routine procedures exist or where new conditions require special procedure. Negotiation may be simply a matter of springing the right officials loose for a particular task or it may represent basic disagreement over the location of authority and power. Thus everywhere in government the decisions on who will decide are extremely important.

In the case of complex governmental institutions in which a great many activities and a great many officials are involved, often the unit may be created by default. That is, the unit is constituted empirically by the actors, who, in effect, select themselves into it.

The Unit and the Setting

Every group of decision-makers functions in a larger setting. Setting is felt, analytically, to be more satisfactory than environment, which has certain explicit connotations in psychology and has ambiguous connotations otherwise. Setting refers to a set of categories of *potentially relevant factors and conditions* which may affect the action of decision-makers. Relevance of particular kinds of factors and conditions *in general* and *in particular situations* will depend on the attitudes, perceptions, judgments, and purposes of particular groups of decision-makers, *i.e.*, how they react to various stimuli. Setting thus is an analytical device to suggest certain enduring kinds of relevance and to limit the number of nongovernmental factors with which the student of politics must be concerned. The setting, empirically, is constantly changing and will be composed of *what the decision-makers decide is important* or *what is "given" as important*.

Two aspects of the setting of any decisional unit deserve mention: the social setting and the political institutional setting. Normally and familiarly, social setting designates public opinion, including the possible reactions of veto-groups. For bureaucracy, this means the general public *and* the specific clientele—either for regular government services or "attentive publics"[19] or an ad hoc interest grouping based on particular issues. However, an adequate concept of decision-making will include in the social setting much more fundamental categories: major common-value orientations, major characteristics of social organization, group structures and function, major institutional patterns, basic social processes (adult socialization and opinion formation), and social differentiation and specialization. From these can be derived conditions and forces of immediate impact on decision-makers.

Several of these can be noted briefly. First, every action taken by the decision-makers has consequences in the society at large. One kind of feedback is that the society experiences its own decisions. Possible effects can range from redistribution of social power to specific complaints, from puzzlement to understanding, from acceptance to rejection. Second, policies are usually accompanied by official interpretations which may or may not agree with nongovernmental interpretations. The strategies of *legitimation* chosen by decision-makers have a very crucial effect on the way policy results are viewed. Third, the society provides decision-makers with a wide

---

[19] The phrase is Gabriel Almond's, *The American People and Foreign Policy*, 1st ed. (New York: Harcourt, Brace, 1950), Chap. 1.

range of means—technical services in which government must rely on private sources. Fourth, the social system has an important bearing on *who* gets recruited into decision-making posts and *how*. This raises the whole question of support for the governmental structure and this question leads to the internal adjustments in response to the social setting. In particular, this point subsumes the number of private agencies and individuals which can hold the decision-makers responsible.

The political institutional setting is perhaps a much more immediate factor. This consists of what might be called the total organizational reservoir from which the constituent elements are drawn, including constitutional prerogatives, rules of the game, responsibility equations, general purposes, concrete membership groups, roles, functions, pools of information, communication links, and so on. These are the items of traditional concern in government—government in general. Basically, the institutional setting viewed in this light is a vast pool of rules, personnel, and information for the decisional units. Within this pool, certain specialized activities—*not* concerned directly with decision-making and execution—are carried on day by day. The decision-making approach does not ignore or render unnecessary structural institutional analysis. On the contrary, it requires more and more thorough analyses of this sort, and, hopefully, it can add to their usefulness.

Unless the particular substructure is very small (*e.g.*, a village or town) any decisional unit is likely to exist simultaneously with other units. These units will be analytically connected because of the following kinds of factors: (*a*) overlapping membership; (*b*) a common set of givens—rules and precedents; (*c*) common objectives throughout the total system; (*d*) overlapping jurisdictions; (*e*) reciprocal impact of courses of action adopted.

To return to the notion of definition of the situation: the line between what is included in the definition and what is not is not just a boundary between relevance and nonrelevance. Two types of relationship appear within the defined situation. On the one hand, there will be relationships among factors within the social setting and the institutional setting and between these two aspects of the setting. On the other hand, there will be relationships between the setting *and* the plans, purposes, and programs of the decision-makers.

Limitations on Decision-Making

The concept of limitations constitutes a set of assumptions about *any* decisional system. The assumptions concern the factors or conditions which limit (*a*) alternative objectives; (*b*) alternative techniques; (*c*) the combination of *a* plus *b* into strategies or projects; (*d*) decision-making resources—time, energy, skills, information; and (*e*) degree of control of external setting. In accordance with our general phenomenological approach, we feel that the range and impact of limitations should be considered from the decision-maker's point of view, although many such assessments will be objectively verifiable. The main categories of limitations in terms of their sources are: those arising from *outside* the decisional system, those arising from the nature and functioning of the decisional system, and those arising from a combination of both of these.

Limitations Internal to the Decision-Making System

For purposes of illustration, let us list briefly some major limitations of this kind. It must be emphasized that the limitations traceable to bureaucratic pathology are perhaps the most dramatic but certainly not the only ones.

1) *Information.*—The decision-makers may lack information or may act on inaccurate information; in either case, the range of alternatives considered may be affected. It would appear to be a permanent liability of the decision-making process that relevant information is almost never completely adequate and testable. The necessity to adopt and employ interpretive schemes and

compensatory devices such as simplification of phenomena provide a related source of limitation.

2) *Communication Failures.*—Reasonably full information may be present in the decisional unit but not circulate to all the decision-makers who need it to perform their roles satisfactorily. A decisional unit may be resistant to *new* information or the significance of new information may be lost because of the way messages are labeled and stored.

3) *Precedent.*—Previous actions and policy rules (the givens for any unit) may automatically narrow the deliberations of the decision-makers. Previous action may prohibit serious consideration of a whole range of projects. Reversal of policies is difficult in a vast organization.

4) *Perception.*—The selective discrimination of the setting may effectively limit action. What the decision-makers "see" is what they act upon. Through perception—and judgment—external limitations gain their significance.

5) *Scarce Resources.*—The fact that any unit is limited in the time, energy, and skills (and sometimes money) at its disposal also tends to limit the thoroughness of deliberation and the effectiveness with which certain related functions are performed. Time pressures may seriously restrict the number of possible courses of action which can be explored.

Lucian W. Pye

The Concept of Political Development

The language of public policy is always in flux, for new concerns produce new terminologies. Yet in the language of politics, in which sloganeering is the common currency of presumed dialogues, fluency in innovation rarely signals advancement in thought. At times fresh terms herald the awareness of novel problems, but more often they indicate merely frustration with intractable circumstance. When the language of politics seeks to define in broadest terms the contemporary human condition, it tends to be sensitive mainly to the emotions of hope, anxiety, or frustration which are inherent in the mind's erratic ability to either race ahead or fall far behind the tempo of substantive change. The political analyst in seeking the neutral ground of the observer inevitably faces the dilemma of being able neither to ignore popular terminology nor to use it as the hard currency of disciplined intellectual exchange. . . .

All of this is of great relevance in trying to find meaning in current discussion of what is or should be happening in the poor and weak countries of the world. During the last decade the worldwide interest in the plight of these societies has produced a Babel of terms. Some of these express the aspirations of statesmen; others are the pompous pretensions of calculating politicians; and still others are merely the euphemisms of people who think that they may be talking about delicate matters. The result is that the study of the problems of these societies is so cluttered up with loosely used terms that clear and disciplined communication has become difficult. Observe how it has now become necessary to employ such optimistic and promiseful expressions as "developing" and "emergent" when discussing

Reprinted from *The Annals of the American Academy of Political and Social Science,* Vol. 358 (March, 1965), 2–13, where it appeared under the title, "The Concept of Political Development." By permission of the author and the publisher.

the gloomy cases of countries that are barely holding themselves together, whose governments are shaky and archaic, and whose peoples are growing faster in numbers than in well-being. The very terms of analysis suggest forecasts that may conflict with the predictions that objective analysis is seeking to make. . . .

From the perspective of intellectual history it is striking that the issue of development in its economic, social, and political guises arose to challenge the social scientists just at the time when we thought we had buried the presumably old-fashioned and innocent concept of "progress." Although earlier social theorists had certainly given support to the notion of human progress and social evolution, modern social scientists have generally been somewhat embarrassed by this popular Western and peculiarly American article of faith. With the rise of the dictators and the holocaust of World War II, the mood of social science was at best agnostic and skeptical to any suggestion about either the inevitability or even the desirability of progress. With this as background, the social sciences were hardly ready to embrace enthusiastically the concept of "development" as applied to the non-Western world. Consequently, we have had to go through a period of adjustment during which there has been some suspicion that the presumably discredited notion of progress was again appearing through a back door. . . .

In a nearer perspective of intellectual history, the question of development caught the social sciences at the high point in our belief in cultural relativism. Although World War II had raised some question about the validity of dispensing tolerance towards all cultures, certainly the mainstream of social science favored the spirit of accepting the propriety of cultural differences and of respecting the realities of contemporary life in every society. In countering the evils of ethnocentrism, rather strong taboos were erected against even implying that some societies might be more "advanced" or more "developed" than others. This meant that general standards of social and political performance were out, and behavior in one society should not be judged against performance in another.

In the light of this experience, it came as a shock to doctrinaire champions of cultural relativism to discover that their doctrines could be cruelly degrading precisely to those to whom it was intended to give respectability. For, when crudely put, the concept of cultural relativism could be read to mean that it was in the nature of some societies to be rich and powerful and for others to be poor and ineffectual. The doctrine could easily be misunderstood as a balm to the poor to make it possible for them to rationalize their lot.

Aside from this, the tendency to misunderstand the ethic of cultural relativism has impeded thought about the problems of development because it has left social scientists unsure as to whether they should properly be concerned with assisting others to change their ways and deviate from their heritages. To be concerned with development can all too often seem the same as trying to make others over into the image of ourselves, as long as it is accepted that we are somehow more developed than they are. The very legitimacy of development is thus brought into question by the spirit of cultural relativism. . . .

The emergence of the question of development also caught political science at a time when the discipline thought that it was successfully breaking from its earlier and strongly normative tradition. Modern political science, in seeking to become an empirical discipline, has been anxious to be highly realistic and to deal with conditions and processes as they actually occur in life. This fundamental trend again seems in some respects to conflict with the orientations necessary for working on the problems of development; for if development means anything it means a rejection of current realities in favor of hoped-for eventualities. The spirit of empiricism, in replacing interest in utopias and in more ideal arrangements, gave a certain sense of legitimacy to the ongoing workings of any political process, which in turn had left political scientists with the feeling that reformism is slightly naive and that change and improvement can only be incremental. This outlook on history was hardly calculated to be of help and encouragement to the leaders and intellectuals of new states impatient for dramatic change.

In addition to dominant trends in the philosophic orientations of the social sciences, the recently fashionable operating procedures and methodologies have also affected our ability to deal with the problems of development. Briefly, after World War II the social sciences felt that they were coming of age as sciences, and thus they tended to place a high value on precision, rigor, and exactness of measurement, qualities which are all more compatible with systematic but essentially static modes of analysis. Our awareness of the possibilities of sophisticated techniques of investigation has made us uncomfortable with loose and broad generalizations. With our methodological sophistication we have also come to appreciate fully the intellectual reasons why dynamic modes of analysis, so essential for understanding the development process, are inherently more difficult, and to some degree beyond our current capabilities, if the highest standards of rigor are to be maintained. Although fortunately many social scientists have been prepared to meet the challenge of work in the imperfect research environments of the new states, they have had to risk criticism of their work being not up to the levels of exactness now expected of studies in our own society.

For all of these and numerous other reasons, Western social science was peculiarly unprepared for providing ready intellectual guidance on the problems of political and social development. Indeed, the very stress of contemporary social science that knowledge must be well grounded in empirical investigation caused many social scientists to feel excessively ill-equipped to pass judgments on the prospects of development in strange and unknown societies; thus, paradoxically, men who considered themselves realists above all else often felt it appropriate to drift along with the almost euphorically optimistic view of the possibilities for rapid development in the new state which were so common a few seasons ago. . . .

DIVERSITY OF DEFINITIONS

It may . . . be helpful to elaborate some of the confusing meanings which are frequently associated with the expression political development. Our purpose in doing so is not to establish or reject any particular definitions, but rather to illuminate a situation of semantic confusion which cannot but impede the development of theory and becloud the purposes of public policy.

(1) *Political Development as the Political Prerequisite of Economic Development*

When attention was first fixed on the problem of economic growth and the need to transform stagnant economies into dynamic ones with self-sustaining growth, the economists were quick to point out that political and social conditions could play a decisive role in impeding or facilitating advance in per capita income, and thus it was appropriate to conceive of political development as the state of the polity which might facilitate economic growth.

. . . .

(2) *Political Development as the Politics Typical of Industrial Societies*

A second common concept of political development, which is also closely tied to economic considerations, involves an abstract view of the typical kind of politics basic to already industrialized and economically highly advanced societies. The assumption is that industrial life produces a more-or-less common and generic type of political life which any society can seek to approximate whether it is in fact industrialized or not. In this view the industrial societies, whether democratic or not, set certain standards of political behavior and performance which constitute the state of political development and which represent the appropriate goals of development for all other systems. . . .

(3) Political Development as Political Modernization

The view that political development is the typical or idealized politics of industrial socetics merges easily with the view that political development is synonymous with political modernization. The advanced industrial nations are the fashion-makers and pace-setters in most phases of social and economic life, and it is understandable that many people expect the same to be true in the political sphere.

. . . .

The question immediately arises as to what constitutes form and what is substance in this view of political development. Is the test of development the capacity of a country to equip itself with such modern cultural artifacts as political parties, civil and rational administrations, and legislative bodies? If so, then the matter of ethnocentrism may be of great relevance, for most of these institutions do have a peculiarly Western character. If, on the other hand, importance is attached only to the performance of certain substantive functions, then another difficulty arises in that all political systems have, historically, in one fashion or another, performed the essential functions expected of these modern *and* Western institutions. Thus, what is to distinguish between what is more and what is less "developed"? Clearly the problem of political development—when thought of as being simply political "modernization"—runs into the difficulty of differentiating between what is "Western" and what is "modern." Some additional criteria seem to be necessary if such a distinction is to be made.

(4) Political Development as the Operations of a Nation-State

To some degree these objections are met by the view that political development consists of the organization of political life and the performance of political functions in accordance with the standards expected of a modern nation-state. In this point of view there is an assumption that, historically, there have been many types of political systems and that all communities have had their form of politics, but that with the emergence of the modern nation-state a specific set of requirements about politics came into existence. Thus, if a society is to perform as a modern state, its political institutions and practices must adjust to these requirements of state performance. The politics of historic empires, of tribe and ethnic community, or of colony must give way to the politics necessary to produce an effective nation-state which can operate successfully in a system of other nation-states.

Political development thus becomes the process by which communities that are nation-states only in form and by international courtesy become nation-states in reality. . . .

It is important to stress that from this point of view nationalism is only a necessary but far from sufficient condition to ensure political development. Development entails the translation of diffuse and unorganized sentiments of nationalism into a spirit of citizenship and, equally, the creation of state institutions which can translate into policy and programs the aspirations of nationalism and citizenship. In brief, political development is nation-building.

(5) Political Development as Administrative and Legal Development

If we divide nation-building into institution-building and citizenship development we have two very common concepts of political development. Indeed, the concept of political development as organization-building has a long history, and it underlies the philosophy of much of the more enlightened colonial practices.

Historically, when the Western nations came in contact with the societies of the rest of the world, one of the principal sources of tension was the discovery that such societies did not share the same Western concepts about law and the nature of public authority in the adjudication of private disputes. Wherever the European went one of his first revealing queries was: "Who is in charge here?" According to the logic of the European mind, every territory should fall under some sovereignty, and all people in the same geographic location should have a common loyalty and the same legal obligations. . . .

In time, however, it was discovered that the smooth operation of an explicit and formalized legal system depended upon the existence of an orderly administrative system. The realization of law and order thus called for bureaucratic structures and the development of public administration, and throughout the colonial period the concept of development was closely associated with the introduction of rationalized institutions of administration.

. . . .

(6) Political Development as Mass Mobilization and Participation

Another aspect of political development involves primarily the role of the citizenry and new standards of loyalty and involvement. Quite understandably, in some former colonial countries the dominant view of what constitutes political development is a form of politcal awakening whereby former subjects become active and committed citizens. In some countries this view is carried to such an extreme that the affective and mass demonstrational aspect of popular politics becomes an end in itself, and leaders and citizens feel that they are advancing national development by the intensity and frequency of demonstrations of mass political passion. Conversely, some countries which are making orderly and effective progress may, nevertheless, be dissatisfied, for they feel that their more demonstrative neighbors are experiencing greater "development".

. . . .

(7) Political Development as the Building of Democracy

This brings us to the view that political development is or should be synonymous with the establishment of democratic institutions and practices. Certainly implicit in many people's views is the assumption that the only form of political development worthy of the name is the building of democracies. Indeed, there are those who would make explicit this connection and suggest that development can only have meaning in terms of some form of ideology, whether democracy, communism, or totalitarianism. According to this view, development only has meaning in terms of the strengthening of some set of values, and to try to pretend that this is not the case is self-deceiving.

. . . .

(8) Political Development as Stability and Orderly Change

Many of those who feel that democracy is inconsistent with rapid development conceive of development almost entirely in economic or social order terms. The political component of such a view usually centers on the concept of political stability based on a capacity for purposeful and orderly change. Stability that is merely stagnation and an arbitrary support of the *status quo* is clearly not development, except when its alternative is manifestly a worse state of affairs. Stability is, however, legitimately linked with the concept of development in that any form of economic and social advancement does generally depend upon an environment in which uncertainty has been reduced and planning based on reasonably safe predictions is possible.

. . . .

(9) Political Development as Mobilization and Power

The recognition that political systems should meet some test of performance and be of some utility to society leads us to the concept of political development as the capabilities of a system. When it is argued that democracy may reduce the efficiency of a system there is an implied assumption that it is possible to measure political efficiency; and in turn the notion of efficiency suggests theoritical or idealized models against which reality can be tested.

This point of view leads to the concept that political systems can be evaluated in terms of the level or degree of absolute power which the system is able to mobilize. Some systems which may or may not be stable seem to operate with a very low margin of power, and the authoritative decision-makers are close to being impotent in their capacity to initiate and consummate policy objectives. In other societies such decision-makers have at their command substantial power, and the society can therefore achieve a wider range of common goals. States naturally differ according to their inherent resource base, but the measure of development is the degree to which they are able to maximize and realize the full potential of their given resources.

. . . .

(10) Political Development as One Aspect of a Multidimensional Process of Social Change

The obvious need for theoretical assumptions to guide the selection of the items that should appear in any index for measuring development leads us to the view that political development is somehow intimately associated with other aspects of social and economic change. This is true because any item which may be relevant in explaining the power potential of a country must also reflect the state of the economy and the social order. The argument can be advanced that it is unnecessary and inappropriate to try to isolate political development too completely from other forms of development. Although to a limited extent the political sphere may be autonomous from the rest of society, for sustained political development to take place it can only be within the context of a multidimensional process of social change in which no segment or dimension of the society can long lag behind.

According to this point of view, all forms of development are related, development is much the same as modernization, and it takes place within a historical context in which influences from outside the society impinge on the processes of social change just as change in the different aspects of a society—the economy, the polity and social order—all impinge on each other.

THE DEVELOPMENT SYNDROME

Without trying to assert any particular philosophical orientation or theoretical framework, it may be useful to scan the various definitions or points of view which we have just reviewed in order to isolate those characteristics of political development which seem to be most widely held and most fundamental in general thinking about the problems of development.[1]

The first broadly shared characteristic which we would note is a general spirit or attitude toward equality. In most views on the subject political development does involve mass participation and popular involvement in political activities. Participation may be either democratic or a form of totalitarian mobilization, but the key consideration is that subjects should become active citizens and at least the pretenses of popular rule are necessary.

Equality also means that laws should be of a universalistic nature, applicable to all and more or less impersonal in their operations. Finally, equality means that recruitment to political office should reflect achievement standards of performance and not the ascriptive considerations of a traditional social system.

A second major theme which we find in most concepts of political development deals with the capacity of a political system. In a sense capacity is related to the outputs of a political system and the extent to which the political system can affect the rest of the society and economy. Capacity is also

---

[1] The themes basic to the concept of political development which follow reflect the work of the Committee on Comparative Politics of the Social Science Research Council and will be developed in much greater detail in a forthcoming volume, *The Political System and Political Development*, to be published in the series, "Studies in Political Development," by the Princeton University Press.

closely associated to governmental performance and the conditions which affect such performance. More specifically, capacity entails first of all the sheer magnitude, scope and scale of political and governmental performance. Developed systems are presumed to be able to do a lot more and touch upon a far wider variety of social life than less developed systems can. Secondly, capacity means effectiveness and efficiency in the execution of public policy. Developed systems presumably not only do more than others but perform faster and with much greater thoroughness. Finally, capacity is related to rationality in administration and a secular orientation toward policy.

A third theme which runs through much of the discussion of political development is that of differentiation and specialization. This is particularly true in the analysis of institutions and structures. Thus, this aspect of development involves first of all the differentiation and specialization of structures. Offices and agencies tend to have their distinct and limited functions, and there is an equivalent of a division of labor within the realm of government. With differentiation there is also, of course, increased functional specificity of the various political roles within the system. And, finally, differentiation also involves the integration of complex structures and processes. That is, differentiation is not fragmentation and the isolation of the different parts of the political system but specialization based on an ultimate sense of integration.

In recognizing these three dimensions of equality, capacity, and differentiation as lying at the heart of the development process, we do not mean to suggest that they necessarily fit easily together. On the contrary, historically, the tendency has usually been that there are acute tensions between the demands for equality, the requirements for capacity, and the processes of greater differentiation. Pressure for greater equality can challenge the capacity of the system, and differentiation can reduce equality by stressing the importance of quality and specialized knowledge.

Indeed, it may, in fact, be possible to distinguish different patterns of development according to the sequential order in which different societies have dealt with the different aspects of the development syndrome. In this sense development is clearly not unilinear, nor is it governed by sharp and distinct stages, but rather by a range of problems that may arise separately or concurrently. In seeking to pattern these different courses of development and to analyze the different types of problems, it is useful to note that the problems of equality are generally related to the political culture and sentiments about legitimacy and commitment to the system; the problems of capacity are generally related to the performance of the authoritative structures of government, and the questions of differentiation touch mainly on the performance of the nonauthoritative structures and the general political process in the society at large. This suggests that in the last analysis the problems of political development revolve around the relationships between the political culture, the authoritative structures, and the general political process.

William Flanigan and *Edwin Fogelman*

Functionalism in Political Science

As it applies to contemporary political science, "functionalism" can refer to several rather disparate types of political analysis; there is no broad, distinctive functional approach to analysis which can be contrasted, say, to an institutional, historical, or legal approach. At no time has functionalism been a prevalent mode of analysis in political science, and political scientists have never borrowed extensively from the functionalists in anthropology and sociology. Functional analysis has come to political science only recently, and relatively few major works in the discipline have explicitly developed a functional analytic scheme. In order to appraise the contributions of functionalism to contemporary political science, we must first clarify the variety of types of analysis that are included within this ambiguous term.

Eclectic Functionalism

In its widest usage functionalism means simply that in analyzing some phenomena the political scientist will be concerned with, among other things, the functions or purposes served by the phenomena. Here function is treated as one—and not necessarily a more significant one—among many relevant considerations that together

Reprinted from Don Martindale, ed., *Functionalism in the Social Sciences: The Strengths and Limitations of Functionalism in Anthropology, Economics, Political Science and Sociology*, Monograph 5 (Philadelphia: The American Academy of Political and Social Science, 1965), pp. 111–26, where it appeared under the title, "Functionalism in Political Science." By permission of the authors and the publisher.

comprise a comprehensive political analysis. In addition to function, the analyst may be equally concerned with the structure, history, ideology, and other aspects of the phenomena. We can call this type of analysis "eclectic functionalism," and in this sense it is not too much to say that we are all functionalists now, although the implications of the commitment differ greatly from scholar to scholar. Eclectic functionalists are found in all branches of contemporary political science; they can be identified by their tendency to ask the question: "What functions does X perform?" There may be reference to the functions of individuals—the functions of the president; groups—the functions of political parties; institutions—the functions of the International Court of Justice; or ideas—the functions of Communist ideology. Depending on the analyst, this simple form of functionalism may only provide a list of activities in which X is engaged, or it may provide answers to more or less explicit questions with respect to how X contributes to the performance of certain purposes of activities.

Eclectic functionalism involves no commitment to a distinctive functional approach and the theoretical implications of including the concept of "function" among the categories of analysis are quite limited. Function is not the focus for analysis, but only one aspect of the analysis; nor is the functional aspect considered in any way primary or exceptionally significant. Eclectic functionalism is by far the most widespread and at the same time the least developed theoretically of current types of functional analysis.

Empirical Functionalism

The attempt to analyze politics from a more consistently functional standpoint, without, however, basing the analysis on a general functional theory, has produced a second type of functional analysis, which we can call "empirical functionalism." Empirical functionalism was given its greatest impetus and its most convincing justification in Robert K. Merton's well-known study of the political machine, contained in his

"Latent and Manifest Functions."[1] Unlike the eclectic functionalists, Merton does not merely consider function as one among a number of equally significant aspects of a political machine, but rather, he concentrates upon function as the "most promising orientation." Moreover, for Merton, function is not the commonsensical concept that it is for eclectic functionalists. Functional analysis requires an elucidation not only of manifest functions, the obvious and intended purposes and consequences, but also of latent functions, the more covert and unintended consequences that are equally important and enlightening as subjects for analysis.

In making the functions of any phenomena the primary focus for his analysis and in enlarging the concept of function to include a variety of significant relationships, Merton gives to functional analysis considerable range and subtlety. For example, Merton argues that the political machine serves several social functions including the provision of welfare assistance in a personal manner. While comparable aid is available through governmental channels, the machine provides assistance without loss of self-respect. This is a thoroughly plausible and typical form of argument in what we are calling "empirical functionalism."

It is important to notice what Merton does and does not do in his functional analysis. First, he does not make reference to functions which must somehow be served in this or all social systems. He simply notes that there is a demand for welfare assistance, and meeting this demand—serving this purpose—is what Merton calls fulfilling a social function. He does not in any sense treat the political machine as a social subsystem with functional requisites of its own fulfilled in various ways. There is so little concern with the functions as such that he does not attempt to assess the extent to which political machines engage in welfare activities. The main aspects of the analysis are straightforward empirical statements of relationship.

Empirical functionalists remain self-consciously limited in their use of a functional perspective. For one thing, they show no concern with functional requisites at the level of the system as a whole. They isolate particular elements within the total system and treat them as discrete units without any presumptions about the significance of these units for the system as a whole. Moreover, empirical functionalists find in functionalism a framework for analysis with limited theoretical implications. It is upon the validity of their empirical findings rather than the analytic power of a possible functional theory that they rest the case for functional analysis. Justification for the restricted perspective adopted by empirical functionalists is provided by Merton in his argument for "middle range" theory. He asserts that whether or not a general functional theory to explain the social system as a whole is ultimately possible, the most advisable course is to deal with more limited units in terms of reasonably precise concepts.

STRUCTURAL FUNCTIONAL ANALYSIS

The most ambitious attempts to introduce a functional approach into contemporary political science have come from those scholars who have applied in political analysis the structural functional framework developed by Parsons[2] and

---

[1] Robert K. Merton, *Social Theory and Social Structure* (Glencoe, Ill.: Free Press, 1957), Chap. 1, especially pp. 72–82.

[2] Talcott Parsons, *The Social System* (Glencoe, Ill.: Free Press, 1951), and with Edward Shils (eds.), *Toward a General Theory of Action* (Cambridge: Harvard University Press, 1951), and with Robert F. Bales and Edward A. Shils, *Working Papers in the Theory of Action* (Glencoe, Ill.: Free Press, 1953) are the main theoretical works on Parsons' structural functionalism. Political scientists will find his "'Voting' and the Equilibrium of the American Political System," *American Voting Behavior*, Eugene Burdick and Arthur Brodbeck (eds.), (Glencoe, Ill.: Free Press, 1959) more readable and more concerned with familiar subject matter. His more recent study of one of the four subsystems, the economy—with Neil Smelser, *Economy and Society* (Glencoe, Ill.: Free Press, 1956)—suggests how he might go about studying other subsystems like the polity, and also touches on more matters of traditional interest to political scientists.

Levy.[3] Here functionalism assumes a theoretical significance potentially far greater than in either eclectic functionalism or empirical functionalism. The promise of structural functionalism is nothing less than to provide a consistent and integrated theory from which can be derived explanatory hypotheses relevant to all aspects of a political system. As William Mitchell explains in his structural functional analysis of *The American Polity*: "I have chosen to use the 'structural functional' approach largely because it seems to offer the best possibilities for eventually developing a general theory . . . of political systems."[4]

Actually, structural functionalists do not all use the same terminology and formulations, and the divergences among such structural functionalists as Almond,[5] Apter,[6] and Mitchell are by no means superficial.

Analytic Framework

Mitchell like other structural functionalists concedes that the present stage of development falls short of scientific social theory. Despite variations in terminology and some confusion within the approach itself, structural functional analysis does embody certain characteristic features: first, an emphasis on the whole system as the unit of analysis; second, postulation of particular functions as requisite to the maintenance of the whole system; third, concern to demonstrate the functional interdependence of diverse structures within the whole system.

Although all the structural functional frameworks in political analysis are more or less related to Parsons' work, some analysts like Almond have restated the scheme so drastically that they have an influence independent of Parsons. For political analysis Almond proposes two categories of functions, the political and the governmental, and in the major study employing this framework *The Politics of Developing Areas*,[7] he emphasizes the political functions. Almond's five political functions are: political socialization, political recruitment, interest articulation, interest aggregation, and political communication. The governmental functions with their obvious parallel to the three branches of government are: rule making, rule application, and rule adjudication. Rather than offer an elaborate rationale for these functions, Almond simply observes that all political systems appear to perform these functions in some way or another, and while this may be true enough, it is not a firm theoretical footing. It seems that Almond is merely making a recommendation that political scientists ought to concern themselves with these activities if they hope to understand politics, and particularly politics under conditions where governments are not highly developed or stable.

The paucity of theoretical structure around these concepts leaves us without detailed hypotheses of the reciprocal relationships among functions, the relationships of groups and institutions to the functions, or the relative significance of the functions. Almond's attempt to present the main analytic interests of political analysis in a system-wide framework leads him to introduce some incongruent elements. The political communication function—an opinion leader proposition basically—is unlike the other functions in that it is more of a mechanism or process, a means of performing functions, but Almond is unwilling to omit an area of investigation which he believes important simply because it does not fit neatly into a structural functional framework. The initial elaboration of Almond's scheme did not suggest many reciprocal interrelationships between structures and functions, nor is there a theoretical concern with mechanisms by which

---

[3] Marion Levy, *The Structure of Society* (Princeton, N.J.: Princeton University Press, 1951).

[4] William C. Mitchell, *The American Polity* (Glencoe, Ill.: Free Press, 1962), p. vii.

[5] Gabriel A. Almond, "A Functional Approach to Comparative Politics," in Gabriel A. Almond and James S. Coleman (eds.), *The Politics of the Developing Areas* (Princeton, N.J.: Princeton University Press, 1960).

[6] David E. Apter, *The Gold Coast in Transition* (Princeton, N.J.: Princeton University Press, 1955).

[7] Almond, "A Functional Approach. . . ."

structures perform functions. The area specialists who contributed the chapters on the developing areas were left with a collection of apparently unrelated categories which served as little more than a basis for organizing the material in each chapter. This level of analysis may be a necessary first step, but the remaining tasks of reformulation and theoretical specification are enormous.

. . . .

One concern of structural functionalism in political science has been simply to advocate the approach. Karl Deutsch persuasively argues that Parsons' scheme is widely applicable in political science and an improvement on most of the conceptualizing in political science as a step toward general systems theory.[8] Deutsch emphasizes the conceptual advantages of viewing Parsons' four fundamental functions as competitive and the interchanges between the structures performing the four functions. Deutsch concedes that there are bound to be difficulties in implementing Parsons' scheme for empirical research, especially in the attribution of observable reality to the categories. He manages to create the impression of considerable dynamism within the structural functional framework, a dynamism lacking in other schemes. To some extent Parsons and Deutsch are coming to discuss structural functionalism in terms of capabilities. The study of the subsystem polity is the study of the capacity of the society to attain its system goals. The analytic framework can be narrowed to a study of the varying capabilities of the structures in the system to perform different functions with available resources. It is too soon to guess what analysis oriented in this way would produce, and while it would omit much of interest to political scientists, it might provide a limited but sophisticated method for studying the survival and disintegration of political systems.

Some Irrelevant Criticisms

Before indicating our own reservations about the structural functional approach, we may notice two criticisms of structural functionalism which seem to us unwarranted. One is an ideological criticism, the second a logical criticism. The ideological criticism holds that structural functionalism is implicitly conservative and biased against social change; the logical criticism holds that the form of the structural functional argument is fallacious.

Although it is true that in the works of Parsons and Levy analysis of the conditions for the stability and survival of a society appears at the very center of the structural functional approach, it is also significant that most political scientists who have found a use for structural functional analysis have been explicitly concerned with the study of political change, and political change of a profound sort. Almond, Apter, and Binder[9] have all drawn upon structural functionalism for the light it could throw on the process of political modernization, and none of these scholars was led to a repudiation of modernization in the name of conservatism. Despite its emphasis upon the conditions for stability, structural functionalism does not lead necessarily to a defense of the *status quo* or to a disregard for processes of change.

An argument against the logic of structural functionalism has been offered by Carl Hempel,[10] and we must differ from his interpretation. In our view, the structural functional argument is not, as Hempel maintains, illogical but rather tautological, and some of its limitations for political analysis are related to its tautological character. Hempel has construed the logic of functional analysis in such a way that the main conclusions prove the existence of a given trait or a given activity. Explanation of a structure in functional

[8]Karl W. Deutsch, "Integration and the Social System: Implications of Functional Analysis," in *The Integration of Political Communities*, Philip E. Jacob and James V. Toscano (eds.), (Philadelphia: Lippincott, 1964).

[9]Leonard Binder, *Iran* (Berkeley and Los Angeles: University of California Press, 1962).
[10]Carl G. Hempel, "The Logic of Functional Analysis," in *Symposium on Sociological Theory*, Llewellyn Gross (ed.), (Evanston, Ill.: Row, Peterson & Company, 1959).

analysis does not refer to predicting the existence of the structure, but rather refers with inappropriate terminology to the existence of a particular activity performed by the structure. Although some functionalists may have argued more or less teleologically that structure X exists because it—and perhaps it alone—performs function F, to reject functionalism in this form is to knock down a straw man. The most interesting statements in functional analysis at the present time, it seems to us, are about structures performing functions, and Hempel treats these statements as suspect premises. What is explained—predicted—by functionalism ideally is the performance of functions and inferentially the maintenance of the system.

CRITICAL EVALUATION

The basic form of the structural functional argument can be presented in two syllogisms:

> I. (1) If system s is to be maintained adequately under conditions c, then requisite functions f_1, f_2 . . . f_n must be performed.
> (2) System s is being maintained adequately.
>
> ∴ Requisite functions f_1, f_2, . . . f_n are being performed.
>
> II. (1) If requisite functions f_1, f_2 . . . f_n are being performed, this will be accomplished by existing structures.
> (2) Requisite functions f_1, f_2 . . . f_n are being performed.
>
> ∴ Requisite functions are being performed by existing structures.

Armed with this argument, the structural functionalist sets out in search of the particular structures which perform the requisite functions. In the argument itself, however, there is no guidance concerning where to look in this quest and no basis for asserting the extent to which any particular structure does in fact perform a specified function. Nor are there any grounds for supposing that one set and only one set of functions is requisite. In short, the analyst can define his "requisite functions" as he pleases, and he can be equally imaginative in locating which structures perform what functions. There is nothing illogical about his quest; the difficulty is rather that his findings may consist of many discrete observations which do no more than illustrate again and again that structures perform functions.

As a basis for analysis, the structural functional argument leaves important problems unresolved. First of all, it is difficult to say when a system is being "adequately maintained." We must have some objective criteria for determining when a system is adequately maintained. But these criteria have not been provided, and as a result the analyst cannot tell whether the system he is observing is flourishing or declining. Thus, he can never be sure when the minor premise in syllogism I is fulfilled, and the same is true of the minor premise in syllogism II, since it actually depends on the former syllogism.

Another difficulty is the failure to elaborate and specify the nature of the interdependence of particular structures. Changes in any structure—which must involve changes in how a function is being performed—must have repercussions throughout the system, but the nature of these effects cannot be determined from the argument. What happens when a structure changes? With its emphasis on structural alternatives in different societies, functional political analysis gives few leads concerning what to expect within a single system over time. This difficulty is related to the failure to classify the conditions of the system. A scheme which indicates when a system is adequately maintained and when it is not would almost necessarily deal with structural continuity and change. Such a taxonomy of systems would introduce new demands for theoretical clarification. This might well entail a categorization of social systems based on the various conditions in which we find them, such as the various schemes suggested for categorization of stages of modernization.

Finally, when we attempt to spell out the functional requisites mentioned in the first major premise, upon which the entire argument rests, we encounter such amorphous concepts that the connection with reality becomes dangerously strained. From the diversity of requisite functions employed in political analysis to date—very nearly a unique set for each study—it is clear that a major weakness of political functional analysis will persist until precise criteria are established for the identification of functions and a theoretically sophisticated argument is made for a particular set of functions. Eventually a set of functions will be selected on the basis of performance in empirical research in comparison with other sets of functions, but for the time being the rationale for a given set of functions cannot be ignored. We need some discussion in political analysis for the rationale of particular functional requisites in order tentatively to arbitrate the differences between incongruent analytic frameworks.

Political scientists have been attracted to functionalism for a number of reasons. Eclectic functionalists find in functionalism an important additional dimension for their analysis, a dimension which brings the analysis into closer touch with the actual consequences of political activity. Empirical functionalists find in functionalism not simply an additional dimension for their analysis but a central organizing concept which serves to illuminate unexpected aspects of important though limited political phenomena. Structural functionalists find in functionalism the promise of a scientific theory of politics, and this is certainly the most ambitious claim that has been made for functionalism by political scientists.

We have already indicated that the promise of structural functionalism has not yet been fulfilled. But the claim of structural functionalists must be taken seriously, and we will attempt to specify some of the qualifications which functionalism must meet in order to justify this claim.

A broad, systematic functional political theory must aspire to a set of statements from which refutable hypotheses may be deduced. The set of statements must have this capacity, although any given analyst may arrive at the hypotheses he tests by induction, deduction, retroduction, or inspiration. The tight, logical qualities of this set of statements, the "theory," may have been exaggerated in the epistemological analysis of scientific knowledge, but nevertheless the requirement persists for a summary of accumulated knowledge and broad guide to "theoretically" critical investigations. Beyond this, useful theory must inspire imaginative hypothesizing, retroduction as some social theorists would now call it, and like much of political and social theory, functionalism appears richly suggestive of a wide range of relationships. The crucial criterion then becomes the capacity of functionalism to generate imaginative hypotheses which can be tested empirically and refuted. The endless proliferation of "interesting and suggestive hypotheses" must be disciplined by empirical tests.

If it is true, as we have suggested, that structural functionalists have usually operated at a level of analysis which did not permit empirical testing of interesting hypotheses, we must inquire whether or not there are elements in structural functional analysis which make it unsuited for empirical hypothesizing. In order to generate refutable hypotheses, functional theory must have the capacity to operationalize its terms. "Integration" or "pattern maintenance," for example, must have precise definitions, definitions which relate them appropriately to social reality. Furthermore, the operational definitions of basic terms must retain the richness and complexity of the referents in the theoretical discussion. (One of the purposes served by Mitchell and Almond for political science is to generate a body of ideas with sophisticated insights and observations which will serve as a guide to the more precise defining and restating of political relationships.) To operationalize without retaining at least most of the meaning of the concepts would lead to unfair testing of the generated hypotheses, although tentatively we might have to accept some quite arbitrary definitions to get on with investigations.

Structural functionalists have not taken the

enormously difficult step of refining, operationalizing, and testing hypotheses. At the same time, they have been well aware of the problems involved in doing so. As yet, however, no convincing solutions have been proposed. It remains for the critic to ask why the remedies have been lacking. Why has no scholar succeeded in presenting a structural functional formulation which meets the requirements of empirical analysis? Although it may be both reckless and presumptuous to pose this question so soon after the introduction of functionalism into political science, nevertheless we can recognize at least three possible explanations for the failings of structural functional analysis: first, limitations of the scholars; second, unavoidable stages in the development of any scientific theory; and third, deficiencies of functionalism itself.

Insofar as the deficiencies of functional analysis derive from the limitations of present-day scholars, there is always the chance that tomorrow may see the appearance of a work of genius which will in fact solve the problems of functionalism and vindicate the hopes of its advocates. It is obvious, however, that structural functionalism has attracted some of the ablest political scientists; it would be unreasonable to attribute the defects of the analysis to the limitations of these scholars.

A second explanation for the present inadequacies of functional analysis lies in the possible requirements of a pattern through which systems of ideas develop. It may be that the contributions being made now provide a necessary part of the foundation for subsequent achievements. In time the weaknesses of contemporary analysis may be remedied through subsequent theoretical developments. There is, of course, no way for us to be sure whether this will be the case, but the contemporary critic must hazard an evaluation before the verdict of history has been rendered.

This brings us, then, to a third explanation for the deficiencies of functional analysis: the defects of the intellectual framework itself. During any active period in the development of an intellectual discipline, we find a variety of interesting formulations being proposed and explored as a basis for progress. From among these varied proposals, some will prove fruitful within the discipline and others will be remembered only as abortive bypaths. Although there is no certain test for determining at the time which formulations hold the greatest promise, one of the most significant characteristics of any proposed framework for social analysis is the nature of the phenomena which is selected as the focus for examination. No framework can encompass all of reality; it is, indeed, a major purpose of any framework to exclude as much of irrelevant reality as possible and to emphasize only selected features. With regard to structural functionalism, we raise the question whether the emphasis upon "functions" as the focus of analysis is likely to prove fruitful.

The contention that an emphasis upon functions may prove abortive, as the focus for analysis at the level of the political system as a whole is based on the difficulties encountered first in defining functional requisites operationally, and second in specifying the indefinite range of activities which fulfill these functions. In attempting to define functional requisites operationally, the scholar is likely either to impoverish his concepts with arbitrarily narrow definitions which deprive the analysis of its characteristic advantages, or to complicate his concepts to the point that they are unresearchable. In attempting to specify in advance the range of activities which fulfill functions, the scholar is faced with the impossibility of anticipating which activities will prove to be relevant to his analysis, so that each study becomes an exploration of particular relationships of structures to functions. To study the few critical factors which determine the survival or disintegration of the political system is undeniably compelling, but it may be that the discovery of these factors is only the culmination of extensive, more sophisticated empirical investigations. To attempt to speculate what these factors may be involves one at this stage in highly abstracted formulations far removed from the realities of political activity which must, after all, comprise the material of political analysis. While these difficulties may not be in principle insurmountable, they bode ill for structural functionalism.

Conclusion

Even if structural functionalism will not provide a scientific theory of politics, it has, nonetheless, enriched the discipline of political science. First, there are several heuristic contributions to political analysis: (1) sensitizing analysis to the complexity of interrelationships among social and political phenomena; (2) drawing attention to a whole social system as a setting for political phenomena; (3) forcing consideration of functions served—particularly latent functions—by political actors groups as something of an antidote for moralizing and "rational" analysis. In addition, structural functionalists have employed a number of frameworks for political analysis which have provided some opportunity to assess the applicability of these frameworks to one's own interests.

In comparison to structural functionalism, the promise of empirical functionalism is more modest. Yet, in departing from analysis of the system as a whole and in restricting the focus of attention to more manageable problems, empirical functionalism suggests opportunities for functional research which have not been fully explored in political science.

The functionalists have not as yet offered more than a loose analytic approach to the study of political systems. At this early stage of functional analysis in political science, most functionalists are reinterpreting other studies through secondary analysis, so there is little to point to as the benefits of original research conducted with a functional scheme. Although it is not insignificant to restate the findings of others in a different way, this is not enough to qualify functionalism for a major role in political analysis.